Energy Efficiency

For Engineers and Technologists

Energy Efficiency

For Engineers and Technologists

T D Eastop

Honorary Research Fellow, Wolverhampton Polytechnic

D R Croft

Principal Lecturer in Energy Studies, Sheffield City Polytechnic

Longman Scientific & Technical

Copublished in the United States with
John Wiley & Sons, Inc., New York

Longman Scientific & Technical,
Longman Group UK Ltd,
Longman House, Burnt Mill, Harlow,
Essex CM20 2JE, England
and Associated Companies throughout the world.

Copublished in the United States with
John Wiley & Sons, Inc., 605 Third Avenue, New York NY 10158

© Longman Group UK Limited 1990

First published 1990

British Library Cataloguing in Publication Data
Eastop, T.D. (Thomas Deas), *1931–*
  Energy efficiency.
  1. Great Britain. Energy. Conservation
  I. Title.   II. Croft, D.R. (David R.)
  621.042

    ISBN 0–582–03184–2

Library of Congress Cataloging-in-Publication Data
Eastop, T.D. (Thomas D.)
  Energy efficiency/T.D. Eastop, D.R. Croft.
    p.   cm.
  Includes bibliographical references.
  ISBN 0–470–21645–X
  1. Energy conservation.   I. Croft, D.R., 1939–   II. Title.
TJ163.3.E2   1990
333.7916—dc20   89–77355
                                                          CIP

Set in 10/12 pt Lasercomp Times
Printed in Great Britain by The Bath Press, Avon

CONTENTS

PREFACE

There is an ever-increasing world-wide demand for energy, usually in the form of electricity or heat. Most of the demand arises from domestic, commercial, industrial and transport users, and virtually all the processes involved in supplying this energy have one thing in common: they are quite inefficient in their conversion of primary energy to an end use. The basic reason for this inefficiency, and what can be done about it, form the content of this book.

In society generally, the driving force for saving energy is usually the prospect of saving money; sometimes the money savings can be used to finance the purchase of capital equipment that will achieve further savings. There are other driving forces, perhaps more significant in the long term than money savings, which are to do with reducing atmospheric pollution and reducing the rate at which non-renewable energy resources are being depleted.

These issues about the effect of large-scale energy conversion and usage are a complicated mixture of technical, economic and political factors with which national governments are having to become increasingly concerned. The so-called 'greenhouse effect' is now a cause of international concern, as is the depletion of the ozone layer, the formation of 'acid rain', and a large number of other environmental side effects arising from the use of energy in industrial processes. These factors become moral issues when the quality of life for future generations is likely to be affected.

While we are aware of these issues, we have restricted the coverage of this book to the technical and economic techniques and implications of improving the conversion and usage of energy. Energy is our most valuable natural commodity and it is vitally important that it is used more efficiently.

The study of the efficient use of energy set in a financial context is becoming increasingly important for the reasons given above; every student on an engineering or technology course should have energy studies as part of their curriculum. Existing engineers and scientists in industry must become 'energy conscious' by continuing their studies through post-graduate short courses; many firms already employ Energy Managers and this trend will continue. This book is aimed at all undergraduate and technician courses that include energy studies as part of their curriculum, and should therefore be suitable for most engineering courses as well as some technology or combined science courses.

The emphasis throughout the book is on cost saving through the efficient use of energy and the theme is continuously illustrated through numerical Examples and Case Studies. It has therefore been necessary to use actual fuel costs, capital costs of equipment etc. and to assume values of interest rates, depreciation, and inflation. The figures used are based on those current in the UK in early 1989. The general principles are of course unaffected and the relative costs in most cases will remain the same giving the same pay-back period for example. Users of the book can check on current prices by contacting the suppliers of primary fuel and energy, and the manufacturers of relevant equipment.

Finally we would like to acknowledge the assistance we have received in writing the book. The following individuals have given permission to use information: Mr J Crombie, Energy Consultant, Chipchase Crombie; Mr A Ford, Chief Engineer, Gunstones Ltd, Sheffield; Mr J Smith, Chief Engineer, ACME-GERRARD Ltd, Sheffield; Dr N Gwyther, Teesside Polytechnic. Our thanks are also due to the following colleagues for their help and encouragement: Dr G Drummond, Teesside Polytechnic; Dr P W Foss, Dr M J Denman, Dr P D Williams, and Dr H C Biggin, all of Sheffield City Polytechnic. Especial thanks to Carolyn Jackson.

TDE
DRC
1989

ACKNOWLEDGEMENTS

We are indebted to the following for permission to reproduce copyright material:

The Building Services Research & Information Association for table 4.13 from Forecast of UK demand for Heat Pumps for 1985 & 1990 from *Statistics Bulletin*, vol. 11:2, 1986, insert 'Product Profile: Heat Pumps, May 1986' and *Statistics Bulletin*, vol. 13:4, insert 'Product Profile: Heat Pumps, December, 1988'; The Chartered Institution of Building Services Engineers for fig. 4.47 from a reduced scale version of the Psychrometric chart from Guide C1–4, (CIBSE), table 4.1 adapted from table C5.1 of CIBSE *Guide to Current Practice* C5, 1986, tables 7.1 & 7.3 adapted from tables A3.17 & A1.3 of CIBSE *Guide to Current Practice*. A, 1986, table 7.7 adapted from tables A7.1 & A7.9 and table 7.14 adapted from CIBSE *Code for Interior Lighting*, 1984; Council of the Institution of Mechanical Engineers for figs 4.19 & 4.20 from figs 4 & 5 (Smart, 1986); The Institution of Chemical Engineers for fig. 6.1 from fig. 1 of the *User Guide* on Process Integration for the Efficient Use of Energy; Longman Group UK Ltd. for fig. 5.7 from fig. 17.28 of *Applied Thermodynamics for Engineering Technologists* 1986, 4th ed. by Eastop & McConkey.

NOMENCLATURE

A	air–fuel ratio; area
ARR	Accounting Rate of Return
BDL	Blow Down Loss
C	thermal capacity; capacitance; fluid velocity; constant
C'	constant
CHP	Combined Heat and Power
CL	Casing Loss
COP	Coefficient of Performance
C_p	molar heat at constant pressure
C_v	molar heat at constant volume; ventilation conductance
c	specific heat
c_p	specific heat at constant pressure
c_v	specific heat at constant volume
d	diameter
D	degree days
DCF	Discounted Cash Flow
DH	District Heating
E	emissive power; heat exchanger effectiveness; energy
e	fraction of heat input at the environmental point
F	geometric factor; fouling factor; fuel consumption
F_U, F_V	factors defined by Eq. [7.16]
F_Y	factor defined by Eq. [7.21]
FGL	Flue Gas Loss
f	frequency; decrement factor
f_r	response factor
Gr	Grashof number
GCV	Gross Calorific Value
GLS	General Light Standard
g	gravitational acceleration
H	enthalpy
ΔH_o	enthalpy of combustion
h	specific enthalpy; heat transfer coefficient
h_{ac}, h_{ec}	equivalent heat transfer coefficients defined by Fig. 7.3
h_{ae}	equivalent heat transfer coefficient defined by Fig. 7.1
I	current; mean solar irradiance; depreciation rate
I_R	real rate of interest

I'	peak solar irradiance
$\tilde{I}$	swing of solar irradiance about mean
IRR	Internal Rate of Return
i	current
K	constant
k	thermal conductivity; constant
L	length; inductance
$LMTD$	Logarithmic Mean Temperature Difference
LPG	Liquefied Petroleum Gas
M	molar mass; mass of matrix; figure of merit
m	mass
$\dot{m}$	mass flow rate
N	rotational speed; number of years; number of heat exchangers
NCV	Net Calorific Value
NPV	Net Present Value
NTU	Number of Transfer Units
Nu	Nusselt number
n	number of kilomoles; number of tubes; number of air changes; number of hours of plant operation
P	power
Pr	Prandtl number
PV	Present Value
p	absolute pressure; number of pole pairs; number of tube passes; pitch
Q	heat; rate of heat transfer
$\tilde{Q}$	swing about the mean heat transfer rate
q	heat transfer per unit mass
R	specific gas constant; thermal resistance; electrical resistance; ratio of thermal capacities; discount rate
R_o	molar gas constant
Re	Reynolds number
RMS	Root Mean Square
RPI	Retail Price Index
r	radius; correlation coefficient
S	solar gain factor; number of streams
$\tilde{S}$	fluctuating solar gain factor
SON	high pressure sodium
SOX	low pressure sodium
s	specific entropy; slip
T	absolute temperature
t	temperature; thickness
$\tilde{t}$	swing about the mean temperature
Δt	temperature difference
U	internal energy; overall heat transfer coefficient; thermal transmittance
ΔU_o	internal energy of combustion
u	specific internal energy
V	volume; voltage
$\dot{V}$	rate of volume flow
v	specific volume; mass of water vapour in unit mass of fuel
W	work; rate of work transfer; Wobbe number
X	reactance

x	dryness fraction; distance from datum; general variable
Y	thermal admittance
Z	impedance

Greek Symbols

α	absorptivity; thermal diffusivity; angle
β	coefficient of cubical expansion
γ	ratio of specific heats, c_p/c_v, proportion of heat input due to direct radiation
δ	proportion of heat input due to convective heating
ε	emissivity
η	efficiency
θ	angle
μ	dynamic viscosity
ρ	density; reflectivity
σ	Stefan–Boltzmann constant; surface tension
τ	time; transmissivity
ϕ	relative humidity; phase angle
ω	specific humidity

Subscripts

A	air; apparatus dew point
AC	air conditioning
a	ambient; air point
ai	inside air
ao	outside air
B	boiler
b	black body; base surface
C	cold; capacitative; casing; condenser; casual internal gain
c	condensate; convective; dry resultant
DB	dry bulb
E	evaporator
e	exit; environmental point
ei	inside environmental
F	fuel; fin; fabric
f	saturated liquid; fuel
fg	change of phase at constant pressure
G	gas; glazing
g	saturated vapour
H	hot
HP	heat pump
i	inlet; a constituent in a mixture; inside surface; input
L	inductive; line
M	mechanical; matrix material
m	mean
mr	mean radiant
ms	mean surface
max	maximum
min	minimum
o	overall; outside surface; reference condition

P	product of combustion; phase; plant; process steam
p	constant pressure
R	reactant; real
r	radiation
ref	refrigerator
S	secondary fluid; steam
s	vapour; synchronous; isentropic; surface; stoichiometric
T	total
t	tube
V	ventilation
v	constant volume
WB	wet bulb
w	water; wall
wv	superheated water vapour

1 THE ENERGY PROBLEM

As you read these words you are probably sitting quietly in a chair, your body operating at near its minimum rate of power consumption, enough to keep your temperature at 37 °C. You will be using about 120 W of power derived from the food you have eaten which is being converted continuously into a form of energy used on demand for bodily motion. If you look around you will see that other forms of energy are being used; the room you are in may be heated, the lights may be on, traffic may be passing outside, and of course all the material objects around you, including the clothes you wear, required energy in their manufacture.

As an illustration of the different types of energy we consume, Table 1.1 shows the annual consumption of a typical family in the UK. The metered units we pay for from the Electricity Board are measured in kW h (i.e. a 1 kW fire burning for 1 hour consumes one unit), and the metered units we pay for from British Gas are measured in therms. Both these units of energy are non-standard units in the SI system; accurate conversion factors are as follows:

$$1 \text{ kW h} = 3.6 \text{ MJ}$$
$$1 \text{ therm} = 29.307 \text{ kW h} = 105.506 \text{ MJ}$$

In the case of gas consumption the metered value is a volume flow and the units charged for are converted to an energy quantity using a Calorific Value which is required to be kept above a certain minimum value by British Gas (currently, 38.7 MJ/m^3).

Table 1.1 does not show one important aspect, that is how much energy has been used to produce the electricity, gas, petrol and food. Gas is a fuel which needs very little processing from the form it takes deep in the earth. Nevertheless

Table 1.1 Annual energy consumption of a typical UK family

Fuel	Consumption in units in normal use		Consumption (kW h)	Consumption (MJ)
Electricity	5 620	kW h	5 620	20 232
Gas	810	therm	23 739	85 460
Petrol	2 500	litre	24 375*	87 750
Food	5.5 × 10⁶	kcal	6 397	23 029

* Taking the specific gravity of petrol as 0.75 and the Gross Calorific Value (GCV) as 46.8 MJ/kg.

the gas has to be extracted and pumped to the point of use; these activities require the expenditure of energy. About 95 % of the initial energy content of the gas is available for use. Electricity is the product of a large number of conversion processes and only about 30 % of the initial energy of the primary fuel eventually appears as useful electrical energy. This is a very low figure and partly explains the high cost of electricity compared with other fuels.

The understanding of energy and energy conversion processes is a central theme of this book. The energy available to us is in two basic categories: (i) *non-renewable* sources such as the chemical energy of the fossil fuels, coal, natural gas and oil, or the nuclear energy available from uranium; (ii) *renewable* sources such as hydro-electric schemes, solar radiation, wind energy, wave energy, tidal energy and the chemical energy in crop and wood residues and animal waste. The conversion process from chemical energy to mechanical or electrical energy is inefficient mainly due to the limitations of the Second Law of Thermodynamics. For example, when you switch on a 1 kW electric fire then the rate of energy extracted from the fossil fuel in the earth to produce the power is about 4 kW.

Similarly a car engine converts only about 20 % of the primary energy of the refined oil from the oil well into useful mechanical work; most of the energy used is dissipated as heat in the cooling water and the exhaust gases. The problem with this inefficient conversion of primary energy is that the supply of primary fuel is being depleted at an alarmingly rapid rate.

In the early 1970s an oil crisis caused the developed countries to take serious stock of their use of energy; various pessimistic forecasts were made of the rate of energy usage and the depletion of natural resources. At that time the world's energy consumption was doubling every 14 years and oil consumption was doubling every seven years. Due to the sudden large increase in oil prices in the late 1970s a world-wide recession occurred and the world's energy consumption rose at a much slower rate than predicted. Also, the predictions of the world's energy resources made in the 1970s were shown to be pessimistic; for example, natural gas is now being found more quickly than it is being consumed. Nevertheless, the earth's resource of fossil fuels is finite and the problem of diminishing supplies will raise its head again in the 21st century. This also applies to metals which in some cases have a shorter predicted life; reclamation and re-cycling of materials, as well as the development of new materials (not derived from oil if possible!), is therefore vitally necessary.

A significant influence on world-wide energy policy in the late 1980s has been the recognition that the so-called *greenhouse effect* is leading to an increase in the average temperature of the earth and hence to potentially serious and damaging changes to the world's climate and ocean levels. The greenhouse effect is due to the fact that the sun's short wave radiation passes freely through the gases which make up the earth's atmosphere but the long wave radiation from the earth's surface back to space is partially absorbed by the atmosphere. One of the gases which is an efficient absorber of long-wave radiation is carbon dioxide, the main product of combustion of any fossil fuel. Therefore as the consumption of fossil fuels has increased over the years the greenhouse effect has become more marked and the rate of the earth's temperature rise is now a serious cause for concern.

The energy consumed by a nation is directly proportional to its standard of living or Gross National Product; for example the under-developed nations use about 0.5 kW per person compared with a rate in the developed world of between 5 and 10 kW. Since the under-developed nations urgently need to raise their standard of living the task of reducing the rate of increase of world-wide energy consumption becomes even more difficult.

Assuming that standards of living are to be continuously improved at the present rate, there are two major ways in which the energy problem should be tackled. Firstly, alternative renewable energy sources must be exploited to the full; solar radiation, wind power, wave power and tidal power must be developed alongside existing hydro-electric schemes. (Note: the combustion of waste in the form of domestic and industrial refuse, and crop and wood residues is also a renewable energy source that must be exploited but in this case the products of combustion will exacerbate the greenhouse effect.) Secondly, more efficient energy conversion methods must be found, existing processes must be operated more efficiently, energy recovery methods must be used whenever possible, and consumers including the general public must be encouraged to conserve energy; financial incentive is the most effective motivating factor.

What scale of change will make a significant difference to the demand for primary energy? One way to answer this is to work out how we use energy at the present time and then see what the effect is of introducing more efficient methods including conservation. Ideally we should take as wide a view as possible and look at the use of energy world-wide. Not an easy task!

We can look at a particular country because a number of studies have been carried out and published, principally for industrialized countries in Western Europe and North America. In the UK, a report was published in 1979[1.1] which detailed a UK-based model for calculating

'what supplies of primary energy sources would be required to enable the demands for energy, arising from the nation's underlying needs, to be satisfied given a set of social and economic conditions.'

In simple terms the model splits the demand for energy into four main sectors of the economy: domestic, commercial, industrial, and transport. Each of these sectors has requirements for particular fuels at the present time, but the model allows the user to say, for example: what happens if the UK starts to use less nuclear energy and more renewable energy sources? what happens if people improve the thermal insulation of their houses? what happens if industry uses less energy-intensive processes for manufacturing products? In other words the model is a 'what if' model; it calculates *what* happens to the demand for primary energy *if* the pattern of demand, conversion efficiency, and supply, change over a period of time. As an example of the type of situation the model can simulate let us consider how the demand for primary energy changes from the time just after the report was published (i.e. 1980) to the year 2025, not too far into the future if we are looking at the depletion of the earth's natural resources. If we specify that during this time period we would like to determine the effect of improvements in conversion efficiency (including the use of renewable energy sources) on the levels of primary demand, then the calculations will show the type of trends in Figs 1.1(a) and (b). These graphs are based on identical demands by the users in the domestic, commercial, industrial, and transport sectors. The reasons for the different demands for primary energy lie in the increasing use of more efficient energy conversion processes (Fig. 1.1(b)), as opposed to the use of current technology over the same period (Fig. 1.1(a)). Most of the ways

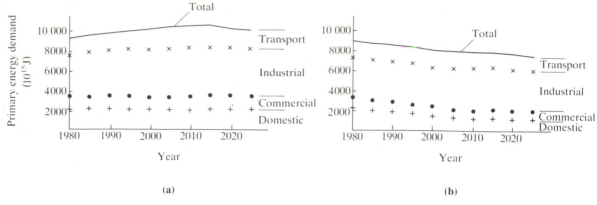

Figure 1.1 UK primary energy supply based on improving conversion efficiencies

of improving the conversion process are known but they require time and capital investment to bring them into prominence in the 'energy economy'. In the UK many are already receiving more attention than before: the use of land-based wind-mills, passive solar design of buildings, the application of new materials in industry, the introduction of combined heat and power in the commercial and domestic sectors, the increasing use of heat pumps, the awareness of the importance of energy recovery etc.

The theme of this book is simply the efficient use of energy. To understand how a device uses energy will enable us to assess its efficiency and hopefully improve this efficiency. There are basic laws of nature which govern the conversion process and these will need to be mastered before analysis of the processes can be undertaken. Chapter 3 gives some of the basic theoretical background and Chapter 4 deals with methods of energy conversion.

Using energy more efficiently should lead to a financial gain whether this is achieved by an individual reducing heating bills through better thermal insulation, say, or a large industrial or commercial concern increasing profits by spending less on primary fuel and electricity. A major feature of this book is therefore the financial aspect of all energy saving measures; Chapter 2 gives some basic theoretical background to methods of calculating whether a particular energy-saving measure will be financially viable, and these methods are applied in later chapters on energy recovery, energy in buildings, and total energy schemes.

Larger organizations have come to realize that the saving of money through more effective energy use justifies the appointment of an Energy Manager. The last chapter of the book discusses the job of the Energy Manager and outlines some of the problems with which he may be faced and some possible solutions.

The future of mankind depends on finding the solution to the problems of depletion of fossil fuel resources, the rise in temperatures due to the greenhouse effect, and other related problems such as the damage to the ozone layer and the general increase of environmental pollution. World-wide interacting economic, social and political factors make the position more complex. This book makes no attempt to analyse energy use and demand on a world-wide, or even a national scale, but instead concentrates on the use of energy within the existing national framework and discusses ways in which processes may be made more cost-effective.

Reference

1.1   Energy Power Number 39, Energy Technologies for the UK, Department of Energy, HMSO 1979

Bibliography

Chapman P 1979 *Fuel's Paradise* 2nd edn Penguin Press

Dorf R C 1978 *Energy, Resources and Policy* Addison-Wesley

Dorf R C 1981 *The Energy Facts Book* McGraw-Hill, New York

Energy Paper Number 54, Energy Technologies for the UK, Department of Energy, HMSO 1986

Energy 2000 – A Global Strategy for Sustainable Development A report for the world commision on environment and development, ZED Books, London 1987

Hedley D 1981 *World Energy, the Facts and the Future* Euromonitor

Hoyle F 1979 *Energy or Extinction* 2nd edn Heinemann

Leach G, Lewis C, Romig F, Van Buren A, Foley G 1979 *A Low Energy Strategy for the United Kingdom* The International Institute for Environment and Development, Science Reviews

Stobaugh R, Yergin S 1979 *Energy Futures* Ballentine Books, New York

2 THE ECONOMICS OF ENERGY-SAVING SCHEMES

Very few energy-saving schemes are implemented for the sake of extending the life of the world's fuel resources; they are adopted because they save money. Any organization which is thinking about investing money in an energy-saving project will ask the question: how much money will be saved and how long will it take to get a return on the initial investment? Unfortunately, it is not usually a simple matter to answer such a question because of the engineering complexity of a project and the difficulties in accurately evaluating savings, particularly when fuel prices are changing over a short period of time. Nevertheless, the question must be answered, and for this reason, a number of techniques have been developed to help the investor in his planning.

Basically, each technique involves the comparison of one scheme with another in a structured manner. A typical situation would be where a company is concerned about its energy costs and decides to invest in some energy-saving measures. Undoubtedly the saving of energy will save money, but this saving may be offset by the cost of the equipment which is needed to achieve the saving. The Energy Manager thus has to use some method of analysis to decide whether or not to invest in a new scheme. Another typical situation is where there are at least two new schemes to consider; one may be expensive to run but have a low capital cost, the other more expensive to purchase but having lower running costs. Again, the Energy Manager requires some systematic method by which to evaluate the two schemes and come to a conclusion about his investment.

This chapter introduces several methods for assessing the economic viability of energy-saving projects. The methods will be used in later chapters where specific costs are used to evaluate particular methods and schemes of energy-saving. At this stage, only the typical source of costs will be considered.

2.1 COSTS

The types of costs associated with energy usage and the calculation of the value of energy savings are:

(a) initial investment (i.e. the capital costs of the project);
(b) fuel costs (e.g. gas, oil, coal, electricity);
(c) other operating costs (i.e. maintenance, materials, labour, service utilities, storage, handling, etc.).

Capital Investment

Capital investment can be pictured as expenditure from which benefits can be expected in the long term. Most energy-saving schemes require an initial investment for new equipment to achieve energy savings. The monetary value of the savings, which will invariably show as reduced fuel costs, must recover the initial costs in as short a time as possible.

If the initial capital is borrowed money then interest will be charged; interest charges are one of two types: simple or compound.

Simple Interest

Simple interest charges are a fixed percentage of the borrowed capital. The charges are calculated for an agreed period of time (say one year) and a total interest charge found by multiplying the charge per year by the number of years given to repay the loan.

Example 2.1

A company is investing in a computerized energy management system which will cost £30 000 to purchase and install. The money is loaned by a bank which charges simple interest at the annual rate of 12 % for a period of three years. Calculate the total repayment required by the bank.

Solution

Total repayment is given by

$$\text{initial loan} + (\text{interest charge/year}) \times \text{number of years}$$

Interest charge/year is 12 % of £30 000 = £3600. Therefore total repayment is equal to

$$£30\,000 + £(3600 \times 3) = £40\,800$$

The advantage to the company of this type of repayment is that there is no incentive to repay the loan in less than three years because the total repayment cost is fixed at £40 800. It is not perceived by the bank as an advantage and therefore a system of compound interest is most commonly used.

Compound Interest

In this system, interest is charged at the end of each time period based on the total (capital plus interest charge up to that point). In Example 2.1:

$$\text{total repayment at end of year 1} = \text{capital} + \text{interest charge}$$
$$= £30\,000 + £3600$$
$$= £33\,600$$
$$\text{total repayment at end of year 2} = \text{year 1 repayment} + \text{interest charge on year 1 repayment}$$
$$= £33\,600 + £(0.12 \times 33\,600)$$
$$= £37\,632$$

Similarly,

$$\text{total repayment at end of year 3} = £42\,148$$

This amount is greater than that accrued by simple interest. The company now has an incentive to repay the loan at regular intervals. The interval may be every year to every month by agreement.

In the above example, the rate of interest was said to be 12 %; a rate agreed between the company and the bank. In one sense, this figure of 12 % per annum represents the cost of borrowing the capital from a bank. As to whether or not the rate is appropriate depends on the prevailing economic climate in the country. The *minimum* rate of interest which any bank may charge is specified, in the UK, by the Bank of England and the Treasury and may change several times per year.

The basic idea of interest charges increasing with time is used later in the economic appraisal of project proposals. When capital is invested in a project then the profits shown by the project can be compared with the income derived from capital which is simply left to accumulate compound interest in a deposit account.

Depreciation

From time to time, a company may choose to sell a piece of equipment after using it for a number of years. The value of the equipment at this time is said to be its depreciated or salvage value, the amount being related to the initial capital cost. Most commonly the depreciation is expressed either as a rate at which the value decreases or as a fixed amount per year.

Example 2.2
A pump is installed to supply cooling water to a heat recovery scheme. After 4 years, the plant layout is changed and the pump is no longer required. The initial cost of the installed pump is £60 000. Calculate the salvage value of the pump set if:

(i) the depreciation is set at £10 000 per year;
(ii) the depreciation is set at the rate of 20 % drop in value per year.

Solution
(i) After four years, total depreciation is

$$4 \times £10\,000 = £40\,000$$

Salvage value is given by

$$\text{initial value} - \text{depreciation}$$
$$= £60\,000 - £40\,000 = £20\,000$$

(ii) In tabular form, the salvage value may be calculated:

Year	Value at start of year (£)	20% depreciation (£)	Residual value (£)
1	60 000	12 000	48 000
2	48 000	9 600	38 400
3	38 400	7 680	30 720
4	30 720	6 144	24 576

Therefore salvage value = £24 576

The salvage value could be found more quickly by applying the following relation:

$$\text{salvage value} = \text{initial capital} \times (1 - I/100)^N \qquad [2.1]$$

(where I = depreciation rate (%) and N = the number of years of use).

The salvage value is a factor which will be used later in the financial assessment of project proposals.

Fuel Costs

The specific value of the cost of a fuel is difficult to state for a number of reasons.

(1) The industrial user is in a powerful position as a large consumer of a particular fuel because he is able to negotiate a price for a fuel which is considerably below that charged to the domestic consumer. Natural gas tariffs are an example of this. A more detailed discussion of fuel tariffs is presented in Chapter 9.

(2) The method of extraction of fuels changes their price. This is particularly true of coal and gas. For example, the price of open-cast coal is much less that of the British deep-mined coal.

(3) The market demand for fuels alters their price, instance the considerable fall in fuel oil prices in the later 1980s. This had a knock-on effect on the price of natural gas for reasons described in (1) above and detailed in Chapter 9.

Notwithstanding the previous comments, Table 2.1 shows typical fuel prices which apply to manufacturing industry at the time of publication. The prices are tabulated in the units in which they are normally sold (e.g. p/kW h for electricity, p/therm for gas, p/l for fuel oils and £/tonne for coal). For comparison purposes, the prices are converted to cost per unit energy. The prices (tariffs) of fuels are an important aspect of energy management and are discussed later in Chapter 9. At this stage it is sufficient to say that a variety of tariffs are available and discussion with the appropriate supplier is likely to lead to an adjustment of the tariff.

Fuel costs are a major factor in calculating the running costs of schemes and the viability of proposed schemes. In the following example two schemes are compared on the basis of fuel costs only.

Table 2.1 Typical fuel prices for manufacturing industry

Fuel	Price (tariff) In units as sold	Per unit energy (£/GW)
Electricity	5 p/kW h	13.89
Natural gas	36 p/therm	3.41
Fuel oil (class DCI)	11 p/litre	2.66
Coal (semi-bituminous)	£65/tonne	1.89

Example 2.3

Gearin plc uses 20 000 MW h of electricity and 900 000 therms of gas per year to satisfy the power and heating requirements of the manufacturing processes. At the present time, the company purchases the electricity and uses a gas-fired boiler to generate the heat. The company is contemplating installing a diesel engine to generate both the power and the heat. It is estimated that the diesel engine will consume 5 million litres of light fuel oil per year. Calculate the fuel costs of the two systems, using the prices in Table 2.1.

Solution

PRESENT SYSTEM: ANNUAL COSTS

$$\text{electricity cost} = \text{number of kW h} \times (\text{cost/kW h})$$
$$= (20\,000 \times 1000) \times 0.05 = £1.0 \text{ M}$$

$$\text{gas cost} = \text{number of therms} \times (\text{cost/therm})$$
$$= 900\,000 \times 0.36 = £0.324 \text{ M}$$

Therefore

$$\text{total fuel cost} = £1.324 \text{ M}$$

DIESEL ENGINE SCHEME: ANNUAL COST

$$\text{fuel oil cost} = \text{number of litres} \times (\text{cost/litre})$$
$$= 5 \times 10^6 \times 0.11 = £0.55 \text{ M}$$

Therefore

$$\text{total savings in fuel cost} = £(1.324 - 0.55) \text{ M}$$
$$= £0.774 \text{ M}$$

These figures indicate that the proposed scheme will save money on fuel costs. Many other costs, particularly maintenance, have not been included. However, it can be imagined that a calculation of this type will be undertaken by the company when new schemes are being planned; the savings will be compared to the initial investment required. More detailed analyses of how this comparison is carried out are shown later in this chapter.

Other Costs

There are a number of other costs relating to manufacturing industries which bear on the operational costs of projects. In terms of manufacturing costs, there are direct costs due to raw materials, labour, energy costs other than fuels, and maintenance etc. Indirect costs such as storage, rates and rent etc. also apply. There are also general overhead costs due to administration and distribution activities. These are not considered in any detail here as they rarely affect the major costings of energy-saving schemes.

General Comments About Costs

It is traditional to classify costs as either *fixed* costs or *variable* costs. Fixed costs are those which do not change with the output of the plant; for example, depreciation and interest charges on plant, site and plant costs such as rent, rates and insurance. Variable costs are those which vary directly with the output of the plant; for example, fuel, maintenance and labour costs.

The total cost of operating a scheme will be the sum of fixed and variable costs. This total will vary with the output of the plant. It is sometimes possible to make a comparison of, for example, two energy supply schemes by calculating how the total costs vary with the level of supply.

Example 2.4

A company requires 1000 kW of electrical power and is contemplating the purchase of a diesel engine as an alternative to its present system of buying from the Electricity Board. The depreciation cost of the engine is £56 000 per year and the generating cost of 1 kW h of energy is 2.8 p. The electricity tariff for bought-in power is 4.2 p/kW h. Calculate the minimum number of hours per year that are required to make the diesel engine a worthwhile investment.

Solution

FOR BOUGHT-IN POWER

$$\text{annual cost} = \text{power(kW)} \times \text{tariff } (\pounds/\text{kW h}) \times \text{hours per year (h)}$$
$$= \pounds(1000 \times 0.042 \times n) \qquad [1]$$

(where n is the number of hours per year).

FOR DIESEL ENGINE SCHEME

$$\text{annual cost} = \text{depreciation(fixed cost)} + \text{operating cost(variable cost)}$$
$$= \pounds 56\,000$$
$$+ \text{power required (kW)}$$
$$\times \text{operating cost}(\pounds/\text{kW h}) \times n$$
$$= \pounds 56\,000 + \pounds(1000 \times 0.028 \times n) \qquad [2]$$

By equating the costs of Eqs [1] and [2], the minimum number of hours required to make the diesel engine a worthwhile investment can be found:

i.e. $\qquad 42n = (56\,000 + 28n)$

giving $\qquad n = 4000$ hours

Since there are 8760 hours in a year this level of operation represents approximately an 11 hour operation for every day of the year.

It is of value to show how Eqs [1] and [2] appear in graphical form by plotting costs against hours per year of operation for the two schemes. Figure 2.1 shows the relationships and confirms the minimum operating level required. This level is represented by what is termed the *break-even point*; below this point it is cheaper to buy the power from the Electricity Board.

Another important point illustrated by Fig. 2.1 is that any future energy-saving measure applied to the diesel engine scheme could only be directed at the variable costs and so must be considered in relation to the unchanging fixed costs. This point is discussed in more detail in the next section.

Figure 2.1 Cost comparison of two schemes

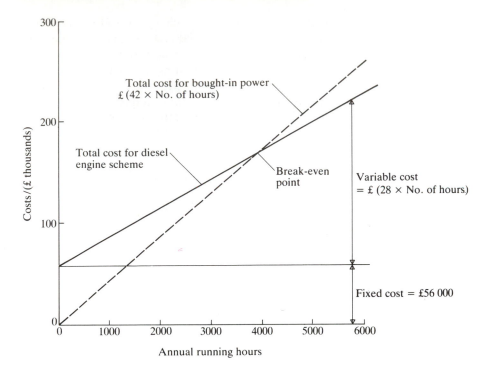

2.2 INVESTING IN NEW ENERGY-SAVING PROJECTS

The remainder of this chapter is concerned with one question: what is the most effective method of investing capital in energy-saving projects?

The basic requirement for such a method is that it should summarize all the cost estimates relating to a project and ideally give a single figure which measures its acceptability, so that the Energy Manager can select a course of action. There are, in fact, a large number of methods which claim to give consistent and meaningful results. The methods in common use are the traditional Accounting Rate of Return (ARR), the payback time, and two Discounted Cash Flow (DCF) methods based on interest rates: the Net Present Value (NPV) and the Internal Rate of Return (IRR).

Accounting Rate of Return (ARR)

This technique is often called the Rate of Return (RR) method. ARR is defined as follows:

$$ARR = \frac{\text{average net annual savings (after depreciation)}}{\text{capital cost}} \qquad [2.2]$$

The basic idea of this method is that it should indicate to the Energy Manager what level of return will be obtained by the investment of capital into energy-saving schemes.

Example 2.5

A company is considering investing £12 000 in energy-saving measures. The Energy Manager has prepared details of three schemes which require £12 000 of capital and which generate the patterns of net savings (after depreciation) shown in the table. (In this discussion, net savings will be taken as the monetary value of savings less depreciation and interest charges.) Calculate the value of ARR for each scheme.

	Project 1	Project 2	Project 3
Capital cost (£)	12 000	12 000	12 000
Net savings after depreciation (£)			
for Year 1	1000	1600	400
Year 2	1000	1400	600
Year 3	1000	1200	800
Year 4	1000	800	1200
Year 5	1000	600	1400
Year 6	1000	400	1600
Total net savings	6000	6000	6000
Average (6 years)	1000	1000	1000
ARR (as a %)	8.3	8.3	8.3

Solution

It is not the intention here to discuss whether or not the value of 8.3 % is an acceptable rate of return. Clearly the higher the rate of return, the more attractive the investment. The example details the basic method and hence illustrates the advantages and disadvantages of the method.

(1) The method is simple to apply and gives a quick indication of the scale of profit.

(2) From the figures it can be seen that the ARR is the same value for all three projects, in spite of the fact that the timing of savings is quite different. Project 2, for example, generates higher savings than the others in the early years of the project; this would be an advantage if further investment was needed for other energy-saving measures.

(3) The fact that the ARR ignores the timing of net savings is a definite flaw in the method and suggests that ARR should be used as a rough indicator only, preparatory to using a more detailed technique.

Payback

The definition of the payback period is: the length of time required for the running total of net savings before depreciation to equal the capital cost of the project. The basic idea is that the shorter the payback time the more attractive the investment.

Example 2.6

Compare the payback times of three projects each with a capital outlay of £24 000, which is invested at the start of the project, and net savings as shown in the table below.

		Project 1		Project 2		Project 3	
Capital cost (£)		24 000		24 000		24 000	
Year		Net annual savings (£)	Running total (£)	Net annual savings (£)	Running total (£)	Net annual savings (£)	Running total (£)
0		−24 000	−24 000	−24 000	−24 000	−24 000	−24 000
1		+6 000	−18 000	+7 200	−16 800	+4 800	−19 200
2		+6 000	−12 000	+6 800	−10 000	+5 200	−14 000
3		+6 000	−6 000	+6 400	−3 600	+5 600	−8 400
4		+6 000	0	+5 600	+2 000	+6 400	−2 000
5		—	—	—	—	+6 800	+4 800
6		—	—	—	—	—	—

Solution

In the table above, year 0 refers to the time of starting the project, that is the present time. For each project, the table has a column of figures showing net savings and a column showing the running total; the latter is formed by adding the net savings for a given year to the previous year's running total. From the above table it can be seen that the payback times are:

exactly 4 years for Project 1
about 3.6 years (by interpolation) for Project 2
about 4.3 years (by interpolation) for Project 3

The example details the basic technique and hence illustrates the advantages and disadvantages of the method.

(1) The method is simple to apply and favours projects with a short payback time, which reduces the uncertainty of calculating savings for periods a long time in the future. The effects of changing technology and fuel prices for example are then reduced.

(2) By its nature, the method does not consider savings produced after the payback time, and therefore does not assess the overall value of the project.

(3) The payback method does not indicate a rate of return on the money invested.

(4) Unlike the ARR method, the payback method does take some account of the timing of the net savings, although the same payback time would be obtained for Project 2 if the years 1 to 3 savings were interchanged.

Discounted Cash Flow (DCF) Methods

A feature of the ARR and payback methods of appraisal is that they both fail to allow sufficiently for the timing of savings. It is worth repeating that a project which generates higher savings in the early years facilitates further investment in other schemes; the DCF methods try to 'weight' the value of savings to reflect this point.

DCF methods are based on interest rates; for a rate of 8 % for example, a deposit of £100 in a bank will accrue £8 of interest in the first year and the account will therefore be worth £108 at the end of year 1. Another way of looking at the figures would be to say that at an interest rate of 8 %, a figure of £108 in one year's time would be worth £100 now, at the present time.

Using the same logic, if the £108 had been left in the bank to accrue another year of 8 % interest, the account would be worth £116.64 at the end of year 2. Again, it could be said that at an interest rate of 8 %, a figure of £116.64 in two years' time would have a present value of £100. Expressing this idea in a general way; the value of an initial deposit D in N years' time will be:

$$S = D \times (1 + I/100)^N \qquad [2.3]$$

where I = interest rate.

It therefore follows that the Present Value (PV) of an amount of money S saved in year N will be:

$$PV = S/(1 + I/100)^N \qquad [2.4]$$

Example 2.7
A project generates cash flows of £1500 in both years 2 and 5 of its operation. Calculate the present value of these savings for an interest rate of 15 %.

Solution
Using Eq. 2.4

$$PV = £1500/(1 + 0.15)^2 = £1134.22$$

for year 2 and

$$PV = £1500/(1 + 0.15)^5 = £745.77$$

for year 5.

If we extend this idea to a complete project which generates savings over a sequence of years then for each year there will be a factor $1/(1 + \text{interest rate}/100)^N$ which relates the savings at year N to a present value. This factor is called a *discount factor* and its effect is to reduce the value of savings achieved in the later years of a project life. Table 2.2 gives discount factors for various

Table 2.2 Discount factors

Year	8 %	10 %	12 %	14 %	16 %	18 %	20 %	22 %
	Interest rates (or discount rates)							
0	1.000	1.000	1.000	1.000	1.000	1.000	1.000	1.000
1	0.926	0.909	0.893	0.877	0.862	0.847	0.833	0.820
2	0.857	0.826	0.797	0.769	0.743	0.718	0.694	0.672
3	0.794	0.751	0.712	0.675	0.641	0.609	0.579	0.551
4	0.735	0.683	0.636	0.592	0.552	0.516	0.482	0.451
5	0.681	0.623	0.567	0.519	0.476	0.437	0.402	0.370
6	0.630	0.564	0.507	0.456	0.410	0.370	0.335	0.303
7	0.583	0.513	0.452	0.400	0.354	0.314	0.279	0.249
8	0.540	0.467	0.404	0.351	0.305	0.266	0.233	0.204
9	0.500	0.424	0.361	0.308	0.263	0.225	0.194	0.167
10	0.463	0.386	0.322	0.270	0.227	0.191	0.162	0.137

interest rates (usually called discount rates) for a 10 year span. (Remember that year 0 is the present time and therefore will always have a discount factor of 1.0.) This basic idea of 'discounting' cash flows of future years to a 'present value' leads directly to a technique of project appraisal known as the Net Present Value method, which overcomes the drawbacks of the ARR and payback methods.

Net Present Value (NPV) Method

This method involves calculating the present value of all yearly capital costs and net savings throughout the life of a project. By summing all these present values (costs being represented as negative amounts and net savings as positive) a total will be obtained; this total is called the NPV of the project. If the NPV is negative then the project will be rejected. If the NPV is positive then the project would not automatically be accepted; other factors, which are discussed later in the chapter, also affect the decision. Of course, such a procedure will require a discount rate to be specified to carry out the calculation. In practice, the value of the discount rate is set slightly above that of the interest rate at which capital has been borrowed to fund the project. In this way, the uncertainties in estimating the net savings of the projects are offset by using a slightly pessimistic discount rate.

The following example illustrates the NPV method and how the application of discounting reveals one project to be more attractive than the others.

Example 2.8
The table below shows the capital costs and net savings of three projects. For a discount rate of 12 %, calculate the NPV of each project.

		Project 1	Project 2	Project 3
Capital cost (£)		12 000	12 000	16 000
Year		Net annual savings (£)	Net annual savings (£)	Net annual savings (£)
1		+3000	+3600	+3500
2		+3000	+3400	+3750
3		+3000	+3200	+4000
4		+3000	+2800	+4250
5		+3000	+2600	+4500
6		+3000	+2400	+4750

Solution
The discount factors for a rate of 12% are taken from Table 2.2 on page 15. The columns of present values in the table below are obtained by multiplying the net saving by the appropriate discount factor. The NPV is obtained by subtracting the capital cost from the sum of the discounted net savings.

Year	Discount factor for 12% (A)	Project 1		Project 2		Project 3	
		Net savings (£) (B)	PV (£) (A × B)	Net savings (£) (C)	PV (£) (A × C)	Net savings (£) (D)	PV (£) (A × D)
0	1.000	−12 000	−12 000	−12 000	−12 000	−16 000	−16 000
1	0.893	+3 000	+2 679	+3 600	+3 215	+3 500	+3 126
2	0.797	+3 000	+2 391	+3 400	+2 710	+3 750	+2 989
3	0.712	+3 000	+2 136	+3 200	+2 278	+4 000	+2 848
4	0.636	+3 000	+1 908	+2 800	+1 781	+4 250	+2 703
5	0.567	+3 000	+1 701	+2 600	+1 474	+4 500	+2 552
6	0.507	+3 000	+1 521	+2 400	+1 217	+4 750	+2 408
			NPV = +336		NPV = +675		NPV = +626

In ranking order the projects appear:

Project 2   NPV = £675
Project 3   NPV = £626
Project 1   NPV = £336

Project 2 has the highest NPV and would therefore be selected from the three schemes under consideration.

The NPV of a project should be positive for the project to be accepted. The magnitude of the NPV represents extra money made available by the project. For example with Project 2, the NPV of £675 could be regarded as money, available now, if £675 is borrowed against the project surplus. In this sense the NPV is immediately available.

In the above example, two of the projects have the same initial capital investment and the NPV affords a direct comparison of these projects. It is more common for projects to require different capital investments. If two such schemes generate different NPVs how then to decide the more attractive one? Referring back to the table in the solution to Example 2.8, the value of the NPV was obtained by subtracting the capital outlay from the sum of all the discounted net savings. For Project 1, the capital outlay was £12 000, and the sum of all the discounted net savings was £12 336, giving a NPV of £336. The ratio of these two quantities (called the *profitability index*) is useful in comparing projects having different capital costs; the ratio is defined as:

$$\text{profitability index} = \frac{(\text{sum of discounted net savings})}{(\text{capital costs})} \qquad [2.5]$$

Based on profitability index, the ranking order of the projects is:

Project 2   profitability index = (12 675)/(12 000) = 1.056
Project 3   profitability index = (16 626)/(16 000) = 1.039
Project 1   profitability index = (12 336)/(12 000) = 1.028

The higher the index the more attractive the project.

Internal Rate of Return (IRR) Method

The Internal Rate of Return (IRR) is defined as the discount rate which will make the NPV of a project equal to zero. Basically, the IRR can be imagined as the yearly net return on capital invested. (Note that the IRR is not the same as the ARR discussed previously, which ignored discount factors.)

When a company invests in new equipment it will have its own guidelines, based on experience, on the minimum acceptable value of IRR for a project. When the IRR is greater than the target value, then the project is acceptable.

The value of IRR produced by a project can be found by direct calculation or graphical means. The calculation method is identical to that of the NPV method described by systematically repeating the calculation with different discount rates until the NPV is zero. The graphical method requires less calculation and is illustrated by the example below.

Example 2.9

Using the cash flow data of Project 2 in Example 2.8, calculate the basic NPV for discount rates of 10, 12, and 15 %. Draw a graph of NPV against discount rate and hence deduce the IRR for the project.

Solution

The discount factors for the rates of 10, 12 and 15 % are taken from Table 2.2. The columns of present values in the table below are obtained by multiplying the net saving by the appropriate discount factor. The NPV is obtained by subtracting the capital cost from the sum of the discounted net savings.

Project 2

Year	Cash flow (£)	10 % discount rate		12 % discount rate		15 % discount rate	
		Discount factor	Present value (£)	Discount factor	Present value (£)	Discount factor	Present value (£)
0	−12 000	1.000	−12 000	1.000	−12 000	1.000	−12 000
1	+3 600	0.909	+3 272	0.893	+3 215	0.870	+3 132
2	+3 400	0.826	+2 808	0.797	+2 710	0.756	+2 570
3	+3 200	0.751	+2 403	0.712	+2 278	0.658	+2 106
4	+2 800	0.683	+1 912	0.636	+1 781	0.572	+1 602
5	+2 600	0.623	+1 620	0.567	+1 474	0.497	+1 292
6	+2 400	0.564	+1 354	0.507	+1 217	0.432	+1 037
			NPV = +1 369		NPV = +675		NPV = −261

Plotting these points on a graph of NPV against discount rate as shown in Fig. 2.2 reveals that the NPV is zero at a discount rate of 14.1 %.

If this calculation is repeated for the other two projects the following IRR values are obtained:

Project 2　14.1 %
Project 3　13.3 %
Project 1　13.0 %

This is the same ranking order as the NPV calculation. In general the NPV and IRR methods lead to the same acceptance or rejection decision. If two

Figure 2.2 NPV variation with discount rate

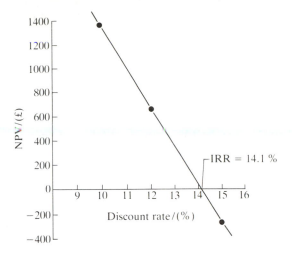

projects have the same values for both the NPV and IRR then the profitability index will indicate the better investment.

The example illustrates the basic method. The viability of the project depends on the target value of IRR set by the Energy Manager. If the calculated IRR exceeds the target value then the scheme is likely to be accepted. Nevertheless, the technique requires assumptions to be made that are not always true:

(a) that the target IRR is always known by the Energy Manager;
(b) that inflation can be ignored;
(c) that discount rates are constant over the life of the project.

Other Factors Affecting Project Appraisal

There are a number of practical considerations which affect the investment in energy-saving projects.

(1) There are outside bodies who will contribute towards the investment cost of a scheme; for example, central government and regional development grants. These contributions may not necessarily be given at the start of a project but can be considered as working capital to be used in later years of the project life.

(2) There are also tax matters to consider; not only tax on net savings but tax incentives which will increase the attractiveness of energy-saving schemes.

(3) Energy prices vary, sometimes unpredictably, over the life of the project. A recent example of this was the dramatic fall in the price of fuel oils which occurred in the late 1980s. Such a fall caused many companies simply to change fuel rather than invest in new measures. (This point and the general management of fuel tariffs to reduce energy costs is discussed in detail in Chapter 9.)

(4) Inflation rates have a direct bearing on the discount factor required for NPV and IRR calculations.

The following examples illustrate the inclusion of these factors when the project is assessed by payback, NPV and IRR methods.

Example 2.10

Glancy plc is planning a heat-recovery scheme for its brick-making kilns. The initial investment is estimated at £30 000. Further investment, depending on the proven success of the scheme, is planned for year 2 of the project; the amount is £20 000 which will be contributed by a regional development grant. The company has a target of 23 % for the IRR.

By comparing the running costs of the present system with those using heat recovery the following savings are estimated for a project life of five years:

Year	1	2	3	4	5
Savings (£)	9 000	6 000	12 000	8 500	5 500

After discussions with the Inland Revenue, the taxation for the proposed scheme would have the following pattern:

(i) Tax = 45 % of net savings. (This is called Corporation Tax.) The tax is to be paid at the end of each year of operation.
(ii) The tax incentive will be a 50 % first year allowance. This means that 50 % of the tax on the amount of the capital investment can be offset against tax paid on the project. The amount is calculated as follows: at 45 % corporation tax, tax on the capital investment amount is

$$0.45 \times £30\,000 = £13\,500$$

Therefore tax incentive is

$$50\,\% \times £13\,500 = £6750$$

This amount is considered as money generated by the heat recovery scheme, because the incentive will not be allowed unless the heat recovery scheme is actually realized. In practice, the £6750 will be an incentive which will equally apply to any of the company's activities in that it represents a general allowance against tax.

Depreciation will be considered over the five-year project life as £3500/year and the salvage value in year 5 will therefore be £2500. This will be considered as a saving in year 5.

NPV ASSESSMENT The NPV is calculated in the table at the top of page 21. The discount rate for the calculation is taken as 12 %, based on the prevailing rates of interest, and no inflation.

IRR ASSESSMENT The value for the IRR is found by graphical/iterative means to be 21.2 % using the method of Example 2.9. Since this value is somewhat below the target figure 23 % then the project is likely to be rejected by the company.

(The payback time can be found by making a running total of the 'net after tax' column and finding the year at which the total becomes zero. By the above figures the payback time is just under two years.)

Year	Capital investment (£) (A)	Net savings (£) (B)	Tax incentive (£) (C)	Tax at 45% (£) (D)	Additional grant (£) (E)	Net after tax (£) (F)	Discount factor (12%) (G)	Present value (£) (H)
0	-30000					-30000	1.000	-30000
1		9000	6750	-1013		7987	0.893	7132
2		6000		-2700	20000	23300	0.797	18570
3		12000		-5400		6600	0.712	4699
4		8500		-3825		4675	0.636	2973
5	2500	5500		-3600		4400	0.567	2495
								5869

∴   Net present value = £5 869

The column items A to H signify the following:

A   £ − 30 000 represents the initial capital investment. £2500 is the salvage value
B   Net savings on new scheme compared to existing scheme
C   Tax incentive (or allowance) for year 1 only
D   Tax for year 1 onwards = $-(A + B - C) \times 0.45$
E   Additional untaxable grant of £20 000 for year 2 only
F   The residual after tax, which equals $(A + B + D + E)$
G   A column of discount factors (for 12% in this case)
H   The PV, which equals $(F \times G)$

It is quite common for companies to arrange tax to be paid one year in arrears. If this situation is applied to this project then the figures in the column 'tax at 45%' will be displaced downwards in the table by one year and a year 6 row will appear; all other capital and savings amounts will be the same. The table will appear as below.

NEW IRR ASSESSMENT   The value for the IRR based on tax being paid in arrears is found to be 24.3%. Since the value is in excess of the target figure of 23% the project is likely to be accepted by the company. It is worth noting that the tax incentive is an important contribution to the viability of the scheme.

Year	Capital investment (£)	Net savings (£)	Tax incentive (£)	Tax at 45% (£)	Additional grant (£)	Net after tax (£)	Discount factor (12%)	Present value (£)
0	-30000					-30000	1.000	-30000
1		9000	6750			9000	0.893	8037
2		6000		-1013	20000	24987	0.797	19915
3		12000		-2700		9300	0.712	6622
4		8500		-5400		3100	0.636	1972
5	2500	5500		-3825		4175	0.567	2367
6				-3600		-3600	0.507	-1825
								7088

∴   Net present value = £7 088

Inflation

Another factor which has a direct bearing on the calculations of project viability is the rate of inflation. For application to energy-saving projects the rate of inflation is defined as the rate of increase in the *average* price of goods and services (for example: fuel prices, wages and maintenance costs, equipment costs and installation costs). The rate of inflation is formally expressed as the *Retail Price Index* (RPI), which in the UK is evaluated and published monthly by central government.

Inflation causes the value of money to decrease with time. For example, when prices are rising by about 10 % a year and the money rate of interest is 15 %, then the real rate of interest is 5 %. In other words, the *real* rate of interest is the rate of interest that would be paid if prices were constant.

The object of expressing something in real terms is, then, to adjust for the effect of rising prices throughout the economic system. To differentiate between actual money and money with constant purchasing power, the economists refer to the latter as money in *real terms*. When the annual rate of inflation is 10 % the *real* purchasing power of £100 received in one year's time is £90.91. This is equivalent to another form of discounting which must be clearly differentiated from the time value of money effect. There is no universally accepted way of accommodating the effects of inflation when discounting the money value of the net savings, but most current practice uses the RPI as the 'discount factor for inflation'. The *real* value of a sum of money S realized N years hence is given by the expression:

$$\text{real value} = S/(1 + \text{RPI})^N \tag{2.6}$$

To summarize: money which is saved by the implementation of energy-saving measures has a certain cash value, but if this money is converted as defined in Eq. [2.6] then the cash value is in *real* terms.

This concept can now be used to assess the effects of inflation on the economic viability of a scheme requiring capital investment by:

(a) discounting the *real* values of the net savings using a discount factor which is based on the *real* rate of interest;
(b) discounting the actual values of the net savings with a discount factor which is based on an interest rate which includes an allowance for inflation. In this case the discount rate R is the combined effect of the real rate of interest I_R and the overall rate of inflation as measured by the RPI, using the relationship:

$$(1 + R) = (1 + I_R)(1 + \text{RPI}) \tag{2.7}$$

Example 2.11

A company plans to invest £30 000 in an energy-saving project whose life will be taken as four years. It is estimated that the savings and costs of the project will be as follows:

energy savings = £30 000 in the first year escalating by 5 % per year
labour costs = £10 000 in the first year escalating by 8 % per year
material and other costs = £8000 in the first year escalating by 10 % per year

Take the RPI as currently 7 %, and the *real* rate of interest as 8 %.

Solution
First evaluate the net savings of the project by subtracting the costs from the energy savings, allowing for the escalation in value of savings and costs.

Year	Capital costs (£) (A)	Energy savings (£) (B)	Labour costs (£) (C)	Material and other costs (£) (D)	Actual money value of savings (£) (E)
0	− 30 000				− 30 000
1		+ 30 000	− 10 000	− 8 000	+ 12 000
2		+ 31 500	− 10 800	− 8 800	+ 11 900
3		+ 33 075	− 11 664	− 9 680	+ 11 731
4		+ 34 729	− 12 597	− 10 648	+ 11 484

The column items A to E signify the following:

A   £ − 30 000 represents the initial capital investment
B   Energy saving: escalation factor 1.05
C   Labour costs: escalation factor 1.08
D   Material/other costs: escalation factor 1.10
E   Net savings = (A + B + C + D)

METHOD 1 – NPV ANALYSIS IN REAL TERMS   By discounting the *real* values of the net savings using a discount factor which is based on the *real* rate of interest, the following table may be completed.

Year	Capital (£) (A)	Net savings (money terms) (£) (B)	Inflation deflator (£) (C)	Net savings (real terms) (£) (D)	Discount factor (real terms, 8 %) (E)	Present value (real terms) (£) (E)
0	− 30 000		1	− 30 000	1	− 30 000
1		+ 12 000	0.935	+ 11 220	0.926	+ 10 390
2		+ 11 900	0.873	+ 10 389	0.857	+ 8 903
3		+ 11 731	0.816	+ 9 572	0.794	+ 7 600
4		+ 11 484	0.763	+ 8 763	0.735	+ 6 440
						+ 3 333

∴   NPV = £ 3 333

The columns A to F signify the following:

A   £ − 30 000 is the initial capital
B   Column E from previous table
C   The factor = $1/(1 + RPI)^N$, where RPI = 0.07
D   Application of Eq. 2.6. (D = C × A) or (D = C × B)
E   The factor = $1/(1 + \text{real interest rate})^N$, where real interest rate = 0.08; see also Table 2.2, (page 15)
F   F = D × E; F is the present value in real terms

METHOD 2 – NPV ANALYSIS IN MONEY TERMS   This method works by discounting the actual values of the net savings with a discount factor which is based on an interest rate which includes an allowance for inflation. In this case the discount rate R is the combined effect of the real rate of interest I_R, and the overall rate of inflation as measured by the RPI. Using Eq. [2.7]:

$$(1 + R) = (1 + I_R)(1 + RPI)$$
$$= (1 + 0.08)(1 + 0.07) = 1.1556$$

i.e. $R = 0.1556 \, (15.56 \, \%)$

The NPV is calculated in the following table.

Year	Capital (£) (A)	Net savings (money terms) (£) (B)	Discount factor (15.56%) (C)	Present value (money terms) (£) (D)
0	−30 000		1.000	−30 000
1		+12 000	0.865	+10 830
2		+11 900	0.749	+8 913
3		+11 731	0.648	+7 602
4		+11 484	0.561	+6 440
				+3 338

∴   NPV = £3 338

The columns A to D signify the following:

A   £ − 30 000 is the initial capital
B   Column B from previous table
C   Discount Factor from Table 2.2, (page 15)
D   (D = C × A) or (D = C × B)

METHOD 3 – NPV ANALYSIS ASSUMING NO INFLATION   The costs, savings and discount factors will now ignore the effects of inflation. The following figures will apply:

$$\text{energy saving for each year} = £30\,000$$
$$\text{labour cost for each year} = £10\,000$$
$$\text{total material and other costs} = \underline{£8\,000}$$
$$\therefore \quad \text{net savings} = \underline{£12\,000}$$
$$\text{capital} = £30\,000$$
$$\text{discount rate} = 8\,\%$$

The NPV is calculated in the following table.

Year	Capital (£) (A)	Net savings (£) (B)	Discount factor (8.0%) (C)	Present value (£) (D)
0	−30 000		1	−30 000
1		+12 000	0.962	+11 112
2		+12 000	0.857	+10 284
3		+12 000	0.794	+9 528
4		+12 000	0.735	+8 820
				+9 744

∴   NPV = £9 744

The columns A to D signify the following:

A £ − 30 000 is the initial capital
B Data from start of method 3
C Discount Factor from Table 2.2, (page 15)
D (D = C × A) or (D = C × B)

This example shows the application of two methods of allowing for the effect of inflation on capital investment appraisal. The two methods provide the same answer for the NPV (allowing for numerical rounding-up). The example shows how the NPV will be overestimated if the inflation is ignored.

2.3 SUMMARY

This chapter has described the type of costs relating to the economic assessment of energy-saving projects and the methods by which the viability of such projects can be evaluated.

Of the four methods described the following can be said:

(1) ACCOUNTING RATE OF RETURN (ARR)   Simple method which takes no account of the timing of net savings.

(2) PAYBACK   Simple method which takes little account of the timing of net savings but useful as a first estimate. Satisfactory for rating projects of short duration ($<2\frac{1}{2}$ years).

(3) NET PRESENT VALUE (NPV)   More complicated method but it does allow for the timing of savings and provides an assessment for long-term projects.

(4) INTERNAL RATE OF RETURN (IRR)   More complicated method, but it does permit a comparison of a new project with a company's established projects by indicating a 'return' on the capital invested.

Factors such as taxation, tax incentives, capital grants and inflation should be included in the calculations as they can significantly affect the investment appraisal of energy-saving projects.

Since investments in energy-saving equipment are usually expected to produce savings for a number of years, the costs on which these savings were estimated will extend some distance into the future. The farther ahead the estimates are made, the more uncertain they become: the estimates are better considered as rough approximations rather than precise numbers.

Problems

2.1   A company is investing in a computerized energy management system which will cost £50 000 to purchase and install. The money is loaned by a bank which charges simple interest at the annual rate of 15 % for a period of two years. Calculate the total repayment required by the bank.

(£65 000)

2.2   An engine is installed to supply on-site power to a hospital. After five years, the plant layout is changed and the engine is no longer required. The initial cost of the installed engine is £90 000. Calculate the salvage value of the engine set if

(i)   the depreciation set at £15 000 per year;
(ii) the depreciation is set at the rate of 15 % drop in value per year.

((i) £15 000; (ii) £39 933)

2.3   A company uses 30 000 MW h of electricity and 700 000 therms of gas per year to satisfy the power and heating requirements of the manufacturing processes. At the present time, the company purchases the electricity and uses a gas-fired boiler to generate the heat. The company is planning to install a diesel engine to generate both the power and the heat. It is estimated that the diesel engine will consume 8 million litres of fuel oil per year.

Calculate the annual saving in fuel cost of the diesel engine scheme, using the following fuel tariffs:

electricity = 4.6 p/kW h
gas = 28 p/therm
fuel oil = 7 p/litre

(£1.016 million)

2.4   A company purchases 2 MW of electrical power and is contemplating the purchase of a gas turbine as an alternative to its present system of buying from the electricity board. The depreciation cost of the gas turbine is £200 000 per year and the cost of generating 1 kW h of energy is 2.6 p. The electricity tariff for bought-in power is 4.4 p/kW h. Calculate the minimum number of hours per year that are required to make the gas turbine a worthwhile investment.

(5556 hours)

2.5   A company is considering investing £8000 in energy-saving measures. The energy manager has prepared details of three schemes which require £8000 of capital and which generate the patterns of net savings (after depreciation) as shown in the following table.

	Project 1	Project 2	Project 3
Capital cost (£)	8 000	8 000	8 000
Net savings after depreciation (£)			
for Year 1	2000	2500	1500
Year 2	2000	2250	1750
Year 3	2000	2000	2000
Year 4	2000	1750	2250
Year 5	2000	1500	2500

For each scheme, calculate the value of ARR and the payback time.

(Project 1: ARR = 25 %, Payback Time = 4.0 years)
(Project 2: ARR = 25 %, Payback Time = 3.7 years)
(Project 3: ARR = 25 %, Payback Time = 4.2 years)

2.6 The following table shows the capital costs and net savings of two projects.
(i) For a discount rate of 16 %, calculate the NPV and Profitability Index of each project.
(ii) Calculate the IRR of each project.

(Project 1: NPV = £8266, PI = 1.413, IRR = 29.15 %)
(Project 2: NPV = £9444, PI = 1.472, IRR = 32.33 %)

Year	Project 1	Project 2
Capital cost (£)	20 000	20 000
	Net annual savings (£)	Net annual savings (£)
1	7000	8500
2	7000	8000
3	7000	7500
4	7000	7000
5	7000	6500
6	7000	6000
7	7000	5500

2.7 A chemical company is planning a heat-recovery scheme for its solvent-producing plant. The initial investment required is estimated to be £70 000. Further investment, from a regional development grant, is planned for year 3 of the project; the amount is £45 000. Using the following data, calculate the IRR of the scheme.

Data
Salvage value of plant in year 4, £10 000; tax incentives, nil; project life, four years.
    Corporation tax (paid at the end of each year of operation), 40 %.
Projected net savings are shown in the following table.

Year	1	2	3	4
Net Savings (£)	20 000	36 000	22 000	19 500

(18.14 %)

Bibliography

Anthony R N, Welsch G A 1977 *Fundamentals of Management Accounting* Revised edn Richard D Irwin Inc
de la Mare R F 1982 *Manufacturing Systems Economics* Holt, Reinhart and Winston
Digest of United Kingdom Energy Statistics HMSO 1987
Lucey T 1982 *Quantitative Methods* DP Publications
Wright M G 1973 *Discounted Cash Flow* 2nd edn McGraw-Hill

3 THEORETICAL BACKGROUND

The reader of this book should have a good understanding of the broad principles of thermodynamics and electricity; the treatment of these subjects in text-books such as references 3.1 and 3.2 provides an adequate basis.

To be able to analyse complex total energy systems, considered in later chapters, it is necessary to have a working knowledge of the laws of thermodynamics, the use of vapour tables, and the various cycles for steam, gas turbine and refrigeration plant. It is also necessary to understand the laws of gas mixtures and psychrometry. Students should be studying first courses in thermodynamics and electricity alongside a course in energy studies or should have completed such courses before starting their further work on energy. Existing Energy Managers or other readers may find it necessary to do some revision using Eastop and McConkey[3.1] or Hughes[3.2] before using this book. Problems at the end of the chapter may be used to check whether the understanding of the subject matter is of a sufficient level to proceed to a study of the chapters which follow.

The topic of combustion is central to the conversion of the chemical energy of primary fuels into thermal energy; this is not always covered in the required detail in a first course in thermodynamics and hence a section on the first principles of the combustion of fuels is included in this chapter as a basis for further work in Chapter 4.

Similarly, the principles of heat transfer are sometimes left until later in a course; also, specific empirical methods for heat exchanger design are required for energy recovery systems and these are covered concisely in this chapter to be referred to later in Chapter 5.

3.1 COMBUSTION THEORY

Chemical elements consist of atoms which may be defined as the smallest particles able to take part in a chemical change. The smallest particle of a substance that can exist separately is known as a molecule and it consists of one or more atoms. For example, two atoms of hydrogen combine with one atom of oxygen to give water, written as H_2O.

The masses of atoms are exceedingly small and hence an agreed relative atomic mass is chosen using an isotope of carbon. For engineering practice these numbers may be rounded up to whole numbers for all elements; Table 3.1 lists some examples.

Table 3.1 Relative atomic mass of some common elements

Element name	Symbol	Relative atomic mass
Oxygen	O	16
Nitrogen	N	14
Hydrogen	H	1
Carbon	C	12
Sulphur	S	32

Similarly a relative molecular mass is defined for substances which exist as molecules, for example hydrogen, H, exists as hydrogen gas, H_2: a list of common substances and their relative molecular mass is given in Table 3.2.

Table 3.2 Relative molecular mass of some common substances

Name of substance	Symbol	Relative molecular mass
Oxygen	O_2	32
Nitrogen	N_2	28
Hydrogen	H_2	2
Water/steam	H_2O	18
Carbon dioxide	CO_2	44
Carbon monoxide	CO	28
Sulphur dioxide	SO_2	64

The molar mass, M, can be taken to be numerically equal to the relative molecular mass for engineering purposes; the molar mass has the units kg/kmol whereas the relative molecular mass is a ratio and is therefore dimensionless.

When certain substances are mixed with oxygen at the correct temperature and pressure and in the correct proportions a chemical reaction takes place in which energy is transferred to the surroundings in the form of heat. The substances taking part in the chemical process are called the reactants and the substances formed are called the products. The process is known as combustion and substances which release energy in this way are called fuels; the chemical reaction is known as an exothermic reaction.

The mass of each element must remain unchanged during the process and hence a kmol balance can be made. For example for the combustion of hydrogen gas:

$$H_2 + 0.5O_2 \rightarrow H_2O$$

i.e. 2 kmol hydrogen, (H) + 1 kmol oxygen, (O) → 1 kmol water (consisting of 2 kmol hydrogen and 1 kmol oxygen)

$$2 \text{ kg hydrogen} + 16 \text{ kg oxygen} \rightarrow 18 \text{ kg water}$$

A mass balance is used to determine the mass of oxygen required for the combustion of a given mass of fuel.

Example 3.1
Calculate the mass of oxygen required for the combustion of the fuel propane, C_3H_8.

Solution
It is known that carbon, C, will combine with oxygen to form CO_2, and hydrogen will combine with oxygen to form water/steam. The number of kmol of carbon

dioxide and water are shown as the unknown quantities a and b. Since the mass of carbon remains unchanged it follows that $a = 3$, and similarly, $b = 8/2 = 4$. The number of kmol of oxygen in the products is therefore

$$a + b/2 = 3 + 2 = 5$$

and therefore we have

$$C_3H_8 + 5O_2 \rightarrow 3CO_2 + 4H_2O$$

$$\{(3 \times 12) + (8 \times 1)\} \text{ kg } C_3H_8 + (5 \times 32) \text{ kg } O_2$$

$$\rightarrow 3(12 + 32) \text{ kg } CO_2 + 4(2 + 16) \text{ kg } H_2O$$

i.e. mass of oxygen required per kg of fuel is

$$(5 \times 32)/\{(3 \times 12) + 8\} = 3.636 \text{ kg}$$

The most convenient way of supplying oxygen is in the form of air and air can be taken as approximately 79 % nitrogen and 21 % oxygen by volume, or as 76.7 % nitrogen and 23.3 % oxygen by mass. The air–fuel ratio would normally be quoted by volume for gaseous fuels and by mass for solid and liquid fuels. Some fuels can exist in the liquid or gaseous states depending on the temperature.

A mixture of fuel and the minimum amount of air required for the complete combustion is called a *stoichiometric mixture*. In practice, in order to ensure that combustion is complete, an excess of air is usually supplied. Note that for a hydrocarbon fuel (i.e. one consisting entirely of carbon and hydrogen), when there is a deficiency of air supplied the hydrogen will burn completely but the carbon will burn to a mixture of carbon dioxide and carbon monoxide.

The H_2O formed in the combustion of hydrogen will remain in the form of steam unless the temperature of the products is reduced to the saturation temperature corresponding to the partial pressure of the steam. Then we can write for the partial pressure of steam

$$p_S = p(n_S/n) \qquad\qquad [3.1]$$

(where p is the total pressure; n_s is the number of moles of steam in the total volume which contains n moles). From steam tables the corresponding saturation temperature can be found. This means that when the gases are cooled below this saturation temperature the steam will just start to condense. This temperature is known as the *dew point* temperature. When the steam has been completely condensed and drained off the remaining gases are called the *dry flue gases*.

Example 3.2

Gaseous propane is burned with an air–fuel ratio by volume of 25. Calculate:
(i) the volumetric analysis of the products;
(ii) the dew point of the products when the total pressure is 1.08 bar;
(iii) the volumetric analysis of the dry flue gas.

Solution
(i) With an air–fuel ratio of 25 by volume the chemical equation can be written as follows:

$$C_3H_8 + 25(0.21O_2 + 0.79)N_2 \rightarrow 3CO_2 + 4H_2O + aO_2$$
$$+ (25 \times 0.79)N_2$$

The quantity a is the excess oxygen supplied in kmol per kmol of fuel, and can be found from a mass balance.

i.e.
$$25 \times 0.21 = 3 + 4/2 + a$$
$$\therefore \quad a = 0.25$$

To find the volumetric analysis of the gaseous products the mol fraction, n_i/n can be calculated. The total number of kmol of the products is

$$3 + 4 + 0.25 + (25 \times 0.79) = 27 \text{ kmol}$$

Then the volumetric fractions are as follows:

CO_2, $3/27 = 11.11\%$; H_2O, $4/27 = 14.82\%$; O_2, $0.25/27 = 0.93\%$; N_2, $19.75/27 = 73.15\%$

(ii)   Using Eq. [3.1]:

$$p_S = 1.08 \times 0.1482 = 0.16 \text{ bar}$$

From steam tables the corresponding saturation temperature at 0.16 bar is 55.3 °C.

i.e.
$$\text{dew point temperature} = 55.3\,°C$$

(iii)   The volumetric analysis of the dry flue gas is given by:

CO_2, $3/23 = 13.04\%$; O_2, $0.25/23 = 1.09\%$; N_2, $19.75/23 = 85.87\%$

A general equation can be written for combustion problems which makes calculation of air–fuel ratio and product analysis easier. For cases where the fuel is a solid or a liquid hydrocarbon the equation is as follows:

$$1 \text{ kg fuel} + A\{(0.233/32)O_2 + (0.767/28)N_2\}$$
$$\rightarrow B\{aCO_2 + bO_2 + (1 - a - b)N_2\} + cH_2O \qquad [3.2]$$

The air–fuel ratio by mass is given the symbol A and hence the volume of oxygen supplied per kg of fuel is $(A \times 0.233/32)$ kmol, and similarly for nitrogen.

The symbol B represents the kmol dry flue gas per kg of fuel; the symbols a and b represent the volumetric fractions of carbon dioxide and oxygen in the dry flue gas; the symbol c represents the kmol of steam per kg of fuel.

The only combustible element likely to be present in a fuel other than carbon or hydrogen is sulphur which burns to give sulphur dioxide, SO_2. For fuels containing sulphur the above equation can be modified to include a symbol representing the fractional volume of sulphur dioxide in the dry flue gas.

From Eq. [3.2] a mass balance can be drawn up for each element in turn; this will give four equations (or five equations if there is sulphur present). If the composition of the fuel is known there are five unknowns in Eq. [3.2], A, B, a, b, and c, and hence if one of these is known the others can be calculated.

When incomplete combustion occurs then the fraction of carbon monoxide in the dry flue gas introduces an additional unknown.

Example 3.3
The analysis by mass of a dry coal is as follows:

86% C; 3.9% H; 1.4% O; 8.7% incombustible ash

The dry flue gas is found to contain 12.7 % carbon dioxide and 1.4 % carbon monoxide with unknown percentages of oxygen and nitrogen. Calculate:
(i) the percentage excess air;
(ii) the complete analysis by volume of the products.

Solution
(i) Using Eq. [3.2]

$$\{(0.86/12)C + 0.039H + (0.014/16)O\}$$
$$+ A\{(0.233/32)O_2 + (0.767/28)N_2\}$$
$$\rightarrow B\{0.127CO_2 + 0.014CO + aO_2 + (1 - 0.127 - 0.014 - a)N_2\}$$
$$+ bH_2O$$

Then

carbon balance: $(0.86/12) = B(0.127 + 0.014)$
$$\therefore \quad B = 0.5083$$
hydrogen balance: $0.039 = 2b$
$$\therefore \quad b = 0.0195$$
oxygen balance: $(0.014/16) + 2A(0.233/32)$
$$= B\{(2 \times 0.127) + 0.014 + 2a\} + b$$
$$\therefore \quad 0.014\,563A = 1.0166a + 0.154\,85 \qquad [1]$$
nitrogen balance: $A(0.767/28) = B(1 - 0.127 - 0.014 - a)$
$$\therefore \quad 0.027\,39A = 0.436\,63 - 0.5083a \qquad [2]$$

Multiplying Eq. [1] by $(0.5083/1.0166)$ and adding it to Eq. [2] we have:

$$0.007\,28A + 0.027\,39A = 0.077\,43 + 0.436\,63$$
$$\therefore \quad A = 14.83$$

To find the percentage excess air we must first calculate the stoichiometric air–fuel ratio, A_s. Now

$$0.86 \text{ kg carbon require } (0.86 \times 32/12) = 2.2933 \text{ kg of oxygen}$$
$$0.039 \text{ kg hydrogen require } (0.039 \times 0.5 \times 32/3) = 0.312 \text{ kg of oxygen}$$

Therefore, the oxygen required for stoichiometric combustion is

$$(2.2933 + 0.312 - 0.014) = 2.5913 \text{ kg/kg of fuel and}$$
$$A_s = 2.5913/0.233 = 11.12$$

Therefore

$$\text{percentage excess air} = 100(14.83 - 11.12)/11.12 = 33.36\,\%$$

(ii) To obtain the complete volumetric analysis it is necessary to find the number of kmol of each of the products. Substituting the value of A in Eq. [1] we have

$$0.014\,563 \times 14.83 = 1.0166a + 0.154\,85$$
$$\therefore \quad a = 0.0601$$

Then for 1 kg of fuel

CO_2, $0.5083 \times 0.127 = 0.064\,45$ kmol; CO, $0.5083 \times 0.014 = 0.007\,12$ kmol;
O_2, $0.5083 \times 0.0601 = 0.030\,55$ kmol;
N_2, $0.5083(1 - 0.127 - 0.014 - 0.0601) = 0.406\,08$ kmol;
H_2O, 0.0195 kmol

i.e. total kmol of products = 0.5277 kmol/kg of fuel

The complete volumetric analysis is then:

CO_2, 0.064 45/0.5277 = 12.21 %; CO, 0.007 12/0.5277 = 1.35 %;
O_2, 0.030 55/0.5277 = 5.79 %; N_2, 0.406 08/0.5277 = 76.95 %;
H_2O, 0.0195/0.5277 = 3.70 %

For a fuel in the gaseous state Eq. [3.2] can be modified by writing the left-hand side as:

$$1 \text{ kmol of fuel} + A(0.21O_2 + 0.79N_2)$$

where A is now the air–fuel ratio by volume. On the right-hand side of Eq. [3.2], B now represents the number of kmol of the dry flue gas per kmol of fuel, and c is the number of kmol of steam per kmol of fuel.

Internal Energy and Enthalpy of Combustion

The process of combustion of a fuel allows chemical energy to be released as heat to the surroundings. It is found that the energy released during the chemical process is dependent on the temperature of the reactants.

When reactants in a closed vessel at a reference temperature, T_0 are burned at constant volume giving products also at T_0, the heat supplied from the surroundings is given from the non-flow equation as

$$Q = U_{PO} - U_{RO} = \Delta U_0$$

The increase in internal energy, ΔU_0, is called the *internal energy of combustion*. Since heat is transferred *to* the surroundings then the internal energy of combustion is always a negative quantity.

Similarly when combustion takes place in steady flow then if the kinetic energy changes are negligible we have from the flow equation that $Q = \Delta H_0$, where ΔH_0 is known as the *enthalpy of combustion*, and is also negative. Values of the enthalpy of combustion of some of the more common fuels are given in Table 3.3.

Table 3.3 Enthalpy of combustion of some common fuels

Fuel	ΔH_0 at 25 °C (kJ/kmol fuel)
Solid carbon	−393 520
Hydrogen	−241 830 (products in vapour phase)
Solid sulphur	−289 910
Carbon monoxide	−282 990
Methane vapour	−802 310

It is necessary to specify whether the fuel is in a solid, liquid or gaseous state since energy is required to produce the enthalpy of vaporization or fusion. For the same reason it must be stated whether the H_2O in the products is in the liquid or the vapour state.

In practice the reactants will be at T_1, say, and the products will be at T_2, say. Hence the complete process including combustion at the reference

Figure 3.1 $U - T$ diagram for a combustion process

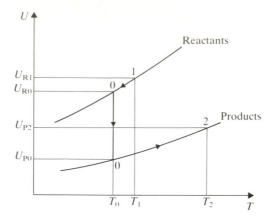

temperature, T_0, is as follows (see Fig. 3.1):

$$Q = U_{P2} - U_{R1} = (U_{P2} - U_{P0}) + (U_{P0} - U_{R0}) + (U_{R0} - U_{R1})$$

i.e.
$$Q = (U_{P2} - U_{P0}) + \Delta U_0 + (U_{R0} - U_{R1})$$

The internal energy of a perfect gas, U, can be expressed as $mc_v T$. The specific heat at constant volume can be expressed per kmol instead of per kg, and the symbol C_v is given to this specific heat which is known as the *molar heat at constant volume*. From the definition it follows that $mc_v = nC_v$.

The reactants and the products can be treated as mixtures of perfect gases and therefore we can write

$$Q = (T_2 - T_0) \sum_{}^{P} m_i c_{vi} + \Delta U_0 + (T_0 - T_1) \sum_{}^{R} m_i c_{vi} \qquad [3.3]$$

where $\sum^{P}$ represents the sum for the products and $\sum^{R}$ represents the sum for the reactants.

If the molar heats are used we have

$$Q = (T_2 - T_0) \sum_{}^{P} n_i C_{vi} + \Delta U_0 + (T_0 - T_1) \sum_{}^{R} n_i C_{vi} \qquad [3.4]$$

For a flow process in a similar way we have

$$Q = (T_2 - T_0) \sum_{}^{P} m_i c_{pi} + \Delta H_0 + (T_0 - T_1) \sum_{}^{R} m_i c_{pi} \qquad [3.5]$$

and
$$Q = (T_2 - T_0) \sum_{}^{P} n_i C_{pi} + \Delta H_0 + (T_0 - T_1) \sum_{}^{R} n_i C_{pi} \qquad [3.6]$$

where C_p is the molar heat at constant pressure.

It is usually a good approximation to take the specific heats of the individual reactants at the arithmetic mean temperature, $(T_1 + T_0)/2$, and to take the specific heats of the individual products at the arithmetic mean temperature, $(T_0 + T_2)/2$.

Example 3.4

A stoichiometric mixture of air and gaseous methane at 54 °C and 2 bar is burned in a rigid vessel of volume 0.1 m³. Assuming complete combustion and using the data below, calculate the heat rejected to the surroundings during combustion.

Data

Internal energy of combustion of gaseous methane at $25\,°C$ with steam in the products, $-802\,310\,kJ/kmol$; molar gas constant, $8.3144\,kJ/kmol\,K$; temperature of products, $1529\,°C$. The relative atomic masses are as given in Table 3.1. The specific heats at constant pressure at the relevant arithmetic mean temperature may be taken from the tables in Rogers and Mayhew.[3.3]

Solution

$$CH_4 + 2O_2 + (2 \times 0.79/0.21)N_2 \rightarrow CO_2 + 2H_2O$$
$$+ (2 \times 0.79/0.21)N_2$$

i.e.
$$CH_4 + 2O_2 + 7.5238N_2 \rightarrow CO_2 + 2H_2O + 7.5238N_2$$

The process is shown on a $U-T$ diagram in Fig. 3.1. The arithmetic mean temperature of the reactants is $(54 + 25)/2 = 39.5\,°C = 312.5\,K$, and the arithmetic mean temperature of the products is $(25 + 1529)/2 = 777\,°C = 1050\,K$.

The problem is most easily solved by using Eq. [3.4], and hence it is necessary to calculate values of the molar heat at constant volume, C_v, for each of the reactants and each of the products at the relevant mean arithmetic temperature. The tables[3.3] give values of c_p, and therefore these must be converted to molar heats at constant volume.

From the definition of molar heat we have:

$$C_p = Mc_p \qquad [3.7]$$

Then since $(c_p - c_v) = R$, it follows that

$$C_p - C_v = MR = R_0 \qquad [3.8]$$

A tabular method can then be used as shown below.

Gas	M (kg/kmol)	c_p (kJ/kg K)	$C_p = Mc_p$ (kJ/kmol K)	$C_v = C_p - R_o$ (kJ/kmol K)
Reactants				
Methane	16	2.2595	36.1520	27.8376
Oxygen	32	0.9205	29.4560	21.1416
Nitrogen	28	1.0400	29.1200	20.8056
Products				
Carbon dioxide	44	1.2470	54.8680	46.5536
Steam	18	2.3230	41.8140	33.4996
Nitrogen	28	1.1770	32.9560	24.6416

(Note: It is a good approximation to treat the steam in the products as a perfect gas.)

Substituting in Eq. [3.4]:

$$Q = (1529 - 25)\{46.5536 + (2 \times 33.4996) + (7.5238 \times 24.6416)\}$$
$$- 802\,310$$
$$+ (25 - 54)\{27.8376 + (2 \times 21.1416) + (7.5238 \times 20.8056)\}$$

i.e.
$$Q = 449\,622.71 - 802\,310 - 6573.08$$
$$= -359\,260\,kJ/kmol\ methane$$

The reactants are contained in a vessel of volume 0.1 m^3 and hence the number of kmol can be found from, $pV = nR_0 T$.

i.e. $\qquad\qquad n = (2 \times 10^5 \times 0.1)/(8.3144 \times 10^3 \times 327) = 0.007\,36 \text{ kmol}$

The total number of kmol of the reactants is also given by,

$$1 + 2 + 7.5238 = 10.5238 \text{ kmol/kmol methane}$$

Hence the number of kmol of methane is $0.007\,36/10.5238 = 699.4 \times 10^{-6}$, and therefore

$$\text{heat rejected to surroundings} = 359\,260 \times 699.4 \times 10^{-6}$$
$$= 251.26 \text{ kJ}$$

When the combustion process is adiabatic the products will attain their maximum temperature. Referring to Fig. 3.1 it can be seen that the temperature of the products is a maximum when point 2 is on a horizontal line through point 1, i.e. when $U_{R1} = U_{P2}$.

For adiabatic combustion under stoichiometric conditions from a reference temperature of 25 °C most fuels have products which reach a temperature in the range 2000 K to 3000 K. The maximum temperature attained can be calculated using the method of Example 3.4; since the temperature of the products is unknown in this case, and since the specific heats are required at the mean temperature of the products, then a trial and error method is necessary.

A phenomenon known as *dissociation* can cause the maximum temperature reached to be lower than that calculated by the method above. Above certain temperatures a gas may break down, or dissociate, and in doing so requires energy which it takes from the bulk of the gases thus reducing their temperature. For example the reaction, $2CO + O_2 \rightleftharpoons 2CO_2$, can proceed in either direction as shown: when CO_2 is formed the reaction is exothermic, i.e. chemical energy is released as heat; when CO_2 dissociates to CO and O_2 the reaction is endothermic, i.e. heat is supplied from the surroundings. If the products of combustion are cooled below about 1500 K then the dissociated gases recombine and hence the energy released in the process is not reduced although the maximum temperature attained in the process is lower. In steam and gas turbine plant the maximum cycle temperature is restricted by the metallurgical limits of the materials used for the turbine blades and hence dissociation does not affect the overall efficiency of the cycle; in reciprocating internal combustion engines dissociation is an important factor since the temperature of the expanding gases is above the temperature at which dissociation occurs (*see Chapter 15*[3.1]).

The maximum combustion temperature attained in practice is reduced because a supply of excess air is always provided to ensure complete combustion. With excess air supplied the mass of products is greater due to the excess nitrogen and hence the temperature rise is less. In some cases it may be necessary to supply excess air in order to restrict the temperature of the products to a value consistent with the metallurgical limit of the materials used; the metallurgical limit may be reduced when the gases are corrosive and in such cases attemperating air can be added after combustion to reduce the temperature.

Calorific Value

In practice the heat transferred on combustion of a fuel is measured by experiment either at constant volume for solid and liquid fuels (e.g. in a bomb calorimeter), or at constant pressure for gaseous fuels (e.g. in a Boys calorimeter). Details of these methods are given in Eastop and McConkey.[3.1]

When the water vapour in the products remains in the vapour phase the heat transferred to the surroundings is called the *Net Calorific Value*, NCV (sometimes called the Lower Calorific Value). When the water vapour condenses the heat transferred is called the *Gross Calorific Value*, GCV (sometimes called the Higher Calorific Value or Higher Heating Value).

The calorific values are approximately the same as the internal energy or enthalpy of combustion but of opposite sign; Table 3.4 gives some typical values at 15 °C and 1.013 bar. The difference between the GCV and the NCV is the energy required to condense the steam at the reference temperature. Therefore, for steady flow of a gaseous fuel

$$\text{NCV} = \text{GCV} - m_c h_{fg} \qquad [3.9]$$

and for non-flow of a solid or liquid fuel

$$\text{NCV} = \text{GCV} - m_c u_{fg} \qquad [3.10]$$

(where m_c is the mass of condensate formed per unit volume of fuel burned for a gaseous fuel and per unit mass of fuel for a solid or liquid fuel); the enthalpy and internal energies of vaporization are read from tables at the reference temperature.

Table 3.4 Calorific values of some common fuels

Fuel	GCV	NCV
Anthracite	34.60 MJ/kg	33.90 MJ/kg
Bituminous coal	33.50 MJ/kg	32.45 MJ/kg
Petrol	46.80 MJ/kg	43.70 MJ/kg
Diesel oil	46.00 MJ/kg	43.25 MJ/kg
Light fuel oil	44.80 MJ/kg	42.10 MJ/kg
Natural gas	38.60 MJ/m^3	34.90 MJ/m^3
Coal gas	20.00 MJ/m^3	17.85 MJ/m^3
Hydrogen	11.85 MJ/m^3	9.97 MJ/m^3

Example 3.5
For hydrogen gas the GCV at 15 °C and 1.013 bar is found to be 11.85 MJ/m^3. Calculate the value of the NCV at the same reference conditions.

Solution
For the combustion of hydrogen gas

$$
\begin{array}{ccccc}
\text{H}_2 & + & 0.5\text{O}_2 & \rightarrow & \text{H}_2\text{O} \\
2 \text{ kg} & & 16 \text{ kg} & & 18 \text{ kg}
\end{array}
$$

i.e. 1 kg of hydrogen gas generates 9 kg of steam/water. From the perfect gas equation the specific volume of the hydrogen is given by

$$
\begin{aligned}
v = RT/p = R_0 T/Mp &= (8314.4 \times 288)/(2 \times 1.013 \times 10^5) \\
&= 11.819 \text{ m}^3/\text{kg}
\end{aligned}
$$

Therefore

$$m_c = 9/11.819 = 0.762 \text{ kg/m}^3 \text{ hydrogen}$$

Then from Eq. [3.9]

$$\text{NCV} = 11.85 - (0.762 \times 2465.5 \times 10^{-3}) = 9.97 \text{ MJ/m}^3$$

(where h_{fg} at $15\,°C$ is 2465.5 kJ/kg).

3.2 HEAT TRANSFER

The transfer of heat is an important subject and specialist text books in the bibliography at the end of the chapter should be consulted for a detailed treatment; Chapter 17 of Eastop and McConkey[3.1] will be found useful as an introduction. In this section a very concise summary is given of topics in heat transfer which are particularly relevant to energy saving.

Conduction

Heat transferred by conduction is proportional to the temperature gradient in the direction of the heat flow and to the area perpendicular to the direction of the heat flow. This law was first postulated by the French mathematician Fourier

i.e.   heat transfer per unit area $(q) = -k(\delta t/\delta x)$

(where t is the temperature in a body at any point, distance x from a datum; k is known as the *thermal conductivity* and is a property of the body).

The temperature decreases in the direction of the heat flow and hence the equation above has a minus sign to give a positive value for q. The partial differential form is used since the temperature may vary in the three space directions x, y and z. For many cases in practice it is possible to assume variation of temperature in only one direction; for example, a plane wall or a cylinder for which the longitudinal heat transfer is negligible. Thermal conductivity can be assumed constant for many solids; in practice there is a variation with temperature but a suitable mean value can usually be taken.

For a plane wall whose thickness is small compared with its width and length we then have:

$$q = -k(\mathrm{d}t/\mathrm{d}x) \qquad \text{or} \qquad q\mathrm{d}x = -k\mathrm{d}t$$

and for a wall of thickness, x, integrating from 0 to x (see Fig. 3.2) we have

$$q = \frac{k(t_1 - t_2)}{x} \tag{3.11}$$

For a plane wall the area perpendicular to the heat flow is constant; this is not so for a cylinder (shown in Fig. 3.3). At any radius within the cylinder, assuming the axial and circumferential heat transfers are negligible, we have:

$$Q = -k\,2\pi r(\mathrm{d}t/\mathrm{d}r) \qquad \text{per unit length}$$

i.e. $Q\,\mathrm{d}r/r = -k\,2\pi\,\mathrm{d}t$

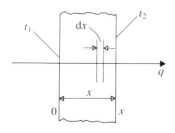

Figure 3.2 Conduction of heat in one dimension

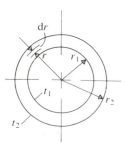

Figure 3.3 Conduction through a cylinder wall

Integrating

$$Q = \frac{2\pi k(t_1 - t_2)}{\ln(r_2/r_1)} \qquad \text{per unit length} \qquad [3.12]$$

Heat transfer by conduction is analogous to the conduction of electricity. In fact, the German physicist Ohm was influenced by Fourier's work on heat transfer in stating that the flow of an electric current is directly proportional to the potential difference and inversely proportional to the resistance. The potential difference, V, is analogous to the temperature difference, the flow of current, I, to the flow of heat, and the electrical resistance, R, to a thermal resistance. Ohm's law can be expressed as, $I = V/R$, and when this equation is compared with Eqs [3.11] and [3.12] above we have, for a plane wall

$$R = x/k \qquad \text{for unit area} \qquad [3.13]$$

and for a cylinder

$$R = \frac{\ln(r_2/r_1)}{2\pi k} \qquad \text{for unit length} \qquad [3.14]$$

As for electric circuits, thermal resistances in series can be added, and the reciprocal of thermal resistances can be added to give the reciprocal of the total resistance when the resistances are in parallel. For any one-dimensional problem we can write:

$$Q = \Delta t / R_T \qquad [3.15]$$

where Δt is the overall temperature difference, and R_T is the total thermal resistance.

At the surface between a solid and a fluid there is a thermal resistance due to a stationary layer of fluid on the surface. The heat transfer rate is governed by the law,

$$q = h\Delta t \qquad \text{per unit area} \qquad \text{or} \qquad Q = hA\Delta t \qquad [3.16]$$

where A is the area of the surface; Δt is the temperature difference between the surface and the bulk of the fluid; h is known as the *heat transfer coefficient*. The heat transfer coefficient is governed by the laws of convection and radiation and will be considered again in more detail later in the chapter. Note that by comparing Eq. [3.16] with Ohm's law we find that for a fluid film

$$R = 1/hA \qquad [3.17]$$

Example 3.6
An insulated pipe carrying wet steam at 300 °C has an internal diameter of 33.5 mm, a wall thickness of 3.4 mm, and an insulation thickness of 25 mm. The thermal conductivity of the pipe material is 50 W/m K and the thermal conductivity of the insulating material is 0.03 W/m K. The inside and outside surface heat transfer coefficients are 10 000 W/m² K and 10 W/m² K. The ambient air temperature is 24 °C. Assuming that the heat transfer along the pipe is negligible, calculate:
(i)   the steady state rate of heat transfer per unit length of pipe;
(ii) the temperature of the outside surface of the insulation;
(iii) the heat loss from an uninsulated pipe assuming that the heat transfer coefficient for the outside surface is unchanged.

Solution

(i) Using Eq. [3.14], thermal resistance, R, for the pipe wall is given by

$$R = \frac{\ln(40.3/33.5)}{2\pi \times 50} = 0.00059 \text{ m K/W}$$

and for the insulation by

$$R = \frac{\ln(90.3/40.3)}{2\pi \times 0.03} = 4.2801 \text{ m K/W}$$

From Eq. [3.17], for the inside fluid film

$$R = 1/(10^4 \times \pi \times 33.5 \times 10^{-3}) = 0.00095 \text{ m K/W}$$

and for the outside fluid film

$$R = 1/(10 \times \pi \times 90.3 \times 10^{-3}) = 0.3525 \text{ m K/W}$$

It can be seen from the above that the resistance of the metal wall is negligible compared with the insulation and the outside surface resistance; also for wet steam the heat transfer coefficient is very high and hence the resistance of the steam film on the inside surface is negligible, i.e. total thermal resistance is given by

$$R_T = 4.2801 + 0.3525 = 4.633 \text{ m K/W}$$

Then, using Eq. [3.15]

$$\text{heat loss from pipe per unit length} = (300 - 24)/4.633$$
$$= 59.6 \text{ W/m}$$

(ii) The temperature at any intermediate point can be found by applying Eq. [3.15] using the resistance up to that point.

i.e.   temperature of outside of lagging $= 24 + (0.3525 \times 59.6) = 45\,°C$

(iii) For the outside surface of an uninsulated pipe

$$R = 1/(10 \times \pi \times 40.3 \times 10^{-3}) = 0.7899 \text{ m K/W}$$

and

$$R_T = 0.7899 + 0.00059 + 0.00095 \text{ m K/W} = 0.7914 \text{ m K/W}$$

(The thermal resistances of the metal wall and the steam film are still negligible; the error in neglecting them is only $0.0015 \times 100/0.7914 = 0.2\,\%$.) Then from Eq. [3.15]

$$\text{heat loss per unit length} = (300 - 24)/0.7914 = 348.8 \text{ W/m}$$

In the above example the insulation reduced the heat loss from 348.8 W to 59.6 W. It may seem obvious that adding insulation to a pipe will decrease the heat loss but in fact under certain circumstances the adding of insulation may actually increase the heat loss. This is because the thermal resistance of the outside surface decreases due to the increased area and this may be more significant than the increase of the thermal resistance caused by the increased thickness of lagging. Figure 3.4 illustrates this effect; when the insulation thickness reaches a value, x, then the pipe loses heat at the same rate as the uninsulated pipe. For values of thickness greater than x the pipe will lose less

Figure 3.4 Heat loss from an insulated pipe

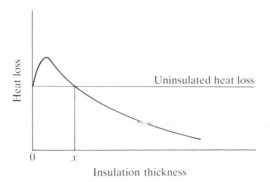

heat than in the uninsulated case. In most practical cases x is very small compared with the pipe diameter.

In Section 5.1 the value of heat insulation in practice is discussed further and the concept of economic thickness of insulation is introduced.

Convection

A proper study of convective heat transfer should be preceded by a study of the fluid mechanics of laminar and turbulent boundary layers; a text book such as Douglas et al[3.4] is recommended for further reading. The heat transfer texts in the Bibliography also give a good basic coverage of the fluid mechanics basis for convection.

The heat transfer coefficient due to convection, h_c, is expressed non-dimensionally as Nusselt number, $Nu = h_c x/k$, where x is a typical dimension of length for the surface and k is the thermal conductivity of the fluid convecting heat to or from the surface.

For flow of a fluid across a surface a *boundary layer* of fluid forms on the surface and this controls the rate of heat transfer to the surface. Where the fluid first impinges on a flat surface the boundary layer is laminar but this laminar layer becomes unstable and eventually is fully turbulent at some distance downstream. At entry to a pipe the boundary layer round the circumference is initially laminar and thickens as the fluid proceeds through the pipe. If the boundary layer meets at the centre line of the pipe before transition to turbulence occurs then the flow becomes *fully developed laminar flow* as shown in Fig. 3.5(a). If on the other hand transition to turbulence takes place before the boundary

Figure 3.5 Boundary layer development in a pipe entry

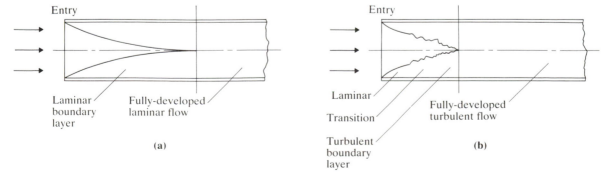

layer meets at the centre line then the flow becomes *fully developed turbulent flow* as shown in Fig. 3.5(b).

Reynolds proved experimentally that the non-dimensional group, known as the Reynolds number, $Re = \rho Cd/\mu$, determines whether a fluid of density ρ, viscosity μ, flowing with velocity C, into a pipe of diameter d, will establish fully developed laminar or turbulent flow. When $Re < 2300$ the flow will always remain laminar.

The section of the pipe up to the point at which the flow becomes fully developed is known as the *entry length*. The entry length is usually small compared with the working length of the pipe but for certain laminar flows, e.g. viscous oils, the entry length is significant.

Since the boundary layer is thin in the entry region as shown in Fig. 3.5, the heat transfer coefficient is high; Figs 3.6(a) and (b) show the variation of heat transfer coefficient for the cases of fully developed turbulent and laminar flow.

Figure 3.6   Heat transfer coefficient in the entry region

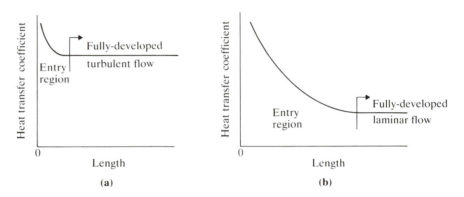

In natural convection the Grashof number, $Gr = \beta g \rho^2 \Delta t x^3/\mu^2$, plays a similar role to that of the Reynolds number in forced convection; the movement of the fluid due to density differences causes a boundary layer to form on the surface and the size of the Grashof number determines whether the boundary layer is laminar or turbulent; β is the coefficient of cubical expansion of the fluid, g is the acceleration due to gravity, and Δt is the temperature difference between the surface and the fluid. For many cases in practice the heat transfer due to natural convection is negligibly small compared with that due to forced convection and radiation.

Many empirical laws have been established for different cases of convective heat transfer and some of the more useful examples are given below.

(1) Fully developed turbulent flow in a pipe:
   (a) fluids of viscosity less than about 0.003 kg/m s and $Re > 2300$

$$Nu = 0.023\ Re^{0.8} Pr^{0.4} \qquad [3.18]$$

   Pr is the Prandtl number, $c\mu/k$, and is therefore a property of the fluid; properties are evaluated at the mean bulk temperature of the fluid.
   (b) fluids of high viscosity and $Re > 2300$

$$Nu = 0.027(\mu/\mu_w)^{0.14} Re^{0.8} Pr^{0.333} \qquad [3.19]$$

   μ_w is the viscosity of the fluid at the wall temperature; other properties are evaluated at the mean bulk temperature.

(2) Laminar flow of viscous fluids in a pipe ($Re < 2300$):

$$Nu = 1.86(\mu/\mu_w)^{0.14}\{(Re\ Pr)d/L\}^{0.333} \qquad [3.20]$$

properties are evaluated at the mean bulk temperature; L is the length of the pipe. Note that the term, d/L, in this equation allows for the entry length.

(3) Flow across a staggered tube bank (see Fig. 3.7(a)):

$$Nu = 0.33\ Re^{0.6}\ Pr^{0.333} \qquad \text{for}\ Re > 2000 \qquad [3.21]$$

For flow across an in-line tube bank (see Fig. 3.7(b)) the same equation can be used with a constant of 0.26.

Figure 3.7 Flow across in-line and staggered tube banks

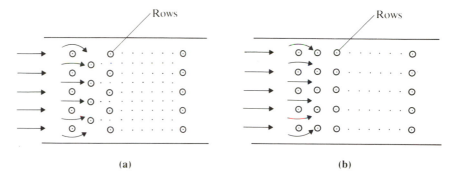

(a) (b)

The value of h_c in the Nusselt number is the average for the tube bank; properties are evaluated at mean film temperature (i.e. midway between the bulk of the fluid and the tube wall temperature); the product ρC in the Reynolds number is taken as the mass flow rate divided by the minimum free flow area; the number of rows must be greater than 10 since after this number of rows the flow pattern has stabilized. Equation [3.21] is a simplified expression; for more accurate evaluation of heat transfer coefficient, variation of tube pitch must also be considered; corrections for a number of rows less than 10 are also available (the average heat transfer coefficient increases as the number of rows increases up to 10).

(4) Natural convection:
(Note: all properties are taken at the mean temperature between the surface and the bulk of the fluid except for β which is taken at the bulk temperature; for a perfect gas, $\beta = 1/T$.)
(a) Vertical surfaces

$$Nu = 0.59(GrPr)^{0.25} \qquad \text{for}\ 10^4 < GrPr < 10^9 \qquad [3.22]$$
$$Nu = 0.13(GrPr)^{0.333} \qquad \text{for}\ 10^9 < GrPr < 10^{12} \qquad [3.23]$$

(b) horizontal surfaces: hot side facing up or cold side facing down

$$Nu = 0.54(GrPr)^{0.25} \qquad \text{for}\ 10^5 < GrPr < 2 \times 10^7 \qquad [3.24]$$
$$Nu = 0.14(GrPr)^{0.333} \qquad \text{for}\ 2 \times 10^7 < GrPr < 3 \times 10^{10} \qquad [3.25]$$

(c) horizontal surfaces: hot side facing down or cold side facing up

$$Nu = 0.27(GrPr)^{0.25} \qquad \text{for}\ 3 \times 10^5 < GrPr < 3 \times 10^{10} \qquad [3.26]$$

Further information on practical expressions for heat transfer coefficients is given in McAdams.[3.5]

Example 3.7

A pipe of 25 mm inside diameter and 0.5 m long transmits 0.030 kg/s of a fluid at a mean temperature of $77\,°C$; the mean pipe wall temperature is $100\,°C$. Using the equations above, taking properties of air and water from the tables[3.3] and the properties of engine oil as given below, calculate the heat transfer coefficient for: (i) dry air; (ii) water; (iii) engine oil. For engine oil at $77\,°C$: $\mu = 0.0376$ kg/m s, $k = 0.138$ W/m K, $Pr = 574$; for engine oil at $100\,°C$: $\mu = 0.0168$ kg/m s.

Solution

(i) For air at $77\,°C$: $\mu = 2.075 \times 10^{-5}$ kg/m s, $k = 0.003$ W/m K, $Pr = 0.697$.

The mass flow rate is 0.030 kg/s and the diameter is 0.025 m. Therefore

$$\rho C = 4 \times 0.030/\pi \times 0.025^2 = 61.12 \text{ kg/m}^2 \text{ s}$$

i.e.
$$Re = 61.12 \times 0.025/2.075 \times 10^{-5} = 7.363 \times 10^4$$

The flow is turbulent ($Re > 2300$) and the entry length is not significant, therefore from Eq. [3.18]

$$Nu = 0.023(7.363 \times 10^4)^{0.8}(0.697)^{0.4} = 155.84$$

and

$$\text{heat transfer coefficient} = 155.84 \times 0.003/0.025 = 18.70 \text{ W/m}^2 \text{ K}$$

(ii) The value of ρC is the same for all fluids since the mass flow rate and the diameter are the same, i.e. $\rho C = 61.12$ kg/m² s. For water

$$Re = 61.12 \times 0.025/360 \times 10^{-6} = 4244.4$$

The flow is turbulent and again the entry length can be ignored. Therefore

$$Nu = 0.023(4244.4)^{0.8}(2.296)^{0.4} = 25.61$$

and

$$\text{heat transfer coefficient} = 25.61 \times 667.6 \times 10^{-3}/0.025$$
$$= 683.8 \text{ W/m}^2 \text{ K}$$

(iii) For oil

$$Re = 61.12 \times 0.025/0.0376 = 40.62$$

i.e. the flow is laminar since $Re < 2300$. Nu can be found from Eq. [3.20] to be

$$1.86(0.0376/0.0168)^{0.14}(40.62 \times 574 \times 0.025/0.5)^{0.333} = 21.86$$

and

$$\text{heat transfer coefficient} = 21.86 \times 0.138/0.025 = 120.7 \text{ W/m}^2 \text{ K}$$

Heat Exchangers

A heat exchanger is a device for transferring heat from one fluid to another. There are three main categories: *recuperative*, in which the two fluids are at all times separated by a solid wall; *regenerative*, in which each fluid transfers heat to or from a matrix of material; *evaporative*, in which the enthalpy of vaporization of one of the fluids is used to provide a cooling effect.

The most commonly used type is the recuperative heat exchanger. In this type the two fluids can flow in counter-flow, in parallel-flow, in cross-flow, or in a combination of these. Figures 3.8(a) and (b) show diagrammatically simple examples of counter-flow and parallel-flow with the temperature variations as shown in Figs 3.9(a), (b) and (c); it can be shown that for all three cases of Fig. 3.9 the true mean temperature difference is the Logarithmic Mean Temperature Difference, $LMTD$, given by:

$$LMTD = \frac{(\Delta t_1 - \Delta t_2)}{\ln(\Delta t_1 / \Delta t_2)} \qquad [3.27]$$

A cross-flow heat exchanger is more difficult to analyse. The flow on either side can be mixed or unmixed as shown in Fig. 3.10. The true mean temperature difference for such a heat exchanger can be obtained using a factor to multiply the $LMTD$ from Eq. [3.27]; such factors are available for a wide range of types of heat exchangers including mixed-flow types such as multi-pass shell-and-tube heat exchangers (see Bowman et al[3.6]).

Figure 3.8 Counter-flow and parallel-flow in an annulus

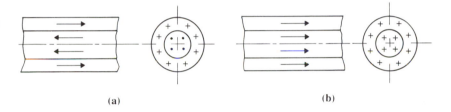

(a)   (b)

Figure 3.9
(a) Temperature variations in counter-flow with $(\dot{m}c)_c > (\dot{m}c)_H$;
(b) temperature variations in counter-flow with $(\dot{m}c)_c < (\dot{m}c)_H$;
(c) temperature variations in parallel-flow

The heat transferred for any recuperative heat exchanger is then given by:

$$Q = \dot{m}_H c_H (t_{H1} - t_{H2}) = \dot{m}_C c_C (t_{C1} - t_{C2}) = U A_o (LMTD) K \quad [3.28]$$

where $\dot{m}_H$ and $\dot{m}_C$ are the mass flow rates of the hot and cold fluids; c_H and c_C are the specific heats of the hot and cold fluids; A_o is the outside area of the wall separating the two fluids; U is the overall heat transfer coefficient based on the outside area; K is a multiplying factor for cross-flow and mixed-flow types ($K = 1$ for counter-flow and parallel-flow).

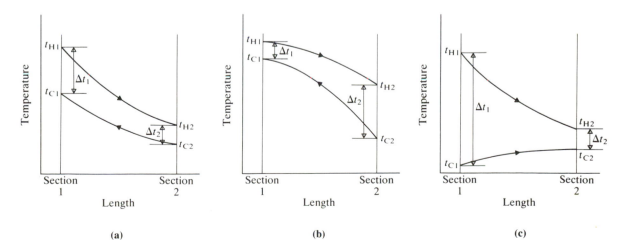

(a)   (b)   (c)

Figure 3.10 Cross-flow
heat exchanger

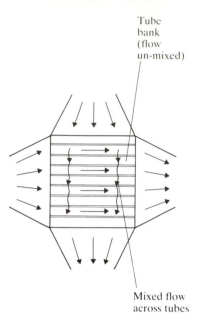

Tube
bank
(flow
un-mixed)

Mixed flow
across tubes

Using the concept of thermal resistances in series we have:

$$\frac{1}{UA_o} = \frac{1}{h_i A_i} + R_w + \frac{1}{h_o A_o} + F_i + F_o \qquad [3.29]$$

where subscripts o and i refer to the outside and inside surfaces of the separating wall; R_w is the thermal resistance of the separating wall; F represents fouling factor (see below).

For most heat exchangers a deposit of salts, oil, or other contaminant will gradually build up on the heat transfer surfaces. This is allowed for in the design by using a *fouling factor*, F, in the form of an additional thermal resistance as in Eq. [3.29]. Cleaning of the heat exchanger takes place when the fouling has reached the design value.

The *effectiveness*, E, of a heat exchanger can be defined as the actual heat transfer divided by the maximum possible.

i.e.
$$E = \frac{Q}{(\dot{m}c)_{min}(t_{Hmax} - t_{Cmin})} \qquad [3.30]$$

The minimum thermal capacity, $(\dot{m}c)_{min}$, is used since this yields the maximum value of temperature difference (see Eq. [3.28]). For the case of a counter-flow heat exchanger for example, we have:

$$E = \frac{\dot{m}_H c_H(t_{H1} - t_{H2})}{\dot{m}_H c_H(t_{H1} - t_{C2})} = \frac{(t_{H1} - t_{H2})}{(t_{H1} - t_{C2})}$$

A useful concept in heat exchanger design is Number of Transfer Units, NTU, which is defined as the ratio of the temperature change of one of the fluids divided by the mean driving force between the fluids.

i.e.
$$NTU_H = \frac{(t_{H1} - t_{H2})}{(LMTD)K} = \frac{UA_o}{(\dot{m}c)_H}$$

and
$$NTU_C = \frac{(t_{C1} - t_{C2})}{(LMTD)K} = \frac{UA_o}{(\dot{m}c)_C}$$

(K is as defined by Eq. [3.28]). A more general definition of NTU is as follows:

$$NTU = \frac{UA_o}{(\dot{m}c)_{min}} \qquad [3.31]$$

The ratio of thermal capacities, R, is usually defined as:

$$R = \frac{(\dot{m}c)_{min}}{(\dot{m}c)_{max}} \qquad [3.32]$$

R varies between 0, when one of the fluids has an infinite thermal capacity (e.g. an evaporating vapour) and 1, when both fluids have the same thermal capacity.

Some designers prefer to define values of R separately for the hot and cold fluids, i.e. $R_H = (\dot{m}c)_H/(\dot{m}c)_C$ and $R_C = (\dot{m}c)_C/(\dot{m}c)_H$. Taking the definition of R given by Eq. [3.32] then when $R = 0$ the temperature variation of the two fluids is as shown in Fig. 3.11(a). When $R = 1$ in counter-flow the temperature variation of both fluids must be linear since the temperature difference between the fluids is the same at any point through the heat exchange as shown in Fig. 3.11(b); in this case the expression for $LMTD$ given by Eq. [3.27] is indeterminate since the numerator and denominator are both zero; the true mean temperature is of course $\Delta t_1 = \Delta t_2$.

Figure 3.11
Temperature variations
for $R = 0$ and $R = 1$

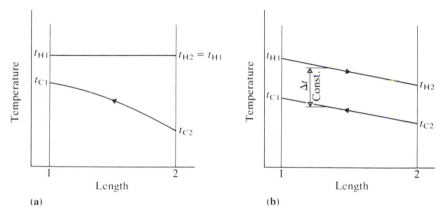

(a)   (b)

Useful relationships can be derived between E, NTU, and R for a given heat exchanger and plotted as shown in Fig. 3.12; the performance of the heat exchanger can therefore be obtained over a range of duties. The mathematical relationship between E, NTU and R for some typical cases is given below.

(1) Counter-flow heat exchanger, general case (except for $R = 1$):

$$E = \frac{1 - e^{-NTU(1-R)}}{1 - Re^{-NTU(1-R)}} \qquad [3.33]$$

(2) Counter-flow heat exchanger with $R = 1$:

$$E = NTU/(1 + NTU) \qquad [3.34]$$

(3) Heat exchanger with condensing vapour or boiling liquid on one side:

$$E = 1 - e^{-NTU} \qquad [3.35]$$

Figure 3.12
Effectiveness against
NTU

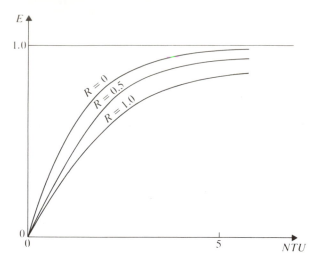

The heat transfer coefficient can be increased by increasing the velocity of flow and by increasing the turbulence of the flow, but friction and other irreversible flow losses increase as the flow velocity and turbulence increase. The pumping power required for both fluids therefore increases as the heat transfer increases and any design is a compromise between the cost of the pumping and the money saved by the increased heat transfer rate.

Radiation

All bodies emit electromagnetic radiation over a wide range of wavelengths. The radiation from a source which strikes a body is either absorbed by, reflected by, or transmitted through, the body. For unit incident radiation we can write:

$$\alpha + \gamma + \tau = 1$$

where α is the fraction of energy absorbed, called the *absorptivity*; γ is the fraction of energy reflected, called the *reflectivity*; τ is the fraction of energy transmitted, called the *transmissivity*.

A *black body* is defined as one which will absorb all the radiation incident on it, i.e. $\alpha = 1$. It can be shown that a black body is also the best possible emitter of radiation. The emissive power of a black body plotted against wavelength is shown in Fig. 3.13; the visible wave band is shown from which it can be seen that light (i.e. solar radiation) is emitted over a small range of wavelengths at a low value of wavelength. The range of temperatures in practice is such that thermal radiation is at much longer wavelengths than solar radiation. When a body is heated and begins to glow red then it has reached a temperature such that a large proportion of the energy emitted is in the red end of the spectrum (i.e. a temperature of about $1000\,^\circ\mathrm{C}$).

The *emissivity*, ε, of a body is defined as the ratio of the energy emitted by the body to that emitted by a black body at the same temperature and wavelength. For most bodies in practice it is a good approximation to assume that the emissivity is equal to the absorptivity and that a total value over all wavelengths can be assumed. Such a body is known as a *grey body*.

Figure 3.13 Emissive power against wavelength

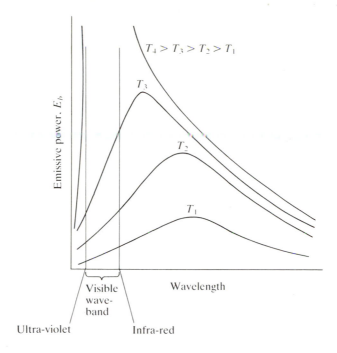

Real surfaces have absorptivities and emissivities which vary with the temperature of the body and the temperature of the source from which they are receiving the radiation. For most cases in practice it is possible to take mean values and to assume that $\alpha = \varepsilon$.

The *Stefan–Boltzmann law* states that the emissive power of a black body is directly proportional to the fourth power of the absolute temperature of the body.

i.e.
$$E_b = \sigma T^4$$

where $\sigma = 5.67 \times 10^{-8}\ \text{W/m}^2\,\text{K}^4$, is the Stefan–Boltzmann constant. The energy emitted by a non-black body is then

$$E = \varepsilon \sigma T^4$$

When a body is in very large surroundings a negligible amount of the radiation reflected from the surroundings is incident on the body; the surroundings are effectively black. The rate of heat transfer from the body to its surroundings is then,

$$Q = A_1 \varepsilon_1 \sigma (T_1^4 - T_2^4) \qquad [3.36]$$

When the surroundings are not large compared with the body, or when there are other bodies within 'sight' of the body, then the heat transferred between the body and any other body depends on the geometrical disposition of the bodies as well as on the temperatures of the bodies and their emissivities. This can be allowed for by using a *geometric factor*, F (sometimes called form factor, or view factor, or angle factor, or configuration factor, or shape factor). The geometric factor for one body with reference to another is defined as the fraction of radiation emitted from the body which is incident on the second body. (For methods of calculating radiation transfer between several bodies at different temperatures see Eastop and McConkey.[3.1])

A heat transfer coefficient for radiation, h_r, can be defined as,

$$Q = h_r A_1 (t_1 - t_2) \qquad [3.37]$$

Introducing a factor f_{1-2} into Eq. [3.36] to allow for the combined effects of emissivities and geometry, and comparing the equation with Eq. [3.37], we have:

$$h_r = f_{1-2}(T_1 + T_2)(T_1^2 + T_2^2) \qquad [3.38]$$

Example 3.8

A flat horizontal surface at $80\,°C$ faces upwards in large surroundings at $20\,°C$; the ambient air temperature is also $20\,°C$. Taking the emissivity of the surface as 0.8 and using an appropriate equation for natural convection, calculate:

(i)   the heat transferred per unit area of the surface;
(ii) the ratio of the heat transferred by radiation to that by convection;
(iii) the heat transfer coefficient for radiation;
(iv) the combined heat transfer coefficient.

Solution

(i)   Using Eq. [3.36] the heat transferred by radiation is

$$5.67 \times 10^{-8} \times 0.8 \times (353^4 - 293^4) = 370 \text{ W/m}^2$$

To find the heat transferred by convection it is first necessary to evaluate the product of the Grashof and Prandtl numbers.

Now $$Gr = \frac{\beta g \rho^2 x^3 \Delta t}{\mu^2}$$

and from the Rogers and Mayhew Tables[3.3] at the mean surface film temperature of $(80 + 20)/2 = 50\,°C$, we have:

$\rho = 1.086 \text{ kg/m}^3$; $\mu = 1.962 \times 10^{-5} \text{ kg/ms}$; $k = 0.028\,16 \text{ W/m K}$;
$Pr = 0.701$

The coefficient of cubical expansion for a gas may be taken at the bulk mean temperature of the fluid, T_a, as $\beta = 1/T_a$

i.e. $$\beta = 1/293 = 0.003\,41 \text{ K}^{-1}$$

Therefore $$Gr = \frac{0.003\,41 \times 9.81 \times 1.086^2 \times 1^3 \times (80 - 20)}{(1.962 \times 10^{-5})^2} = 61.55 \times 10^8$$

and $$(GrPr) = 61.55 \times 10^8 \times 0.701 = 43.15 \times 10^8$$

Hence Eq. [3.25] can be used for the Nusselt number.

i.e. $$Nu = 0.14(43.15 \times 10^8)^{0.33} = 227.9$$

Then the heat transfer coefficient for convection is given by

$$h_c = Nu\, k/x = 227.9 \times 0.028\,16 = 6.418 \text{ W/m}^2 \text{ K}$$

Therefore the heat transferred by convection from the plate

$$6.418 \times (80 - 20) = 385.1 \text{ W/m}^2$$

Then

$$\text{total heat transferred} = 370 + 385.1 = 755.1 \text{ W/m}^2$$

(ii)   The ratio of heat transferred by radiation to that transferred by convection is then $370/385.1 = 0.961$.

(iii)   The heat transfer coefficient for radiation is given by Eq. [3.38]

i.e.
$$h_r = 5.67 \times 10^{-8} \times 0.8(353 + 293)(353^2 + 293^2)$$
$$= 6.17 \text{ W/m}^2 \text{ K}$$

(iv)   The combined heat transfer coefficient is then

$$h = 6.17 + 6.418 = 12.59 \text{ W/m}^2 \text{ K}$$

(This can also be found from $h = 755.1/(80 - 20) = 12.59 \text{ W/m}^2 \text{ K}$)

3.3 ELECTRICITY

An understanding of basic electrical principles is necessary in order to consider the efficient generation of electricity and the most cost effective methods of using it in industry. In this chapter a brief treatment is given of the background theory; further coverage of generators and motors is given in Chapter 4. A reference such as Hughes[3.2] should be consulted for a fuller treatment of basic electrical engineering theory.

The important discovery of *electromagnetic induction*, leading to the electric generator, was made in the 1830s by Faraday who found that if there was relative movement between a conductor and a magnetic field then a flow of electricity was generated in the conductor. The flow of electricity thus caused was said to be produced by an *electromotive force*.

At about the same time Lenz showed that when an electromotive force is induced in a coil the current in the coil always tends to flow such as to oppose the motion or change of flux which induced the electromotive force.

The current available in the days of the first experimenters was steady current, that is both the current and the electromotive force producing it did not vary with time (except for the very short time period on switching on or off); this is now known as direct current (dc).

In Faraday's experiments with two coils of wire wrapped round a common iron core he observed a current instantaneously in the secondary coil when the current in the primary was switched on or off; a phenomenon known as *mutual inductance*. The magnetic field created in the iron core started from zero up to its full strength when the current was switched on but after that instant it was at a constant value. The current induced in the secondary coil is due to the change of magnetic flux with time across the coil.

Oersted demonstrated that a conductor carrying a current has a magnetic field surrounding it. When such a conductor is placed in a magnetic field then the two fields interact with a distortion of the lines of flux causing a force on the conductor which if not resisted will cause the conductor to move at right angles to the lines of flux. This is the basis of conversion of mechanical power to electrical power or vice versa.

If mechanical power is used to cause a relative movement between a conductor and a magnetic field then by Faraday's law a current is induced in the conductor. The magnetic field induced round the conductor interacts with the initial magnetic field causing a force which opposes the applied force; the mechanical power is therefore converted into electrical power through the electrical load across the generator. Conversely if electrical power is applied by supplying current to a conductor placed in a magnetic field the interacting magnetic fields cause a force on the conductor relative to the magnetic field. There is therefore relative motion between the conductor and the magnetic field which can be resisted by a mechanical load, this is the basis of the electric motor.

Alternating Voltage and Current

The most practical way of moving a conductor in a magnetic field in order to generate electricity is by rotating it. When the conductor is rotated it cuts the lines of flux at right angles twice per cycle and passes parallel to the lines of flux twice per cycle. It follows that the electromotive force (emf) generated in the conductor alternates from zero to a maximum positive value, back to zero, down to a maximum negative value, then back to zero as shown by Fig. 3.14. The current induced follows the same sine wave as the emf and hence is known as alternating current (ac); it can be seen that ac is the basic current generated in a rotating generator. The current and voltage waves are repeated with a given *frequency*, f, which is defined as the number of cycles per unit time; the unit used for frequency is the Hertz (Hz), which is defined as one cycle per second. From Fig. 3.14 it can also be seen that the voltage v at any angle θ from the zero position is given by $v_{max} \sin \theta$, where v_{max} is the maximum voltage; note also that the angle of the cycle, 2π, is equivalent to a time of $1/f$, i.e. $\theta = 2\pi f t$, where t is the time from the zero point.

Figure 3.14 Voltage variation in one cycle

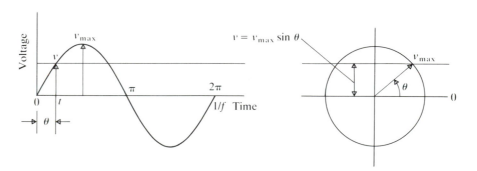

An alternating current can be generated in exactly the same way as the above by maintaining the conductor stationary and rotating a magnetic field to cut the stationary coils; this is the normal way of generating ac. The coils in which the current is generated (known as the secondary windings) are placed on the stationary annulus, known as the *stator*, and the assembly of rotating magnets mounted on the shaft is called the *rotor*; the windings which are used to create the magnets on the rotor are known as the primary windings or field coils.

When the magnetic field consists of one pair of poles (i.e. one magnet) then one complete cycle as shown in Fig. 3.14 is generated for each revolution of the shaft. When two pairs of poles are used there will be two complete cycles of voltage per revolution and hence the frequency will be twice the rotational speed. In general therefore

$$f = p \times N \qquad\qquad [3.39]$$

(where p = number of pairs of poles and N = revolutions of the shaft per unit time).

Faraday's discovery of mutual inductance can be applied using ac current to transfer power from one coil to another (see Fig. 3.15); since the current in the primary coil varies continuously then the magnetic field induced also varies and hence an ac current is induced to flow in the secondary coil; by varying the number of turns of wire on each coil it is possible to transfer a current at one emf to one at a higher or lower emf. The device is known as a *transformer* and is a vital element in the distribution of ac power from generating station to consumer.

Figure 3.15 A simple transformer

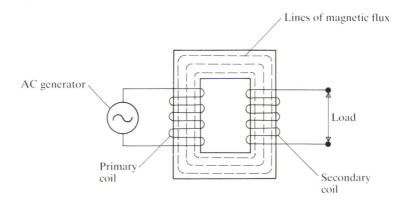

The heating effect of a dc current, I, in a circuit of resistance, R, is given by $I^2 R$, but the heating effect of an ac current varies as shown in Fig. 3.16. A mean current I can be defined such that the mean heating effect is given by, $I^2 R$. For the ac circuit the mean heating effect can be seen from Fig. 3.16 to be given by

$$\int_0^\pi \frac{i^2 R \, d\theta}{\pi} = \int_0^\pi \frac{(i_{max} \sin \theta)^2 R \, d\theta}{\pi} = \frac{(i_{max})^2 R}{2}$$

i.e. $$I = \sqrt{\{(i_{max})^2 R / 2R\}} = 0.707 \, i_{max} \qquad [3.40]$$

This mean value is known as the root mean square value or rms current; in the same way the rms voltage can be defined as 0.707 times the maximum voltage. A voltmeter, or an ammeter, in an ac circuit is designed to measure the rms value of the voltage, or current.

Figure 3.16 Current and heating effect over one cycle

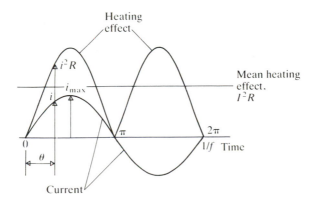

The voltage and current in an ac circuit can be represented by a diagram similar to the right-hand side of Fig. 3.14 but using the rms values; this known as a phasor diagram and the lines representing the rms value are known as phasors. Figure 3.17(a) represents two ac currents of the same frequency but out of phase by an angle α, and Fig. 3.17(b) represents the corresponding phasor diagram; it can be seen that the combined current, OC, is the vector sum of the two currents, OA and OB.

When an electrical load is connected across the terminals of a generator then the relationship between the current in the load and the potential difference, or voltage, across the terminals depends on the type of electrical load. The circuit comprising the load can be resistive, inductive, capacitive, or a combination of any of these. All conductors have resistance to the flow of current and hence in practice no circuit can be purely inductive or capacitive.

An *inductive load* is one due to a coil. When an ac voltage is applied to a coil then the magnetic field generated fluctuates with time thus inducing an alternating emf in the coil which by Lenz's law, (see above), tends to oppose the current. The induced emf is given by Faraday's and Lenz's law as

$$\text{emf} = -L \times \text{rate of change of current with time}$$

where L is known as the *inductance* of the coil and has the units of henry; a circuit with an inductance of one henry will induce an emf of one volt with a current change at the rate of one ampere per second.

The voltage applied to the coil therefore opposes the induced emf and in a purely inductive circuit (i.e. where the resistance in theory is zero) the applied voltage is always exactly equal and opposite to the induced emf as shown in Fig. 3.18(a). When the rate of change of current is at a maximum the induced emf is at its maximum negative value and when the rate of change of current is zero the induced emf is zero.

It can be seen therefore that in a purely inductive circuit the applied voltage is always quarter of a cycle ahead of the current; the current is said to *lag* the voltage by 90° or $\pi/2$ radians in a purely inductive circuit. The phasor diagram is shown in Fig. 3.18(b).

Figure 3.17   Two out-of-phase currents

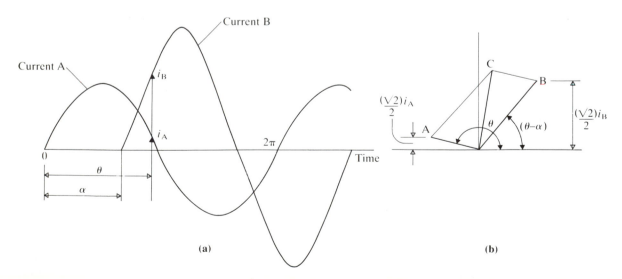

(a)   (b)

Figure 3.18 A purely
inductive circuit

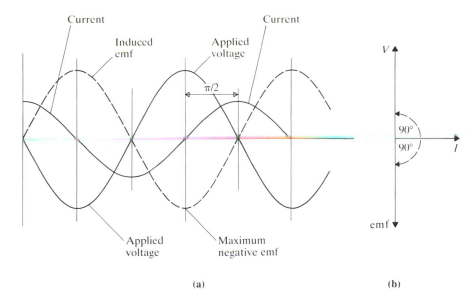

(a)   (b)

It can be shown that the rms value of voltage, V, and the rms value of current, I, for a purely inductive circuit are related by

$$V = I \times 2\pi f L = I X_L \qquad\qquad [3.41]$$

where $X_L = 2\pi f L$ is known as the *inductive reactance.*

A *capacitive load* is one containing what is known as a capacitor (formerly known as a condenser). A capacitor consists basically of two large metal surfaces separated by insulation with one of the metal surfaces charged to a much higher potential than the other; the charge is obtained electrostatically in the same way as the rubbing of silk on glass; a capacitor, therefore, acts as a store or reservoir of electricity. Now the coulomb is defined as the quantity of electricity that passes a given point when a current of one ampere is maintained for one second. The charge of a capacitor, Q, is then given by, $Q = CV$, where V is the applied voltage and C is defined as the capacitance of the capacitor; C has the units of farad, defined as the capacitance that will give the charge of one coulomb when the applied voltage is one volt. When an ac current is supplied to a capacitor then the charging current is given by

$$i = C \times \text{rate of change of potential difference across the capacitor}$$

Figure 3.19(a) shows the applied voltage and resulting current; when the rate of change of voltage is at a maximum the current is at a maximum and when the rate of change of voltage is zero the current is zero. In the case of a purely capacitive load therefore, the current *leads* the voltage by 90° or $\pi/2$; the phasor diagram is shown in Fig. 3.19(b).

It can be shown that for a purely capacitive load then the rms values of the voltage and current are related by

$$V = I \times (1/2\pi f C) = I X_C \qquad\qquad [3.42]$$

where $X_C = 1/2\pi f C$ is known as the *capacitive reactance.*

When a circuit consists of a resistance, an inductance and a capacitance in series then a phasor diagram can be used to find the common current and the

Figure 3.19 A purely capacitive circuit

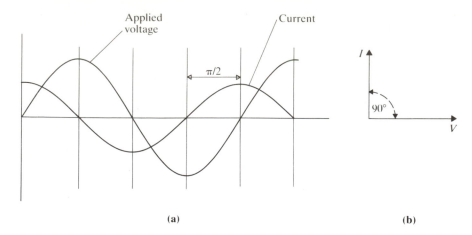

(a)   (b)

phase angle between the current and the applied voltage; similarly if an inductance, capacitance, and resistance are connected in parallel then from the phasor diagram the common voltage and the phase angle between the voltage and the current can be found.

Example 3.9
A supply of 415 V, 50 Hz is connected across a circuit consisting of a coil of 0.2 H, a capacitor of 80 μF, and a total resistance of 10 Ω which includes the resistance of the coil. Calculate the current and the phase angle between the current and the applied voltage.

Solution
The circuit and the phasor diagram are shown in Figs 3.20(a) and (b). In the phasor diagram the current which is common to all three components is drawn as a horizontal line as shown. The voltage drop across the resistance, V_R, is in phase with the current I; the voltage drop across the coil, V_L, leads the current by 90°, and the voltage drop across the capacitor, V_C, lags the current by 90°. Using Eqs [3.41] and [3.42] we have

$$V_L = 2\pi f L I = 2\pi \times 50 \times 0.2 \times I = 62.83I$$

and $\qquad V_C = I/(2\pi f C) = I/(2\pi \times 50 \times 80 \times 10^{-6}) = 39.79I$

Figure 3.20 A simple series ac circuit

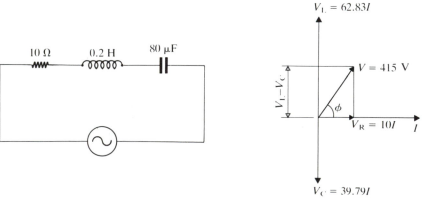

(a)   (b)

Also $\qquad V_R = 10I$

From the phasor diagram it can then be seen that the applied voltage, V, is the vector sum of $(V_L - V_C)$ and V_R

i.e. $\qquad V = I\sqrt{\{(62.83 - 39.79)^2 + 10^2\}} = 415$ (given)

$\qquad \therefore \quad I = 415/25.12 = 16.52$ A

The phase angle, ϕ, is given by

$$\tan \phi = (62.83 - 39.79)/10 = 2.3$$
$$\therefore \quad \phi = 66.5° \text{ lagging}$$

Note that for a circuit such as the above the ratio V/I is known as the *impedance*, Z. It can be seen for the series circuit of the above example we have

$$Z = \sqrt{\{R^2 + (X_L - X_C)^2\}} \qquad\qquad [3.43]$$

Power Factor

An ac circuit having a constant resistance with no inductive or capacitive components is known as a *non-reactive* circuit. It was shown previously how the heating effect, or power, P, varied in one cycle (see Fig. 3.16); the mean power in a non-reactive circuit is given by $P_m = I^2 R = IV$, where I and V are the rms values read from a voltmeter and ammeter in the circuit.

In the case of a purely inductive circuit the power becomes zero every quarter cycle when either the current or the voltage is zero. During one quarter cycle power from the generator is passed to the coil, and during the next quarter cycle power from the coil is passed back to the generator; it follows that the mean power for one cycle is zero.

For a purely capacitive circuit a similar power variation takes place but at a different phase angle. The energy stored in the capacitor during one quarter cycle is returned to the generator in the next quarter cycle; as for the case of a purely inductive circuit the mean power over one cycle is zero.

For the general case of a reactive circuit as shown in Figs 3.21(a) and (b), where the case of a lagging current has been assumed, the mean power can be found by integrating the instantaneous value of the product of current and

Figure 3.21 Typical ac circuit with phasor diagram

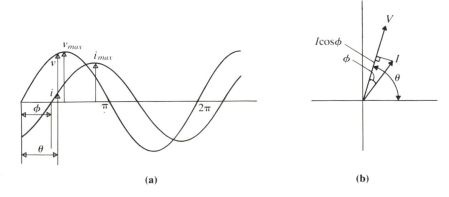

(a)   (b)

voltage over a complete cycle and dividing by 2π.

i.e.

$$P_m = \int_0^{2\pi} \frac{i_{max}\sin(\theta - \phi) \times v_{max}\sin\theta\,d\theta}{2\pi}$$

From Eq. [3.40] we can write the rms value of current, I as $i_{max} \times (\sqrt{2}/2)$ and similarly for the rms voltage, V.

i.e.

$$i_{max}v_{max} = 2VI$$

$$\therefore \quad P_m = \frac{VI}{\pi}\int_0^{2\pi}\sin(\theta - \phi)\sin\theta\,d\theta$$

$$= VI\cos\phi$$

It can be seen from Fig. 3.21(b) that the mean power is in fact the product of the non-reactive component of the current, $I\cos\phi$, and the voltage, V. This is therefore known as the *active* or *power* component of the current; the component $I\sin\phi$ is known as the *reactive* or *wattless* component.

Since $\cos\phi$ is always less than unity the active power is always less than the product, VI; the latter term is known as the apparent power (kilovoltamperes, kVA) to distinguish it from the true mean power in watts (or kilowatts, kW); the reactive or wattless power is known as the kilovoltamperes reactive (kVAr). The *power factor* is defined as

$$\text{power factor} = P_m/VI = \cos\phi \qquad [3.44]$$

It is the convention to refer to a *lagging power factor* when the current lags the voltage, and to a *leading power factor* when the current leads the voltage.

A power triangle is sometimes useful to illustrate the active and reactive power components. Figure 3.22 shows a power triangle for a circuit with resistance, inductance and capacitance; the angle between the active power (kW) and the apparent power (kVA) is the phase angle, ϕ, with the third side of the triangle representing the reactive power (kVAr).

The larger the angle, ϕ, and hence the smaller the value of $\cos\phi$, the lower is the value of the active power (kW) for a given applied voltage, V, and current, I. This means that for a given power with a low power factor the current, I, is much higher than it would be if the power factor were unity. Hence the losses in the generator itself are larger and the switchgear, transformers and cables etc. are all larger and more expensive; heat losses are equal to I^2R in any component. Since the power producer would need to provide much larger generator capacity for a highly reactive load taken by a consumer, then the Electricity Boards charge a penalty for low power factor loads. They do this by making a charge based on the *maximum demand* of the customer in kVA; more details of the way this is done are given in Chapter 9. There is therefore a financial incentive for a customer to increase the power factor of the load.

In general low power factors are caused by induction motors, induction furnaces, and arc and seam welders. Since an inductive load has a lagging power factor and a capacitive load has a leading power factor then a low power factor due to an inductive load can be improved by adding a capacitor to the circuit.

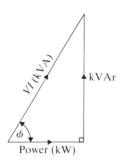

Figure 3.22 Power triangle

Example 3.10

A motor running from a 415 V, 50 Hz supply requires an input power of 15 kW and has a power factor of 0.6 lagging at full load. A capacitor of 250 μF is connected in parallel with the motor to increase the power factor. Calculate:
(i)   the percentage reduction in current taken from the supply after the capacitor is installed;
(ii) the new power factor;
(iii) the kVAr rating of the capacitor.

Solution

(i)   Without the capacitor the power triangle is as shown by Fig. 3.23; the input kVA is given by $15/0.6 = 25$ kVA, and the reactive power by $\sqrt{(25^2 - 15^2)} = 20$ kVAr. The angle, ϕ, is given by $\cos \phi = 0.6$, i.e. $\phi = 53.13°$.
      Also the current taken by the motor is

$$25\,000/415 = 60.24 \text{ A}$$

With the capacitor added the circuit is as shown in Fig. 3.24(a), with the corresponding phasor diagram given in Fig. 3.24(b). The new current from the supply, I, is the vector sum of the motor current (60.24 A at the phase angle of 53.1° lagging), and the current through the capacitor which leads the applied voltage by 90°.
      The current through the capacitor is found from Eq. [3.42].

i.e.
$$I_C = 2\pi f C V = 2\pi \times 50 \times 250 \times 10^{-6} \times 415$$
$$= 32.59 \text{ A}$$

From Fig. 3.24(b) it can be seen that the reactive component of the new supply current is given by $(60.24 \sin 53.1° - 32.59) = 15.60$ A. The power component of I is given by $60.24 \times 0.6 = 36.15$ A. Then the new supply current can be found as

$$I = \sqrt{(15.60^2 + 36.15^2)} = 39.37 \text{ A}$$

Hence

$$\text{percentage decrease in supply current} = \frac{(60.24 - 39.37) \times 100}{60.24}$$
$$= 34.65\%$$

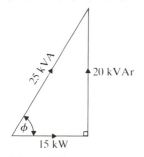

Figure 3.23 Power triangle for Example 3.10

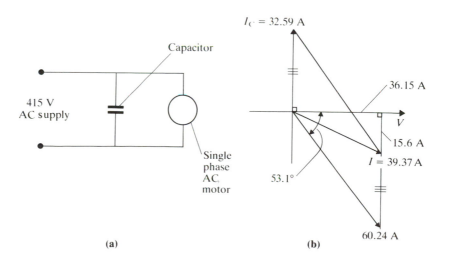

Figure 3.24 Circuit and phasor diagram with capacitor

It follows that the kVA are also reduced by this percentage, since the applied voltage is 415 V in both cases.

(ii)   Equation [3.44] gives the new power factor.

i.e.   power factor $= \cos \phi_2 = 36.15/39.37 = 0.92$ lagging

(iii)   The current through the capacitor is 32.59 A and hence the kVAr rating is given by $32.59 \times 415 \times 10^{-3} = 13.53$ kVAr.

Three-phase Supply

Single-phase ac supply as considered above is adequate for purposes such as heating and lighting but for ac motors it is more satisfactory to use more than one phase; for example, a single-phase induction motor is not self-starting and requires an auxiliary winding whereas one with a three-phase supply is not only self-starting but is also more efficient and has a better power factor.

Windings with more than one phase are referred to as *polyphase windings*; the reason for using three phases as compared with two phases or four phases etc. is that the capacity of a three-phase supply is greater than a two-phase supply for the same number of lines to the load, whereas for a phase number greater than three the capacity increase is not very great and the number of lines is increased.

In a three-phase generator three windings are arranged such that three separate emfs are generated at 120° intervals in the cycle. Now since there is a magnetic field formed round a conductor whenever a current flows in it then each of the coils on the stator has its own magnetic field which will interact with that of the other coils. Each of the stator coils will have a current flowing only when the rotor magnetic flux cuts it at an angle. Hence the magnetic flux created in the stator moves round as the rotor revolves and it can be shown that for a two-pole rotor it rotates through one revolution for every cycle. For a rotor with p poles the magnetic flux rotates at f/p revolutions per unit time. Comparing this with Eq. [3.39] it can be seen that the magnetic flux in the stator rotates at the same speed as the rotor, and this is known as the *synchronous speed*. The magnetic flux in the stator rotates at synchronous speed for all polyphase windings but remains stationary for a single-phase winding. It is for this reason that a three-phase induction motor is self-starting compared with a single-phase motor; this is explained more fully in Chapter 4.

A three-phase system basically requires six conductors to connect the phases to the load but this can be reduced by connecting the phases together in either *delta*, or *star* connection (called wye connection in the USA). Three possible connections are shown in Figs 3.25(a), (b), and (c); Fig. 3.25(a) is a delta

Figure 3.25   Delta and star connections for 3-phase supply

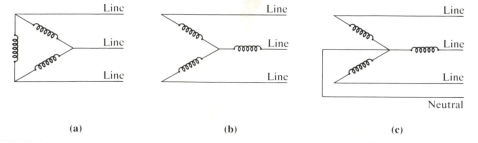

(a)   (b)   (c)

connection and Figs 3.25(b) and (c) are star connections with three and four wires respectively.

It can be seen that for a star connection the line current, I_L, is equal to the phase current, I_P, and it can be proved that for a star connection the line voltage is equal to 2 cos 30^0 times the phase voltage, i.e. $V_L = (\sqrt{3}) V_P$. The line voltage is the voltage between any two of the lines, and the phase voltage is the potential difference across a phase coil.

For a delta connection it can be seen that the line voltage is equal to the phase voltage. If the load is balanced so that the current in each line is equal then it can be shown that for a delta connection $I_L = (\sqrt{3}) I_P$.

For a four-wire system star connection (see Fig. 3.25(c)) with a balanced load there is no current in the neutral and hence a three-wire connection could be used (Fig. 3.25(b)).

The total power with a balanced load in either a star or delta connection is given by

$$P = 3 \times I_P \times V_P \times \text{power factor}$$
$$= (3/\sqrt{3})I_L V_L \times \text{power factor}$$
$$= \sqrt{3} \times I_L \times V_L \times \text{power factor} \qquad [3.45]$$

(The voltage quoted for a three-phase system is the line voltage.)

Power fraction correction calculations can be made for a three-phase circuit in a similar way to that of Example 3.10.

Problems

3.1 A steam plant operates with turbine entry conditions of 20 bar, 350 °C and a condenser pressure of 0.03 bar. The steam mass flow rate is 30 kg/s. Assuming a turbine isentropic efficiency of 0.85 and neglecting all other losses, calculate:
(i) the net power output;
(ii) the cycle efficiency;
(iii) the Carnot cycle efficiency between the same temperature limits.
                                                    (27.34 MW; 30 %; 52.3 %)

3.2 A steam plant operates with turbine entry conditions of 30 bar, 400 °C and a condenser pressure of 0.01 bar. The turbine isentropic efficiency is 0.9. For a net power output of 25 MW, neglecting losses, calculate:
(i) the required mass flow rate of steam;
(ii) the input power to the pump;
(iii) the cycle efficiency;
(iv) the Carnot cycle efficiency between the same temperature limits.
                                        (21.5 kg/s; 64.5 kW; 36.4 %; 58.4 %)

3.3 A heat pump operates on a vapour compression cycle using Refrigerant-12. Dry saturated vapour enters the compressor at 3.086 bar and saturated liquid refrigerant enters the throttle valve at 20.88 bar. The heat output of the plant is 100 kW. Assuming isentropic compression, calculate:
(i) the coefficient of performance of the heat pump;
(ii) the required mass flow rate of refrigerant.
                                                    (3.18; 0.92 kg/s)

3.4 An open cycle gas turbine plant operates with a pressure ratio of 9 with an inlet air temperature of 10 °C. The maximum cycle temperature is 1200 K, and the isentropic efficiencies of the compressor and turbine are 0.82 and 0.86 respectively. Taking the value of γ for air and for gases as 1.4 and 1.333 respectively, and the specific heats of air and gases as 1.005 kJ/kg K and 1.15 kJ/kg K respectively, neglecting all losses, and assuming that the mass flow rate in the turbine is the same as that in the compressor, calculate:
(i) the mass flow rate required for a net power output of 30 MW;
(ii) the plant thermal efficiency;
(iii) the Carnot cycle thermal efficiency between the same temperature limits.

(151.3 kg/s; 28.0 %; 76.4 %)

3.5 An open cycle gas turbine plant operates with an air inlet temperature of 15 °C and a maximum cycle temperature of 900 °C. The pressure ratio in the compressor is 10 and the isentropic efficiencies of the compressor and turbine are 0.8 and 0.85 respectively. The mass flow rate of air entering the compressor is 70 kg/s and the air–fuel ratio in the combustion chamber is 75. Taking the specific heats at constant pressure for the air and combustion gases as 1.005 kJ/kg K and 1.5 kJ/kg K, and γ for air and the gases as 1.4 and 1.333, and neglecting all pressure and heat losses, calculate:
(i) the net power output;
(ii) the cycle efficiency;
(iii) the Carnot cycle efficiency between the temperature limits.

(12 MW; 26.8 %; 75.5 %)

3.6 (a) For gaseous methane, CH_4, calculate the stoichiometric air–fuel ratio by volume.
(b) Gaseous methane is burned with an air–fuel ratio by volume of 15, calculate:
(i) the percentage excess oxygen in the products;
(ii) the volumetric analysis of the dry flue gas;
(iii) the dew point of the products when the total pressure is 1.1 bar.

(9.52; 7.19 %; 7.14 % CO_2, 8.21 % O_2, 84.64 % N_2; 52.1 °C)

3.7 A gas used in a gas engine has the following composition by volume:
93 % CH_4; 3 % C_2H_6; 3 % N_2; 1 % CO
The air–fuel ratio is 10. Calculate:
(i) the percentage by volume of carbon dioxide in the dry exhaust gas assuming there is no carbon monoxide present;
(ii) the dew point of the wet gases when the total pressure of the gases is 1.016 bar.

(11.04 %; 57.8 °C)

3.8 Carbon monoxide, CO, initially at 25 °C is burned at constant volume with air in stoichiometric proportions. The products attain a temperature of 1029 °C. Calculate:
(i) the internal energy of combustion at 25 °C given that the enthalpy of combustion at 25 °C is $-282\,990$ kJ/kmol of CO;
(ii) the heat rejected during the process per kmol of CO burned.

($-281\,751$ kJ/kmol CO; 194 873 kJ/kmol CO)

3.9 Benzene vapour, C_6H_6, at 25 °C is burned adiabatically at constant pressure and the H_2O in the products is in the vapour phase. Using the

information given in the Tables,[3.3] assuming complete combustion, and neglecting dissociation, calculate:
(i) the temperature attained by the products assuming stoichiometric combustion;
(ii) the temperature attained by the products for an air–fuel ratio by volume of 70.

(2203 °C; 1189 °C)

3.10 (a) Prove from first principles that the specific humidity of a mixture of air and water vapour is given by:

$$\omega = 0.622 p_s (p - p_s)$$

(where p_s is the partial pressure of the water vapour; p is the total pressure.)
(b) A rigid vessel of volume 0.2 m^3 contains humid air at a pressure of 1.013 bar, a temperature of 22 °C and a relative humidity of 40 %. The vessel is cooled until condensation just begins to form on the inside surfaces. Calculate, using tables only:
(i) the mass of water vapour in the vessel;
(ii) the percentage saturation of the air initially;
(iii) the heat rejected during cooling.

(0.001 55 kg; 39.37 %; 1.58 kJ)

3.11 A wall consists of 105 mm brickwork, a 25 mm air gap filled with expanded polystyrene, a second layer of 105 mm brickwork, and a 13 mm thickness of lightweight plaster. For an outside temperature of −1 °C and an internal temperature of 20 °C, using the data below, calculate:
(i) the overall heat transfer coefficient for the wall;
(ii) the steady state rate of heat transfer per unit area through the wall;
(iii) the temperature at the inside surface of the room;
(iv) the rate of heat transfer per unit area when the air space is not filled with insulation.

Data
Thermal conductivities: brickwork (outer leaf), 0.84 W/m K; brickwork (inner leaf), 0.62 W/m K; expanded polystyrene, 0.035 W/m K; lightweight plaster, 0.16 W/m K.
Heat transfer coefficients: outer surface, 16.7 W/m^2 K; inner surface, 8.3 W/m^2 K.
Resistance of air space, 0.18 m^2 K/W.

(0.787 W/m^2 K; 16.53 W/m^2; 18 °C; 28.53 W/m^2)

3.12 A steel pipe of 60.8 mm external diameter and 53.0 mm internal diameter carries gases at 300 °C through an air space at 15 °C. The rate of heat loss from the pipe is to be restricted to 150 W per m length. Using the data below, calculate:
(i) the required thickness of glass fibre insulation;
(ii) the temperature of the outside surface of the insulation;
(iii) the temperature at the pipe/insulation interface.

Data
Thermal conductivities: steel, 48 W/m K; glass fibre, 0.07 W/m K. Heat transfer coefficients: outside surface, 10 W/m^2 K; inside surface, 500 W/m^2 K.

(31.9 mm; 30 °C; 136.7 °C)

3.13   Air is heated by passing it across a bank of tubes in a square duct. Using the data below and additional air properties from the tables of Rogers and Mayhew,[3.3] estimate the mean heat transfer coefficient from the tube bank to the air using Eq. [3.21].

Data
Duct size, 550 mm × 550 mm. Number of tube rows, 15. Number of tubes per row across the duct at right angles to the flow, 5. Diameter of tubes, 50 mm. Pitch of tubes across duct, 100 mm. Mean temperature of air through bank, 104 °C. Mean temperature of tube walls, 150 °C. Mass flow rate of air, 5 kg/s.
$$(153.2 \ W/m^2 \ K)$$

3.14   A heated vertical panel 1 m high by 2 m wide is at 84 °C in large surroundings in still air at 20 °C. Taking the emissivity of the surface as 0.85, data for air from the tables of Rogers and Mayhew,[3.3] and using either Eq. [3.22] or Eq. [3.23], calculate:
(i)   the total heat transfer from the panel;
(ii) the ratio of the heat transferred by convection to that by radiation.
(iii) the effective heat transfer coefficient for the panel surface.
$$(3.258 \ kW; \ 0.904; \ 12.73 \ W/m^2 \ K)$$

3.15   A single phase 240 V, 50 Hz supply is connected across a resistor of 100 Ω, a coil with an inductance of 0.2 H and negligible resistance, and a capacitor of 200 μF. Calculate the supply current and the phase angle between the supply current and the applied voltage:
(i)   when each element is connected in series with the supply;
(ii) when each element is connected in parallel with the supply.
$$(3.757 \ A; \ 25.13°; \ 19.901 \ A; \ 77.96°)$$

3.16   A single phase induction motor operates from a supply of 415 V, 50 Hz with a power input of 10 kW, and a lagging power factor of 0.7. Calculate the capacitance and kVAr rating of a capacitor to be connected in parallel with the motor such that the power factor is raised to 0.9 lagging.
$$(99.05 \ \mu F; \ 5.359 \ kVAr)$$

References

3.1   Eastop T D, McConkey A 1986 *Applied Thermodynamics for Engineering Technologists* 4th edn Longman

3.2   Hughes E 1977 *Electrical Technology* 5th edn Longman

3.3   Rogers G F C, Mayhew Y R 1988 *Thermodynamic and Transport Properties of Fluids* 4th edn Basil Blackwell

3.4   Douglas J F, Gasiorek J M, Swaffield J A 1986 *Fluid Mechanics* 2nd edn Longman

3.5   McAdams W H 1954 *Heat Transmission* 3rd edn McGraw-Hill

3.6   Bowman R A, Mueller A E, Nagle W M 1940 Heat temperature difference in design *Tr ASME* **62**: 283

Bibliography

Bayley F J, Owen J M, Turner A B 1972 *Heat Transfer* Nelson
Kays W M, London A L 1984 *Compact Heat Exchangers* 3rd edn McGraw-Hill

Knudsen J G, Katz D L 1958 *Fluid Dynamics and Heat Transfer* McGraw-Hill
Laycock C H 1985 *Applied Electrotechnology for Engineers* Pavic Publications
Shepherd J, Morton A, Spence L 1986 *Higher Electrical Engineering* 2nd edn
  Longman
Sucec J 1985 *Heat Transfer* Wm C Brown

4 ENERGY CONVERSION

A study of the efficient and cost effective use of energy must start with an examination of methods of energy conversion from one form to another. The various possible ways in which this energy conversion may be effected are summarized in Fig. 4.1. The main primary path of energy conversion is from the Chemical Energy of a fossil fuel to the Thermal Energy of steam or hot water in a boiler, or to the thermal energy of hot gases in an Internal Combustion (IC) engine. The thermal energy thus produced can be converted to Mechanical Energy in a rotating turbine or reciprocating engine, and this in turn can be used directly to drive machinery in factories or vehicles on road, sea and air. Alternatively the mechanical energy produced can be further converted into Electrical Energy using a generator. The same energy conversion path can be

Figure 4.1 Energy conversion diagram

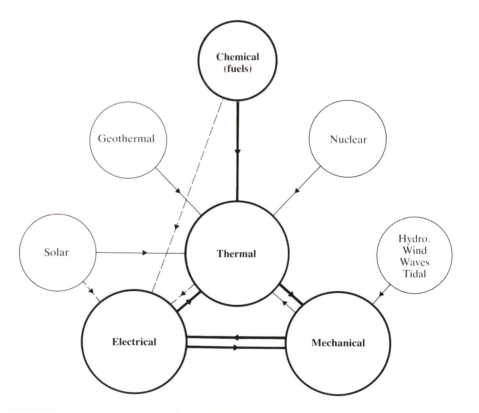

followed starting from Nuclear Energy, Geothermal Energy, or Solar Energy. A more direct conversion path to mechanical energy and hence electrical energy is by using Hydro, Wind, Wave or Tidal Energy as shown in the figure.

The ease with which electricity can be distributed via the national grid and on site makes it practical to convert mechanical energy to electrical energy and then reconvert it to mechanical energy using electric motors, or to thermal energy using electric resistance elements.

The lines shown dotted in Fig. 4.1 represent energy conversion paths which are feasible but not economically viable on a large scale at the current state of development. Direct conversion from chemical energy to electrical energy can be achieved using a fuel cell; direct conversion from solar energy to electrical energy in a photovoltaic cell; direct conversion from thermal energy to electrical energy from the Seebeck effect (or refrigeration using electrical energy from the Peltier effect).

The Department of Energy has set up an Energy Technology Support Unit which commissions research in Renewable Energy and advises on the types of development which should be supported. Renewable energy covers all forms of energy in which the source is continuously replenished. The Energy Technology Support Unit has decided that for the UK the most attractive schemes in the short term are: biofuels including waste, coppice wood crops, and straw; passive solar energy; small scale land-based wind farms (e.g. 600–750 kW). In the longer term they perceive as attractive possibilities: tidal barrages (the promising sites are the Severn and the Mersey); small scale shore-based wave devices; small scale hydro schemes (the potential for large scale hydro has been exploited fully); geothermal hot dry rocks; energy forests (i.e. the growing of large trees as fuel); off-shore wind devices. Schemes which are seen to be less feasible in the UK are: geothermal aquifers; large scale off-shore wave schemes; photovoltaic devices; animal waste and crop waste other than wood and straw.

The main aim of this book is to help the reader to understand and make more cost effective the use of energy in industry and commerce. The detailed coverage is therefore restricted to the main areas of thermal, mechanical and electrical energy with input from chemical energy (taken to include waste as well as fossil fuels), i.e. the routes marked in bold lines in Fig. 4.1. The use of waste as a fuel is covered in Section 4.3 but other forms of renewable energy are not included in this book. For further information on these and on nuclear power the reader is referred to the bibliography at the end of the chapter.

4.1 FUELS AND COMBUSTION

The efficient use of fuel is a detailed subject in its own right and for a comprehensive treatment readers are recommended to the bibliography given at the end of the chapter. This section will cover the main technical factors relevant to the preparation and use of various fuels for the production of thermal energy. A decision on which fuel to use in a particular case must be made in the context of the complete plant and with a knowledge of the current market prices of fuels and the likely trend in prices within the lifetime of the plant. These points will be considered in more detail in later chapters.

Coal

Coal consists basically of carbon, hydrogen, nitrogen, sulphur, oxygen, mineral matter (called ash), and inherent moisture. It is the name given to matter which has formed within the earth over millions of years by the decay of trees and plants. The most basic solid fuel formed in this way is *peat* which has a low carbon content and a high moisture content. Coal can be described by an *ultimate analysis* which gives the percentage by mass of the basic constituents. For example, a typical anthracite has an ultimate analysis of 89.2 % carbon, 2.7 % hydrogen, 1.2 % nitrogen, 1.2 % sulphur, 1.7 % oxygen, 3 % mineral matter, 1 % inherent moisture.

When dry coal is heated at a temperature of 925 °C at atmospheric pressure in an air-free atmosphere the loss of weight gives the *volatile matter* initially present; the carbon which remains is called the *fixed carbon*. The *proximate analysis* of the coal on a moist mineral-matter-free basis is defined as the percentage amounts of volatile matter, fixed carbon, and inherent moisture. It can also be given on a dry mineral-matter-free basis.

Coal is given a *rank* based on the proximate analysis. Anthracite with the highest proportion of fixed carbon is the most highly ranked coal; bituminous coal contains a high percentage of volatile matter and occupies the middle range of the ranking. An international system of coal classification defines low rank coal as that with a Gross (or Higher) Calorific Value, GCV (or Higher Heating Value, HHV), of less than 24 MJ/kg. Such coals are called *brown coals* or *lignite*. Lignite is not found to any significant extent in the UK but it forms a large proportion to the coal mined in Eastern Europe (e.g. Poland, East Germany, and Czechoslavakia); it has a high inherent moisture content as shown by Fig. 4.2 and the ultimate analysis shows a high percentage of mineral matter (10 to 20 %).

The other property of coal which is important is its tendency to *cake* (i.e. swell and bind to form coke). When the volatile matter is below about 35 % by mass the caking tendency is small and approximately constant with changes in volatile matter. Above this figure the caking tendency varies widely and provides an additional factor in the classification. In the *Gray–King test* a sample of coal is heated to 600 °C, the resulting coke is classified visually into categories and the coal given a grade accordingly. The middle ranking bituminous coals are the most strongly caking with the low ranking bituminous and lignite non-caking. At the high end of the ranking, anthracite is also non-caking. Figure 4.2 shows the ranking of coal based on the moist mineral-matter-free proximate analysis.

In the UK, high ranked coals are given a low rank number; for example anthracites start at 100, bituminous coals in the 300–600 rank are the strongly caking coals, and those in the range 600–900 are general purpose coals.

Calorific value varies according to the rank as shown in Fig. 4.3. However, calorific value of a coal in the 'as-fired' state will have lower values due to the surface moisture content, and the adventitious ash.

The CIBSE *Guide*[4.1] gives information for some typical 'as-fired' coals; an extract is given in Table 4.1, (page 70), by permission of CIBSE.

Large coals are graded by sieving to agreed minimum and maximum sizes; for example *singles* (see Table 4.1) have an agreed minimum of 13 to 18 mm, and the next size up, *doubles*, to a minimum size of 25 to 38 mm. *Smalls* (see

Figure 4.2 Classification of coal

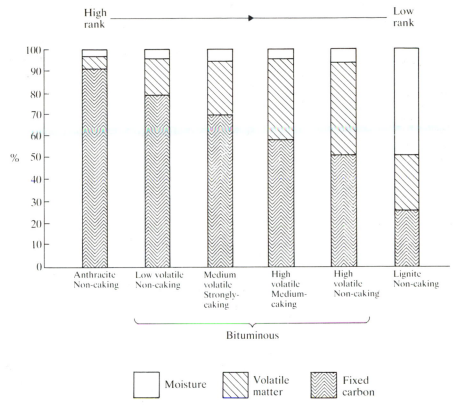

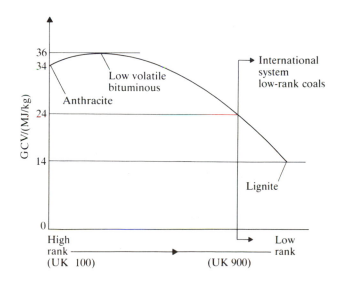

Table 4.1) are defined as coal that is all smaller than a certain sieve size (usually 13 mm); smalls therefore contain fine material down to <3 mm which can affect the flow of the coal and its combustion, particularly if the surface moisture is high.

Coal-fired boilers can have a variety of different designs from simple gravity-fed sectional boilers to the 660 MW power station water-tube type with pulverized fuel firing. In all cases the boiler requires: (a) a means of introducing the coal into the combustion zone; (b) a system for providing air for combustion;

Figure 4.3 GCV against coal rank

Coal	Rank	GCV (MJ/kg)	As-fired analysis						
			C %	H %	N %	S %	O %	Ash %	Moisture %
Washed singles:									
Anthracite	101	32.10	84.7	2.6	1.0	1.1	1.6	5.0	4.0
Bituminous	401	32.40	78.5	4.7	1.7	1.9	4.2	5.0	4.0
Non-caking	902	25.45	63.0	3.9	1.3	1.8	9.0	5.0	16.0
Washed smalls:									
Anthracite	101	29.65	78.2	2.4	0.9	1.0	1.5	8.0	8.0
Bituminous	401	29.55	71.6	4.3	1.6	1.7	3.8	8.0	9.0
Non-caking	902	23.85	59.0	3.7	1.2	1.7	8.4	8.0	18.0
Coke	—	27.90	82.0	0.4	1.7	—	—	7.0	8.9

Table 4.1 Analysis and GCV of some as-fired coals

(c) a controlled combustion region; (d) an efficient flue gas outlet to the chimney; (e) a method of ash collection and disposal.

A system using gravity feed from a hopper with air blown through a grated hearth is suitable for boilers up to 1 MW. Such a system uses anthracite since the volatiles from bituminous coals would be chilled too quickly by the proximity of the heating surface. For sectional or shell boilers up to sizes of about 1.8 MW an underfeed stoker can be used with bituminous coal; in this system an Archimedian screw delivers the coal to a retort to which air is supplied through slots; the combustion starts in the retort forming an incandescent zone and continues in the space above (see Fig. 4.4). Manual removal of the ash is usually required in such boilers, either as fused clinker from the combustion space itself, or via a reciprocating bar grate combined with the retort which allows the ash to be removed by a screw conveyor below the boiler.

For small water-tube boilers up to 10 MW, ram-feed, or coking, stokers can be used; in this system coal from a hopper is ram-fed onto a plate from which it moves down onto a moving bar grate; low caking bituminous coals are used, the volatiles released initially turning the coal into coke before it reaches the moving grate.

Figure 4.4 Fire-tube boiler with screw coal-feed

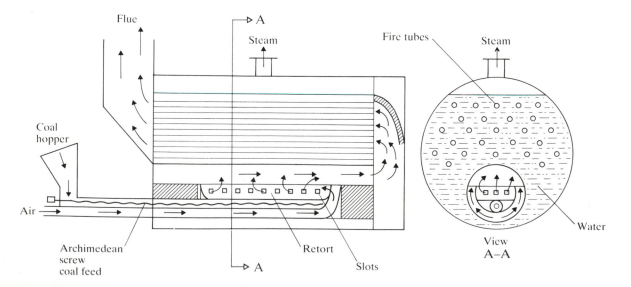

For boilers up to 80 MW the travelling chain grate stoker is in common use; coal from a hopper passes through a guillotine door onto the travelling grate; a refractory arch above the grate entry facilitates the combustion of the volatiles thus turning the coal into coke as it moves through the furnace; the fixed carbon is finally burned and the clinker and ash left at the end of the grate's travel are passed to an ash pit below the grate where a conveyor automatically removes them; the air is supplied by a forced draught fan as primary air from below the grate and as secondary air from above the coal bed; fully automatic control can be achieved by varying the speed of the grate and by using dampers for the air flow. Such boilers are suitable for use with a wide range of coals. Spreader stokers are also used in which the coal is fed to the grate by a rotating impeller which allows the small particles to burn in suspension while the larger pieces fall onto the grate forming the fire-bed (see Fig. 4.5); an advantage of this type is that the grate movement is towards the front of the boiler so that the ash extraction is easier.

Above about 80 MW the size of the combustion space precludes the use of a mechanical stoker and water-tube boilers from 80 MW up to 660 MW and above use pulverized coal; coal is supplied via large hoppers to the pulverizing mills in which it is ground down by large steel balls within upper and lower grooved rings on a track, the track being driven by a vertical spindle; heated primary air from a fan blows the pulverized fuel through an outlet pipe to burners mounted on the side of the boiler combustion chamber; the burner consists of a pipe carrying the primary air/pulverized fuel which runs concentrically through a cylindrical register supplying secondary air through swirl vanes on the periphery; a steam atomized oil burner with peak ignition is located through the centre of each burner to provide a light-up facility. A wide range of coals can be burned using this system and the heated primary

Figure 4.5 Travelling grate stoker

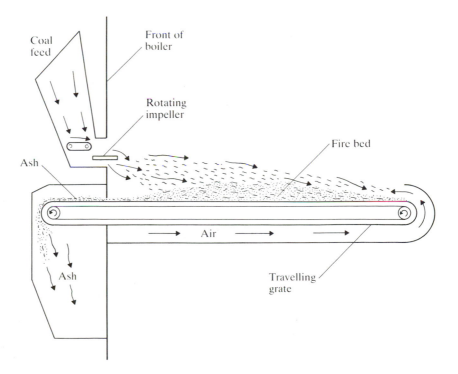

Figure 4.6 Water-tube boiler

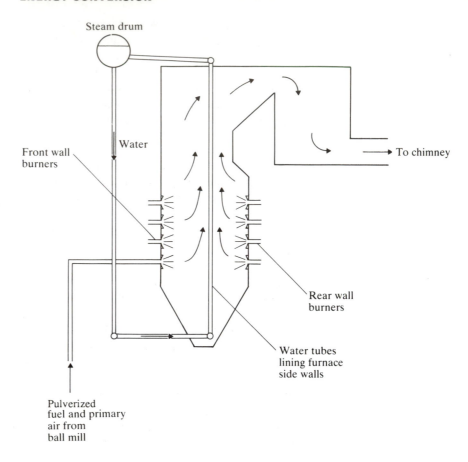

Steam drum

Water

Front wall burners

Rear wall burners

To chimney

Water tubes lining furnace side walls

Pulverized fuel and primary air from ball mill

air makes it possible to handle coal with moisture contents up to 27 %. Figure 4.6 shows a diagrammatic arrangement of a typical boiler of this type.

Fluidized bed combustion is becoming of increasing importance. In this method a bed of inert material is fluidized by blowing air through it until the drag force on each particle is balanced by the gravitational pull; the material is then heated to a high temperature so that coal, or other combustibles such as solid waste, can be fed into the bed; after the coal has ignited the original source of heat to the bed is stopped. Fuels with a very high moisture and ash content can be burned successfully using this method; it is therefore particularly important for those countries in which there is a large resource of lignite. There are other advantages which make it increasingly important in the UK. The main advantage is in the possibility of reducing atmospheric pollution; this is becoming increasingly important as international regulations are tightened. Limestone can be added to the bed causing the retention within the bed of the sulphur in the fuel and hence reducing the possibility of pollution from the formation of SO_3 in the flue gases. Also, the high heat transfer rates within the bed allow combustion to take place at a much lower temperature than usual ($< 1000\,°C$), which minimizes the formation of the oxides of nitrogen. The low combustion temperature also leads to more compact boilers for the same output and an absence of problems due to ash fusion. Fluidized bed combustion has been developed by several boiler manufacturers in the UK for both shell and water tube boilers. A typical example is shown diagrammatically in Fig. 4.7.

Figure 4.7 Fluidized-bed boiler

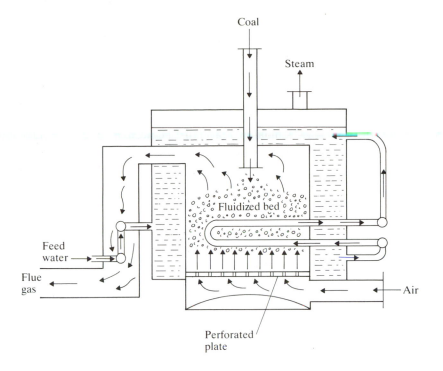

Oil

Crude oil obtained from an oil well goes through a complex refining procedure in which various fractions are distilled and then condensed to form fuels such as gasoline, kerosine and gas oil. The residue from the refining process is blended with the distillates to form what are called residual fuel oils.

The categories used in industry in the UK are given letters: C1 is the paraffin used in domestic burners; C2 is kerosine; D is known as gas oil (also called diesel oil); E, F and G are the residual fuel oils, known as light fuel oil, medium fuel oil and heavy fuel oil. The specific gravities vary from 0.79 for C2 to 0.97 for G. (The lighter fractions known as gasoline and benzine are important as fuels for cars and aircraft, but are not considered here; note that the use of the name 'petrol' as a fuel for cars is restricted to the UK and is a confusing misnomer since the term petroleum is the generic name for the crude oil.) The approximate ultimate analysis of the various grades of oil used in industry is given in Table 4.2. The percentages given under the heading 'other' in Table 4.2 include moisture and matter which is incombustible. The viscosities given are only indicative values for the purpose of comparison (see below).

Table 4.2 Ultimate analysis of oils used as industrial fuel

Fuel	C %	H %	S %	Other %	GCV (MJ/kg)	Viscosity at 16 °C (kg/m s)
C2, kerosine	86.2	13.6	0.2	—	46.4	0.002
D, gas oil	86.2	12.8	1.0	—	45.5	0.004
E, light fuel oil	83.8	12.1	3.5	0.6	43.4	0.205
F, medium fuel oil	83.4	11.7	4.0	0.9	42.9	1.238
C, heavy fuel oil	82.9	11.4	4.5	1.2	42.5	7.275

It can be seen that the calorific value is not significantly different over the whole range of oils. Of more significance are the physical properties of the oils which affect their storage, pumping, firing and burning. The most important single property of an oil is the viscosity since this is directly related to the pumping power required and also affects the oil's ability to be atomized (i.e. broken into small droplets), an essential prerequisite to combustion. The viscosity of an oil decreases rapidly as the temperature is increased and hence the residual fuel oils must be heated, both for storage and for atomization. For example, a medium fuel oil must be stored at 20 °C, and heated to about 100 °C before delivery to the burner (its viscosity will be reduced by a factor of about six in the process).

Burners for the combustion of oil vary with the density and viscosity of the fuel and with the importance or otherwise of being able to modulate the rate of burning. The very small boilers (say domestic up to 40 kW), burn kerosine using a *vaporizing pot burner*; the kerosine supplied from a reservoir is vaporized and burned with primary air in a pot; the boiler is an on/off type with electric ignition.

Industrial boilers for heating plant burn light fuel oil and may use *low or medium pressure air atomizers*. In these types air is supplied under pressure and entrains the oil in a swirl chamber thus atomizing it; further atomization takes place in the secondary air stream; the turn down ratio is only about 2/1 but can be extended by using oil spill ports.

A *rotary or spinning cup atomizer* can be used for medium sized shell boilers burning medium or heavy fuel oils (see Fig. 4.8); burners of this type can have a turn down ratio of 5/1.

For larger boilers using heavy fuel oil a *pressure jet atomizer* is used in which oil at pressures from 30 bar to 80 bar is forced through tangential slots into a swirl chamber from which it escapes through a small orifice to form a hollow conical sheet in the primary air stream. Atomization can be assisted by supplying high pressure air or steam (see Fig. 4.9). An extension of this principle is used in the *twin–fluid atomizer*; oil and steam under pressure mix in a chamber within the burner tip and atomization takes place as the mixture expands into the air stream; turn down ratios of up to 20/1 can be achieved or alternatively a lower oil pressure can be used, or higher viscosity oils used.

For oil engines and gas turbines the fuel used depends on the application: aircraft gas turbines use an aviation turbine fuel (AVTUR) which is a type of kerosine; high speed oil engines for transport and some industrial applications

Figure 4.8 Rotating-cup oil burner

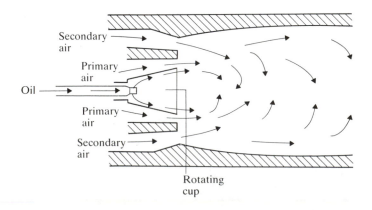

Secondary air
Primary air
Oil
Primary air
Secondary air
Rotating cup

Figure 4.9 Pressure-jet atomizing oil burner

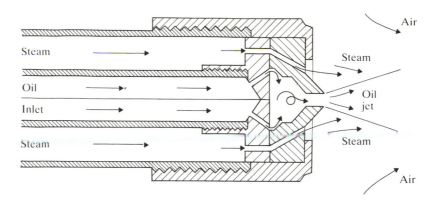

use a gas oil, usually called 'diesel' after Rudolf Diesel, one of the inventors of the compression–ignition engine; most industrial gas turbines use a light fuel oil; large slow-speed oil engines mainly for marine propulsion use one of the heavier fuel oils.

For combustion in an oil engine the fuel is pumped to a pressure as high as 600 bar and injected into the cylinder by a spring controlled needle valve injector; combustion takes place spontaneously when the fine atomized fuel spray encounters the high temperature air in the cylinder; air–fuel ratios of 20 to 25 are commonly used and the maximum cylinder temperatures attained can be as high as 2500 °C. In the combustion chamber of a gas turbine unit the fuel is sprayed into a primary air stream forming a stable flame which is then cooled by secondary air surrounding the flame tube; air–fuel ratios of 60 to 120 are used to keep the gas temperature entering the turbine blades down to the metallurgical limit (about 900 °C, or 1200 °C using blade cooling).

Gas

Since the 1970s the UK has used *natural gas* piped from the North Sea. Before the discovery of this natural resource the gas used was manufactured from coal or light distillates of oil. A feature of all natural gases world wide is their high percentage of methane; an analysis of the natural gas in the UK is given in Table 4.3. At some time in the future the natural gas supply, either indigenous or imported as a Liquefied Natural Gas (LNG), will run down and it will be necessary then to revert to the use of manufactured gas. Research has taken place to produce a *Substitute Natural Gas* (SNG), with properties similar to the natural gas currently in use. For example, British Gas has developed a means of catalytic conversion of gas produced from coal in order to increase the methane content; an approximate analysis of a possible SNG produced in this way is given in Table 4.3, labelled SNG; a town gas manufactured by the formerly used Lurgi process is also shown for comparison.

It is a legal requirement in the UK that the calorific value of the gas supply must be declared and that the value as supplied must be not less than 95 % of the declared value; the current declared value is 38.6 MJ/m^3.

The energy released by a gas of a given calorific value per unit volume, CV, flowing through a burner of given size is proportional to $CV\sqrt{(\Delta p/\rho_G)}$, where Δp is the pressure drop through the orifice of the burner and ρ_G is the density

Table 4.3 Analysis of natural gas, SNG and town gas

	Natural gas (% by vol)	SNG (% by vol)	Town gas (% by vol)
Methane, CH_4	92.6	80.0	27.0
Ethane, C_2H_6	3.6	—	1.2
Propane, C_3H_8	0.8	0.3	—
Butane, C_4H_{10}	0.3	7.0	3.5
Nitrogen, N_2	2.6	—	6.5
Hydrogen, H_2	—	10.0	51.8
Carbon dioxide, CO_2	0.1	2.7	2.0
Carbon monoxide, CO	—	—	8.0

of the gas. In order to compare gases the Wobbe number, W, is defined as follows:

$$W/(MJ/m^3) = \frac{CV/(MJ/m^3)}{\sqrt{(\rho_G/\rho_a)}}$$

(where ρ_a is the density of standard air). The Wobbe number for natural gas is between 48 and 53 MJ/m^3; the SNG in Table 4.3 has approximately the same Wobbe number, but town gas has a Wobbe number of about 27 MJ/m^3.

Liquefied Petroleum Gas (LPG) is the name given to fuels burned as gases which are liquid oils at ambient temperature and a moderate pressure; the oils are in categories C3 and C4 (i.e. between C2 and D in Table 4.2). These are marketed as commercial propane and commercial butane. Propane requires to be pressurized to about 90 bar to transport and store it as a liquid at normal temperatures whereas the equivalent pressure for butane is about 30 bar. When LPG is throttled down through the release value it vaporizes and is burned as a gas. The GCVs of propane and butane on a volume basis are 96 and 122 MJ/m^3 respectively; on a mass basis the corresponding values are 50 and 49.5 MJ/kg.

The simplest type of burner used for natural gas or LPG is the *aerated jet burner*; in this type the gas passes through a perforated pipe entraining air as it flows so that an air–gas mixture is delivered to the flame port. Better control can be achieved with *forced draught*, or *fan-assisted*, burners; air is used to entrain gas to form a combustible mixture as shown diagrammatically in Fig. 4.10. In another form of the fan-assisted type the air and gas are pre-mixed before being supplied to the burner port. A high pressure air stream can be used to form a high velocity jet of burning gas in the furnace which creates turbulence within the furnace and therefore improves heat transfer. It is also

Figure 4.10 Fan-assisted gas burner

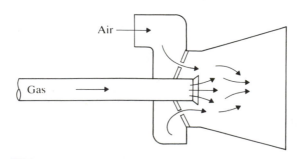

Figure 4.11
Recuperative gas burner

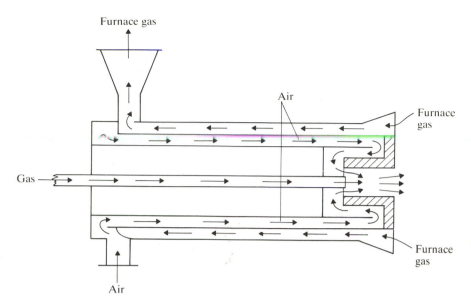

possible to pre-heat the air using furnace gases in a *recuperative burner* (see Fig. 4.11).

The similarity between burners for oil, gas and pulverized fuel will have been noted. Dual fuel burners of the gas/oil type are quite commonly installed in boilers to permit the use of both types of fuel. High pressure gas may be used in some types but in such cases the gas boost system must be carefully designed to prevent blow back of gas into the mains.

4.2 EFFICIENT COMBUSTION

In Chapter 3 it was stated that the highest temperature is attained when a fuel is burned with a stoichiometric supply of air and when combustion is complete. When excess air is supplied it must be heated thus taking useful energy from the process; on the other hand if insufficient air is supplied leading to incomplete combustion the efficiency is even less and pollution may be caused. Therefore in practice excess air is always supplied; in some cases this is also necessary in order to reduce gas temperatures to within metallurgical limits (e.g. in the gas turbine engine).

Equation [3.2] gives a general equation for combustion from which it is possible to calculate the percentage of carbon dioxide and/or oxygen for a given air–fuel ratio. Instruments are available to measure the CO_2 and O_2 in the dry flue gas and hence a continuous check can be made on whether the air–fuel ratio is being maintained as designed. (Note: the instruments dry the flue gas before measurement and hence give the percentage in the dry flue gas.)

Example 4.1
For a bituminous coal, rank 401, supplied as washed singles, using the 'as-fired' analysis given in Table 4.1, construct a graph of the carbon dioxide and oxygen in the dry flue gas plotted against the percentage excess air.

Solution
From Eq. [3.2]:

$$\left(\frac{0.785C}{12} + 0.047H + \frac{0.017N}{14} + \frac{0.019S}{32} + \frac{0.042O}{16}\right)$$

$$+ A\left(\frac{0.233O_2}{32} + \frac{0.767N_2}{28}\right) \rightarrow B\{aCO_2 + bO_2$$

$$+ cSO_2 + (1 - a - b - c)N_2\} + dH_2O$$

From a hydrogen balance, $d = 0.0235$.

Carbon balance: $Ba = 0.785/12 = 0.065\,42$ [1]

Sulphur balance: $Bc = 0.019/32 = 0.000\,59$

Oxygen balance: $\dfrac{0.233A}{32} + \dfrac{0.042}{2 \times 16} = B(a + b + c) + \dfrac{0.0235}{2}$

$\therefore$ $0.007\,28A = Bb + 0.076\,45$ [2]

Nitrogen balance: $\dfrac{0.017}{2 \times 14} + \dfrac{0.767A}{28} = B(1 - a - b - c)$

$\therefore$ $0.027\,39A = B - Bb - 0.066\,62$ [3]

From Eqs [2] and [3] by eliminating B we have:

$$b = \frac{(0.007\,28A - 0.076\,45)}{(0.034\,67A - 0.009\,83)}$$ [4]

A stoichiometric air–fuel ratio, A_s, implies that $b = 0$; this can be found by putting the numerator of Eq. [4] to zero.

i.e. $A_s = 0.076\,45/0.007\,28 = 10.50$

The values of b can then be found from Eq. [4] at excess air percentages of 10 %, 20 %, 30 % etc. by substituting values of $A = 1.1A_s$, $A = 1.2A_s$, $A = 1.3A_s$, etc.; B can be found from either equation [2] or [3]; the fraction of carbon dioxide, a, is then given by Eq. [1]. A graph can then be plotted as shown in Fig. 4.12.

Figure 4.12 CO_2 and O_2 against excess air for bituminous coal

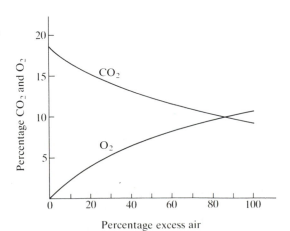

(Note: by giving symbols to the fractions of carbon, hydrogen, sulphur etc. in the fuel and applying Eq. [3.2] the calculation procedure above can be done using a simple computer program and hence graphs can be drawn for coals of any composition.)

A similar method can be applied for oil fuels. For gas fuels the equation is written in terms of volumes throughout; for example, for natural gas we have:

$$(0.926CH_4 + 0.036C_2H_6 + 0.088C_3H_8 + 0.003C_4H_{10}$$
$$+ 0.026N_2 + 0.001CO_2) + A(0.21O_2 + 0.79N_2)$$
$$\rightarrow B\{aCO_2 + bO_2 + (1 - a - b)N_2\} + dH_2O$$

where A is now the air–fuel ratio by volume and B is in kmol dry flue gas per kmol fuel.

Carbon, hydrogen, oxygen and nitrogen balances are written as before; Fig. 4.13 shows the percentages of carbon dioxide and oxygen plotted against percentage excess air for natural gas.

Figure 4.13 CO_2 and O_2 against excess air for natural gas

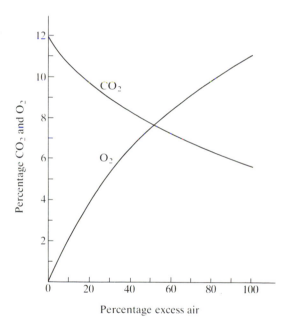

Gas and oil burners should be able to operate down to 15 % excess air with an upper limit of about 20 %, and measures should be taken to adjust the burner if the carbon dioxide in the flue gases is less than about 9 % for natural gas and less than about 13 % for oil. For coal, particularly with mechanical stokers, the range of air–fuel ratios is wider, say 30 to 80 %; a percentage of less than 12 % carbon dioxide is considered to be satisfactory. When the burners are first set up it is important to establish that combustion is complete by monitoring for carbon monoxide in the flue gas. In the case of oil fuel incomplete combustion shows itself as smoke and a visual test is available giving a smoke number. For coal-fired boilers incomplete combustion gives unburnt carbon in the ash; a test for this can be made by heating a representative sample under laboratory conditions and measuring the loss of weight.

Pollution control legislation makes it essential to maintain efficient combustion equipment giving complete combustion and avoiding air–fuel ratios below the stoichiometric value; good energy management makes it economically sensible to maintain the excess air to a minimum while ensuring complete combustion. Monitoring of carbon dioxide and/or oxygen in the dry flue gas is the easiest way of checking that boiler efficiencies are being maintained. Some systems now have automatic control of burner air supply operated from an oxygen sensor in the flue gases.

Boiler efficiency

A boiler is essentially a heat exchanger in which the thermal energy released by the combustion process is transferred from the products of combustion to the coolant fluid (e.g. water or water/steam). The increase in enthalpy of the coolant fluid represents the useful output. An energy balance applied to the system within the dotted line shown in Fig. 4.14 gives the following equation:

$$\dot{m}_F h_F + \dot{m}_A h_A + \dot{m}_F \text{GCV} = \dot{m}(h_2 - h_1) + \dot{m}_G h_G + Q_C \qquad [4.1]$$

The GCV is used in Eq. [4.1] which implies that the water vapour in the products is condensed out. This is unlikely to happen unless a condensing boiler is used (see later) and therefore the NCV is frequently used. Boiler efficiency can then be defined as:

$$\eta_B = \frac{\dot{m}(h_2 - h_1)}{\dot{m}_F \text{GCV}} \qquad \text{or} \qquad \frac{\dot{m}(h_2 - h_1)}{\dot{m}_F \text{NCV}}$$

It must be made clear which calorific value is used when a boiler efficiency is quoted; efficiencies using the NCV will give higher values.

The first term in Eq. [4.1] is the enthalpy of the fuel and is usually negligible compared with the other terms in the equation. The term, Q_C, is the loss from the boiler casing to the atmosphere, usually referred to as the *casing loss*, or *radiation loss*; heat is lost from the casing by both convection and radiation but the latter term is in general use. The casing loss is usually less than 1 % for shell and water tube boilers, but can be as high as 10 % if the boiler is not adequately insulated or if the insulation is allowed to deteriorate.

Figure 4.14 Boiler energy analysis

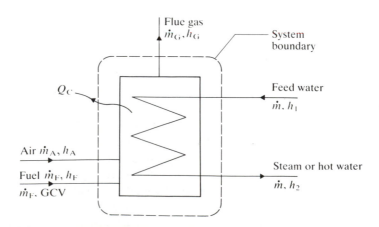

Substituting for $\dot{m}(h_2 - h_1)$ from Eq. [4.1] in the expression for boiler efficiency we have

$$\eta_B = \frac{\dot{m}_F \mathrm{GCV} + m_A h_A - \dot{m}_G h_G - Q_C}{\dot{m}_F \mathrm{GCV}}$$

$$\therefore \qquad \eta_B = 1 - \frac{(\dot{m}_G h_G - \dot{m}_A h_A)}{\dot{m}_F \mathrm{GCV}} - \frac{Q_C}{\dot{m}_F \mathrm{GCV}} \qquad [4.2]$$

The enthalpy of the flue gas should be considered in two parts: the enthalpy of the dry flue gas, and enthalpy of the superheated water vapour. The flue gas loss will then reflect the thermal energy necessary to provide the enthalpy of evaporation for the water vapour formed during combustion, i.e. flue gas loss is given by

$$\{\dot{m}_F + \dot{m}_A - (18d + v)\dot{m}_F\} t_G \Sigma(m_i c_{pi}/m) - \dot{m}_A c_{pA} t_A$$
$$+ (18d + v)\dot{m}_F (h_{wv} - h_f) \qquad [4.3]$$

where d is the number of kmol of water vapour in the flue gas per unit mass of fuel and v is the mass of water vapour in unit mass of fuel supplied; h_{wv} is the specific enthalpy of the superheated water vapour at the partial pressure of the water vapour and the flue gas temperature, t_G; h_f is the specific enthalpy of water at the temperature of the air–fuel mixture at combustion, t_A.

Moisture present in the fuel before combustion, vm_f, is included in the above since it will appear as an increased percentage in the flue gas and has been introduced to the boiler at the approximate temperature, t_A.

For a steam boiler it is necessary to eject some of the water to remove precipitated salts which form a sludge which can foul heat exchange surfaces and can be carried over into the superheater and steam mains. This process is known as boiler *blow-down*. It represents a loss of money (a) in the reduction in boiler efficiency due to the loss of energy of the hot water blown down and (b) in the fraction of the cost of chemically treating this water since the feed water is treated before entering the boiler. The most efficient system of blow down is one which is fully automatic with continuous measurement of the total dissolved solids in the boiler. Blow down can be as high as 10 % or more of the steam flow rate which represents a loss of about 3 % of the fuel energy input. This loss can be reduced by recovering part of the thermal energy of the blow down water by using it to heat the feed water; the most effective way of doing this is to allow the flash steam formed from the blow down to mix direct with the feed water; this also recovers some of the cost of the water treatment.

Equation [4.2] can be modified to include a term for the blow down loss. The useful output of the boiler is, $\dot{m}(h_2 - h_1)$, and the boiler efficiency is given by the useful output divided by the energy from the fuel input, as before, hence Eq. [4.2] still applies with the inclusion of a term to allow for the blow down loss

i.e. $\qquad$ Blow down loss $= x\dot{m}(h_{f2} - h_w) \qquad\qquad$ [4.4]

where x is the fraction of the steam flow which is lost as blow down, and h_w is the enthalpy of the make up water. Note that when calculating boiler efficiency h_w is always equal to h_1; any recovery of the blow down thermal energy affects the efficiency of the steam cycle but not strictly the boiler efficiency since it is the enthalpy of the feed water entering the boiler which is important.

We can now re-write Eq. [4.2] to include the additional terms given by Eqs [4.3] and [4.4], i.e. boiler efficiency is given by

$$\eta_B = 1 - FGL - BDL - CL \qquad [4.5]$$

where *FGL*, *BDL* and *CL* represent the flue gas, blow down, and casing losses divided by the total energy input from the fuel.

It can be seen from Eq. [4.2] that the efficiency can be increased by decreasing the enthalpy and hence the temperature of the flue gases. Now a boiler is designed for a given flow of feed water, a given exit condition of the working fluid and for a given feed water temperature (a steam cycle is more efficient when the condensate is returned or if feed heating is used, see Section 4.4); therefore from the definition of effectiveness given in Section 3.2 it can be seen that for a given combustion temperature and a given feed temperature the flue gas temperature depends on the overall heat transfer coefficient and the heat transfer surface area. The temperature of the flue gas is therefore fixed by the design of the boiler.

An *economizer* may be incorporated into the design of a boiler; this is an additional heat exchanger in the flue passage which uses energy in the flue gas to pre-heat the feed water before it enters the main boiler. Similarly a heat exchanger in the flue passage may be used to pre-heat the air used for the combustion. A boiler should be designed with the minimum flue gas temperature either with or without these additional heat exchangers; adding an economizer or air preheater to an existing boiler is therefore only feasible if the original boiler efficiency was low because of a designed high value of flue gas temperature.

Referring to Fig. 4.15 it can be seen that Eq. [4.2] is still applicable for a boiler with an air preheater and an economizer if the boundary for the energy balance is redrawn as shown. Pre-heating the combustion air also has the advantage of increasing the flame temperature leading to a possible reduction in the excess air required for a given fuel; alternatively with air at a higher temperature a lower grade of fuel may be burned. Air preheaters are either

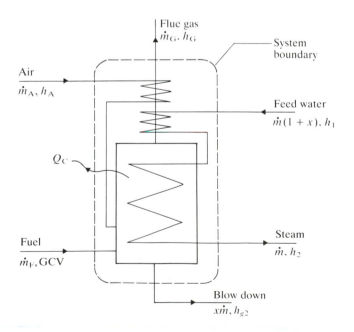

Figure 4.15 Energy analysis including economizer and air pre-heater

recuperative or regenerative; in the latter case a rotary regenerator is commonly used (see Section 5.4). The recuperative gas burner discussed earlier is also a type of air preheater.

Thermal energy recovered from the flue gas can also be used for other purposes such as space-heating, hot water heating etc. and such applications will be considered again in later chapters. When the flue gas temperature is reduced below the dew point of the flue gas then condensation of the water vapour takes place. Dew point was defined in Section 3.1 and Example 3.2 shows how to calculate dew point temperature. For a fuel containing sulphur the sulphur dioxide in the flue gas may be oxidized further to sulphur trioxide which then forms sulphuric acid when combined with water vapour in the reaction

$$SO_3 + H_2O \rightarrow H_2SO_4$$

The *acid dew point* occurs at a higher temperature than the water vapour dew point and since sulphuric acid is highly corrosive it is important not to cool the metal heat exchange surfaces below the acid dew point temperature. It can be seen from Example 3.2 that water vapour dew points are of the order of $60\,°C$; acid dew points for sulphur-based fuels are of the order of $150\,°C$ depending on the sulphur content of the fuel and the excess air percentage. It is found that the corrosion rate is highest at about 40 K below the acid dew point. Metal heat exchange surfaces should be maintained at temperatures above 150 to $160\,°C$ in flue gas containing sulphur dioxide.

(Note: coal and oil fuels contain sulphur and hence sulphur dioxide is present in the flue gas; when the flue gas passes into the atmosphere the interaction with the cloud layer and with ozone can lead to the formation of *acid rain* which poisons the soil, rivers and lakes leading to the destruction of trees and fish stock. Dry sulphur is also deposited where the flue gas plume touches the ground up to 20 km from the chimney, causing damage to crops. Concern expressed by countries throughout Europe has led to a policy of installing plant in power stations to treat the flue gas before releasing it into the atmosphere.)

Example 4.2

An oil-fired boiler burning a medium fuel oil generates steam. The blow down water is flashed into a vessel and the flash steam thus formed is recovered into the feed water with the remaining blow down water passing to waste. Using the data below calculate:

(i)   the percentage flue gas loss;
(ii) the percentage loss due to heat loss from the casing;
(iii) the percentage blow down loss;
(iv) the boiler efficiency;
(v)   the total annual cost of blow down;
(vi) the reduction in fuel consumption and hence annual fuel cost if the boiler is redesigned for a flue gas temperature of $177\,°C$, with all the other data unchanged.

Data

Fuel: medium fuel oil of composition by mass, 83.4 % C, 11.7 % H, 4.0 % S, 0.9 % incombustibles; GCV, 42.9 MJ/kg; mass flow rate of fuel supplied, 175 kg/h; fuel cost, 0.25 p/MJ. Steam produced, 2250 kg/h, dry saturated at 8 bar; temperature of feed water entering boiler, $30\,°C$; blow down, 10 % of steam mass flow rate; blow down flash vessel at 1.1 bar; temperature of make-up water for blow down, 15 °C; cost of chemical treatment of water, 0.02 p/kg. Excess air, 20 %; air temperature at entry, 25 °C. Flue gas temperature at boiler exit, 227 °C; flue gas pressure at boiler exit, 1.027 bar. Heat loss from casing, 80 kW.

The boiler is in use at full load for 6000 h/annum.

Specific heats at constant pressure, c_p, are as given below.

Gas	c_p at 500 K (kJ/kg K)	c_p at 450 K (kJ/kg K)	c_p at 298 K (kJ/kg K)
Air	—	—	1.0048
Carbon dioxide	1.014	0.978	
Oxygen	0.972	0.956	
Sulphur dioxide	0.812	0.800	
Nitrogen	1.056	1.049	

Solution

(i) To find the flue gas loss it is necessary to calculate the air–fuel ratio, A, and the composition of the flue gas in order to calculate the enthalpy. The combustion equations for the three constituents are:

$$C + O_2 \rightarrow CO_2 \qquad H_2 + \tfrac{1}{2}O_2 \rightarrow H_2O \qquad S + O_2 \rightarrow SO_2$$

Hence the oxygen required for complete combustion is

$$\left(0.834 \times \frac{32}{12}\right) + \left(0.117 \times \frac{0.5 \times 32}{2}\right) + \left(0.04 \times \frac{32}{32}\right)$$
$$= 3.2 \text{ kg/kg fuel}$$

i.e. the stoichiometric air–fuel ratio is given by

$$A_s = 3.2/0.233 = 13.734$$

Therefore the actual air–fuel ratio is given by

$$A = 13.734 \times 1.2 = 16.481$$

Then using Eq. [3.2]:

$$\frac{0.834C}{12} + 0.117H + \frac{0.04S}{32} + 16.481\left(\frac{0.233O_2}{32} + \frac{0.767N_2}{28}\right)$$
$$\rightarrow B\{aCO_2 + bO_2 + cSO_2 + (1 - a - b - c)N_2\} + dH_2O$$

Hydrogen balance: $d = 0.117/2 = 0.0585$
Carbon balance: $0.834/12 = Ba$ $\quad \therefore \quad$ $Ba = 0.0695$
Sulphur balance: $0.04/32 = Bc$ $\quad \therefore \quad$ $Bc = 0.001\,25$
Oxygen balance: $16.481 \times 0.233/32 = Ba + Bb + Bc + 0.0585/2$
$\quad \therefore \quad$ $Bb = 0.12 - 0.0695 - 0.001\,25 - 0.029\,25 = 0.02$
Nitrogen balance: $16.481 \times 0.767/28 = B - Ba - Bb - Bc$
$\quad \therefore \quad$ $B = 0.451\,46 + 0.0695 + 0.02 + 0.001\,25 = 0.542\,21$
Then $a = 0.1282$, $b = 0.0369$, $c = 0.0023$, $1 - a - b - c = 0.8326$

Since the analysis by volume of the dry flue gas is known rather than the analysis by mass, it is convenient to find the enthalpy by using molar specific heats (see

Eq. [3.6]).

i.e.
$$C_p \text{ for } CO_2 = 1.014 \times 44 = 44.616 \text{ kJ/kmol K}$$
$$C_p \text{ for } O_2 = 0.972 \times 32 = 31.104 \text{ kJ/kmol K}$$
$$C_p \text{ for } SO_2 = 0.812 \times 64 = 51.968 \text{ kJ/kmol K}$$
$$C_p \text{ for } N_2 = 1.056 \times 28 = 29.568 \text{ kJ/kmol K}$$

Then for the dry flue gas

$$C_p = (0.1282 \times 44.616) + (0.0369 \times 31.104) + (0.0023 \times 51.968)$$
$$+ (0.8326 \times 29.568) = 31.605 \text{ kJ/kmol K}$$

The molar mass of the dry flue gas is given by:

$$M = (0.1282 \times 44) + (0.0369 \times 32) + (0.0023 \times 64)$$
$$+ (0.8236 \times 28) = 30.282 \text{ kg/kmol}$$

Hence, for the dry flue gas:

$$c_p = 31.605/30.282 = 1.044 \text{ kJ/kg K}$$

(Note that since from the previous calculation we have $B = 0.542\,21$ kmol dry flue gas per kg fuel then

$$B/\{A + 1 - (0.0585 \times 18)\} = 0.542\,21/16.428$$
$$= 0.033 \text{ kmol/kg dry flue gas}$$
$$\therefore \quad c_p \text{ for the flue gas} = 31.605 \times 0.033$$
$$= 1.044 \text{ kJ/kg K as above.})$$

To find the state of the water vapour in the flue gas it is necessary to find the partial pressure of the water vapour using Eq. [3.1], i.e. the fraction of water vapour by volume in flue gas is

$$0.0585/(0.542\,21 + 0.0585) = 0.0974$$

Therefore, partial pressure $= 0.0974 \times 1.027 = 0.1$ bar i.e. enthalpy of superheated steam at 0.1 bar and 227 °C from tables $= 2932.4$ kJ/kg.
The enthalpy of the water vapour at combustion is h_f at 25 °C $= 104.8$ kJ/kg.
Then using Eq. [4.3]

$$\text{flue gas loss} = \frac{\{(1 + A - 18d) \times 1.044 \times 227\} - (A \times 1.0048 \times 25)}{42.9 \times 10^3}$$
$$+ \frac{18d(2932.4 - 104.8)}{42.9 \times 10^3}$$
$$= \frac{(16.428 \times 1.044 \times 227) - (16.481 \times 1.0048 \times 25)}{42\,900}$$
$$+ \frac{(18 \times 0.0585 \times 2827.6)}{42\,900}$$
$$= 0.0811 + 0.0694 = 0.1505 \text{ or } 15.05\,\%$$

(Note that in this case there is no water vapour in the fuel supplied and hence v is zero in Eq. [4.3].)

(ii)
$$\text{Casing loss} = \frac{80 \times 3600}{175 \times 42\,900} = 0.0384 \text{ or } 3.84\,\%$$

(iii)   The blow down loss can be found from Eq. [4.4], where the enthalpy of the blow down water is the saturated value at the steam pressure of 8 bar, and the water entering the boiler is at 30 °C.

i.e.   Blow down loss $= \dfrac{0.1 \times 2250 \times (721 - 125.7)}{175 \times 42\,900} = 0.0178$ or 1.78 %

(iv)   The boiler efficiency can be found from Eq. [4.5]:

$$\text{Boiler efficiency} = 1 - 0.1505 - 0.0178 - 0.0384$$
$$= 0.793 \text{ or } 79.3 \%$$

(Note that the boiler efficiency can also be obtained in this example from the energy output of the steam produced divided by the fuel energy input.

i.e. $\eta_B = 2250 \times (2769 - 125.7)/(175 \times 42\,900) = 0.792$)

(v)   To find the total cost of the blow down it is necessary to calculate the actual mass flow rate of water which goes to waste so that the cost of chemical treatment can be included. The hot saturated water at 8 bar is flashed into a vessel at 1.1 bar, therefore

$$721 = 429 + x2251$$

where x is the proportion of dry saturated steam in the total.

i.e. $x = 0.1297 \qquad 1 - x = 0.87$

Hence the mass flow rate of water wasted is $0.87 \times 0.1 \times 2250 = 195.8$ kg/h. Since the make-up water must be treated then the cost of treatment in one year is $195.8 \times 6000 \times 0.02/100 = £234.9$. The feed water is introduced to the boiler at 30 °C and therefore the energy cost can be calculated as

$$\dfrac{0.1 \times 2250 \times (721 - 125.7) \times 6000 \times 0.25}{1000 \times 100 \times 0.793} = £2532.6$$

total annual cost of blow down $= £234.9 + £2532.6 = £2767.5$

(vi)   The flue gas temperature is now 177 °C and hence the specific heat of the dry flue gas must be recalculated; the analysis by volume is unchanged.

i.e. $C_p = (0.1282 \times 44 \times 0.978) + (0.0369 \times 32 \times 0.956)$
$$+ (0.0023 \times 64 \times 0.8) + (0.8326 \times 28 \times 1.049)$$
$$= 31.218 \text{ kJ/kmol K}$$

Therefore for the dry flue gas

$$c_p = 31.218/30.282 = 1.031 \text{ kJ/kg K}$$

(This compares with the values of 1.044 kJ/kg K at a flue gas temperature of 227 °C.)

Using Eq. [4.3] the new flue gas loss can be calculated as

$$FGL = \dfrac{(16.428 \times 177 \times 1.031) - (16.481 \times 1.0048 \times 25)}{42\,900}$$
$$+ \dfrac{18 \times 0.0585 \times (2835.4 - 104.8)}{42\,900}$$
$$= 0.0602 + 0.0670 = 0.1272 \text{ or } 12.72 \%$$

The blow down and casing losses are independent of the mass flow rate of fuel. The casing loss is given as 80 kW, and the blow down loss from (iii) above is $0.1 \times 2250 \times (721 - 125.7)/3600 = 37.06$ kW. The initial boiler output was $0.793 \times 175 \times 42\,900/3600 = 1653.7$ kW, and this is unchanged.

Let the new mass flow rate of fuel be $\dot{m}_F$, then

$$\dot{m}_F \times 42\,900 = 1653.7 + 80 + 37.06 + (0.1272 \times \dot{m}_F \times 42\,900)$$

i.e.

$$\dot{m}_F = 1770.8/(0.8728 \times 42\,900)$$
$$= 0.0473 \text{ kg/s} = 170.26 \text{ kg/h}$$

Therefore

$$\text{decrease in annual fuel cost} = \frac{(175 - 170.26) \times 6000 \times 42.9 \times 0.25}{100}$$
$$= \pounds 3050$$

(Note that the boiler efficiency has increased from 79.3 % to $79.3 \times 175/170.26 = 81.5$ %.)

The savings due to this redesign represent only $(175 - 170.26)/175 = 2.7$ % of the fuel cost.

The reduction in flue gas temperature, as in the above example, may be achieved by adding or modifying baffles within the flue passes thus improving the heat transfer.

When burners are badly adjusted thus giving too high an excess air–fuel ratio the flue gas losses will increase approximately in direct proportion to the increase in the mass flow of gases. In the above example for the flue gas temperature of 227 °C an increase in excess air from 20 % to 30 % would increase the dry flue gas loss from 8.11 % to $(8.11 \times 1.3/1.2) = 8.79$ %. Since the loss due to the water vapour is approximately unchanged then the increase in the flue gas loss is from 15.05 % to 15.73 %. In practice the increased flow through the boiler flue passages will increase the convective heat transfer but the reduced flame temperature will reduce the radiant heat transfer. For the convective heat transfer applying Eq. [3.28] we have:

$$\dot{m}_G c_{pG}(t_1 - t_G) = UA(LMTD)$$

(where t_1 is the temperature of the gases leaving the radiant section and entering the tubes). The thermal resistances of the metal tube wall and the water film on the outside surface of the tube are negligible and hence we can assume that the inside surface of the tube is approximately equal to the saturation temperature of the steam, t_g. Therefore the $LMTD$ is given by $(t_1 - t_G)/\ln\{(t_1 - t_g)/(t_G - t_g)\}$, and then substituting in the equation above we have:

$$\ln\{(t_1 - t_g)/(t_G - t_g)\} = UA/m_G c_{pG}$$

The heat transfer by convection is controlled by the heat transfer coefficient for the inside of the tube and it was shown in Chapter 3 that the heat transfer coefficient varies with the mass flow rate to the power 0.8 (Eq. [3.18]). Hence for the case where only the mass flow rate changes (i.e. neglecting changes in properties due to the change in the temperatures) we have:

$$UA/m_G c_{pG} = k/m_G^{0.2}$$

(where k is a constant). Then

$$t_G = t_g + (t_1 - t_g)/e^{(k/m_G^{0.2})} \qquad [4.6]$$

For the case of an increase in excess air supplied to an existing shell boiler the temperature t_1 will be reduced (see Section 3.4). Then referring to Eq. [4.6] it can be seen that as the mass flow of gases increases $k/m_G^{0.2}$ decreases and hence $e^{k/m_G^{0.2}}$ decreases. This offsets the decrease in t_1.

Approximate formulae

For speed of calculation it is found convenient to use approximations for the dry flue gas loss and the flue gas water vapour loss. A formula for dry flue gas loss due to Siegert is as follows:

$$\text{percentage dry flue gas loss} = \frac{C \times \{(t_G - t_A)/(\text{K})\}}{(\%\ CO_2\ \text{in dry flue gas})} \qquad [4.7]$$

Values of C can be assumed based on either the GCV or the NCV; some examples are given in Table 4.4.

Table 4.4 Values of constant C for Eq. 47

Fuel	C for GCV basis	C for NCV basis
Fuel oil	0.53	0.56
Natural gas	0.35	0.38
Bituminous coal	0.61	0.63

A formula for the water vapour loss is given as follows:

$$\text{percentage loss} = C' \times \{1121 + (t_G - t_A)/(\text{K})\} \qquad [4.8]$$

Some values of C' are given in Table 4.5.

Table 4.5 Values of constant C' for Eq. 48

Fuel	C'
Fuel oil	0.0051
Natural gas	0.0083
Coal	0.0041

Note: Eqs [4.7] and [4.8] are only approximations and the value of C' for coal in particular can vary widely depending on the amount of moisture present in the coal; the equations are very useful none-the-less for estimating the possible effect of excess air and flue gas temperature on the flue gas loss; the flue gas loss is normally between 15 and 25 % of the total losses. Applying Eqs [4.7] and [4.8] to the data of Example 4.2 we have:

$$\text{dry flue gas loss} = 0.53(227 - 25)/12.82 = 8.35\ \%$$
$$\text{water vapour loss} = 0.0051\{1121 + (227 - 25)\} = 6.75\ \%$$

These figures compare with the original answers of 8.11 % and 6.94 %; the total flue gas loss of 15.10 % compares with the original answer of 15.05 %. For part (vi) of Example 4.2 the flue gas temperature is 177 °C and the approximate formulae give values of 6.28 % (compared with 6.02 %), and 6.49 % (compared with 6.70 %) with a total flue gas loss of 12.77 % (compared with 12.72 %).

Boilers at part load

The simplest form of part load working is achieved by using an on–off burner which therefore gives cyclic operation. An improvement on this method is obtained by using a burner with a high and a low setting, and a further improvement in performance is obtained by using a fully modulating burner with as great a turn down ratio as possible.

The *load factor* is defined as the ratio of the rated input at part load to that at full load; for an on–off burner the load factor is therefore the percentage of burner on-time. It is important to make the load factor as high as possible.

A modulating boiler may have an efficiency variation with load of the form shown in Fig. 4.16; note that the part load efficiency is greater over a considerable part of the range.

Figure 4.16 Variation of boiler efficiency with load

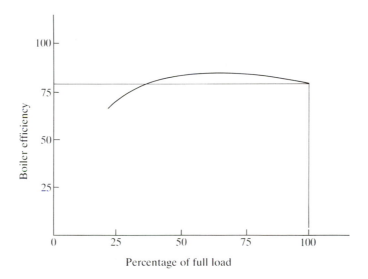

When there is a wide variation of load during the year as with space-heating, for example, then the use of a multiple boiler installation is recommended. The boilers are controlled in sequence with an additional boiler brought into use whenever this would make the system more efficient; this is illustrated in Fig. 4.17.

Condensing boilers

Considerable improvements in efficiency can be obtained if the water vapour in the flue gas is condensed by further cooling; the maximum increase in energy obtainable is approximately equal to the difference between the GCV and NCV. If the flue gas temperature is reduced below the water vapour dew point then condensation takes place which requires the flue gas to reject heat equivalent to the enthalpy of evaporation of the water vapour at its partial pressure. Boilers

Figure 4.17 Boiler efficiency for multi-boiler plant

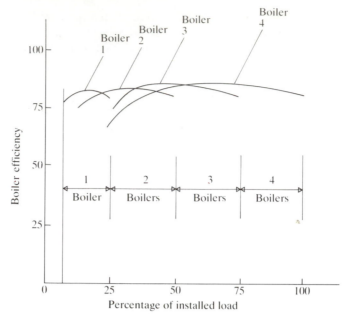

Figure 4.17 Boiler efficiency for multi-boiler plant

which make use of the energy of the water vapour in this way are known as condensing boilers.

From the previous discussion on acid dew point it will be seen that the problems of corrosion rule out the use of condensing boilers for fuels containing sulphur; such boilers are therefore not suitable for coal or oil but can be used for natural gas. Condensing boilers are usually designed with a secondary heat exchanger for the condensing mode with the feed water entering at temperatures between 30 °C and 50 °C depending on the design; efficiencies as high as 95 % can be obtained. Such boilers are now available from a number of manufacturers, mainly for space-heating, in the range from 20 kW up to about 1.5 MW. Typical examples are shown diagrammatically in Figs 4.18(a) and (b).

Figure 4.18 Condensing boilers

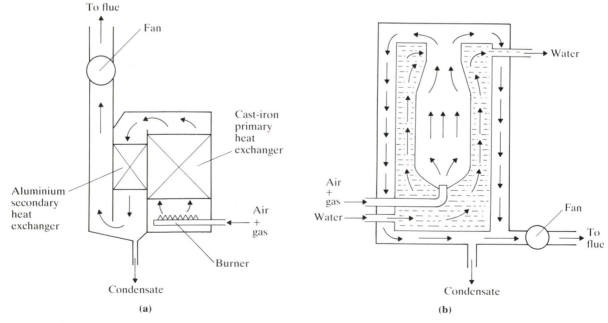

Since condensate is continuously formed a means for its removal is necessary; there is also the possibility of corrosion unless suitable materials are selected for the secondary heat exchanger (e.g. stainless steel or aluminium). With the lower flue gas temperature a fan is needed to ensure adequate dispersal of the products of combustion from the chimney; since the flue gas entering the outside air is at a low temperature, perhaps as low as 35 °C say, then further condensation may take place forming a vapour cloud in the chimney plume.

A condensing boiler, because of the more expensive additional secondary heat exchanger, the condensate draining system, and the additional fan, is about twice as expensive as an equivalent conventional heating boiler (say about £16/kW compared with about £8/kW). Taking an improvement in efficiency of from say 80 % to 90 %, the annual fuel saving could be about £3/kW for a normal heating season length. The boiler would therefore pay for itself in about three years.

Example 4.3

A gas-fired condensing boiler with an output of 1.2 MW operates with 20 % excess air and an exit flue gas temperature of 40 °C. Condensate is collected at a rate of 0.65 kg/m^3 of fuel supplied. Using the data below, calculate:

(i)   the boiler efficiency;
(ii) the mass flow rate of condensate formed;
(iii) the annual saving due to the reduction in gas consumption compared with a conventional boiler with an efficiency of 80 % based on GCV;
(iv) the break-even time if the conventional boiler costs £13 000 and the condensing boiler costs £24 000.

Data
Take the analysis by volume of natural gas given in Table 4.3. GCV, 38.6 MJ/m^3. Price of gas, 37 p/therm. Air inlet at 20 °C and 1.013 bar. Temperature of condensate, 35 °C. Casing loss 1.5 %. Boiler usage at full load, 2000 h/annum. The values of specific heats at constant pressure for the constituents of the dry flue gas at the relevant temperature may be taken from Tables such as those of Rogers and Mayhew.[4.2]

Solution
(i)   It is necessary to find the stoichiometric air–fuel ratio and hence the actual air–fuel ratio so that the actual mass flow rate of products including water vapour may be found. Taking each of the constituents in turn:

$$CH_4 + 2O_2 \rightarrow CO_2 + 2H_2O$$
$$C_2H_6 + 3\tfrac{1}{2}O_2 \rightarrow 2CO_2 + 3H_2O$$
$$C_3H_8 + 5O_2 \rightarrow 3CO_2 + 4H_2O$$
$$C_4H_{10} + 6\tfrac{1}{2}O_2 \rightarrow 4CO_2 + 5H_2O$$

Therefore

$$\begin{aligned}
\text{oxygen required} &= (2 \times 0.926) + (3.5 \times 0.036) + (5 \times 0.008) \\
&\quad + (6.5 \times 0.003) \\
&= 2.0375 \text{ kmol/kmol fuel}
\end{aligned}$$

$$\begin{aligned}
\text{water vapour formed} &= (2 \times 0.926) + (3 \times 0.036) + (4 \times 0.008) \\
&\quad + (5 \times 0.003) \\
&= 2.007 \text{ kmol/kmol fuel}
\end{aligned}$$

$$\text{carbon dioxide formed} = 0.936 + (2 \times 0.036) + (3 \times 0.008)$$
$$+ (4 \times 0.003)$$
$$= 1.044 \text{ kmol/kmol fuel}$$

The stoichiometric air–fuel ratio by volume is then $2.0375/0.21 = 9.702$, and the actual air–fuel ratio is $9.702 \times 1.2 = 11.643$. Therefore the oxygen in the flue gas is $11.643 \times 0.21 = 2.445$ kmol/kmol fuel, and the nitrogen in the flue gas is $0.0026 + (11.643 \times 0.79) = 9.201$ kmol/kmol fuel. The carbon dioxide in the flue gas is $(0.001 + 1.044) = 1.045$ kmol/kmol fuel. The total number of kmol of dry flue gas is therefore $(1.045 + 2.445 + 9.201) = 12.691$ kmol, and the fraction by volume of each gas is

$$1.045/12.691 = 0.0823 \text{ CO}_2 \qquad 2.445/12.691 = 0.1927 \text{ O}_2$$
$$9.201/12.691 = 0.7250 \text{ N}_2$$

Then, for the flue gas

$$C_p = (0.0823 \times 44 \times 0.858) + (0.1927 \times 32 \times 0.920)$$
$$+ (0.725 \times 28 \times 1.04)$$
$$= 29.892 \text{ kJ/kmol K}$$

For the dry flue gas

$$M = (0.0823 \times 44) + (0.1927 \times 32) + (0.725 \times 28)$$
$$= 30.089 \text{ kg/kmol}$$

Hence for the dry flue gas at $30\,^\circ\text{C}$

$$c_p = 29.892/30.089 = 0.994 \text{ kJ/kg K}$$

The mass flow of dry flue gas is given by

$$(\text{number of kmol dry flue gas}) \times (\text{molar mass})$$

which is equal to

$$12.691 \times 30.089 = 381.86 \text{ kg/kmol fuel}$$

For the gaseous fuel at 1.0125 bar and $20\,^\circ\text{C}$ we have

$$n/V = p/R_o T$$
$$\therefore \qquad n/V = 1.013 \times 10^5/(8314.4 \times 293) = 0.04158 \text{ kmol/m}^3$$

Therefore the mass flow of dry flue gas is

$$381.86 \times 0.041\,58 = 15.878 \text{ kg/m}^3 \text{ fuel}$$

Similarly the mass flow of water vapour formed is

$$18 \times 2.007 \times 0.041\,58 = 1.502 \text{ kg/m}^3 \text{ fuel}$$

(Note: this is the maximum theoretical condensation possible.) The mass of condensate formed in this case is given as 0.65 kg/m^3. The enthalpy of the condensate at $35\,^\circ\text{C}$ is approximately equal to h_f at $35\,^\circ\text{C} = 146.6$ kJ/kg. The enthalpy of the superheated water vapour remaining in the flue gas at exit depends on its partial pressure which in turn depends on the fraction by volume of the water vapour. However, at low partial pressures it is a good approximation to assume that the enthalpy of superheated water vapour is equal to the dry saturated value at the same temperature, i.e. enthalpy of water vapour in flue gas at exit is approximately 2573.7 kJ/kg ($= h_g$ at 40°C). We then have:

> flue gas loss = enthalpy of dry flue gas per unit volume of fuel
> + enthalpy of water vapour in flue gas per unit volume of fuel
> + enthalpy of condensate formed per unit volume of fuel
> − enthalpy of inlet air per unit volume of fuel
> − enthalpy of water vapour at inlet air temperature per unit volume of fuel

i.e.

$$\text{flue gas loss} = (15.878 \times 0.994 \times 40) + \{(1.502 - 0.65) \times 2573.7\}$$
$$+ (0.65 \times 146.6)$$
$$- (11.643 \times 0.041\,58 \times 29 \times 1.0042 \times 20)$$
$$- (1.502 \times 83.9)$$

(where the molar mass of air is taken as 29 kg/kmol and the specific heat at constant pressure at 20°C is 1.0042 kJ/kg K)

$$\therefore \quad \text{flue gas loss} = 2511.4 \text{ kJ/m}^3$$

Then

$$\% \, FGL = 100 \times 2511.4/38\,600 = 6.51\,\%$$

and

$$\text{Boiler efficiency} = 100 - 6.51 - 1.5 = 92.0\,\%$$

(ii)   The volume flow rate of gas supplied can be found using the GCV given, i.e. the volume flow rate of fuel is given by

$$\dot{V} = 1.2/(0.92 \times 38.6) = 0.0338 \text{ m}^3/\text{s}$$

Therefore,

$$\text{mass flow rate of condensate} = 0.65 \times 0.0338 \times 3600$$
$$= 79.08 \text{ kg/h}$$

(iii)   For the same output of 1.2 MW with a conventional boiler of 80% efficiency, annual fuel cost is

$$\frac{1.2 \times 37 \times 2000 \times 3600}{0.8 \times 105.5 \times 100} = £37\,877$$

(where 1 therm = 105.5 MJ).

For the condensing boiler annual fuel cost is

$$\frac{0.8}{0.92} \times £37\,877 = £32\,937$$

Therefore

$$\text{annual saving} = £4940$$

(iv)   Not including discount factors and neglecting all other annual costs a simple break-even point can be obtained

i.e.

$$\text{break-even point} = (24\,000 - 13\,000)/4940$$
$$= 2.23 \text{ years}$$

The condensing boiler will therefore start making an annual saving after 2.23 years which makes it an economic proposition.

4.3 WASTE AS A FUEL

The disposal of waste is a major problem for all developed nations with the amount of waste tending to increase in proportion to the affluence of the population. It is becoming increasingly important to conserve scarce resources by reclaiming and recycling them as much as possible from waste; for example it is possible to recycle empty glass bottles and paper, and to reclaim ferrous and non-ferrous metals. Industry can save money by collecting different categories of waste separately and local communities can install bottle banks and waste paper collection systems. Expensive systems for metal separation from solid or even liquid refuse are available but it is cheaper if separation takes place at the source as far as possible.

For any reclamation system to work it must be economically worthwhile for the participants; there must be a market for the reclaimed material. The UK set up a Reclamation Council in 1984 with its main objective to stimulate interest in the need for economic reclamation and recycling of waste. There is also a special unit at Warren Springs Laboratory, Stevenage set up by the Department of Trade and Industry to provide technical advice on recycling of waste.

It has been estimated that the UK produces 400 to 500 million tonnes (Mt) of waste each year in the broad categories shown in Table 4.6. These figures must be taken as approximate estimates only.

Table 4.6 Waste produced in the UK each year

Waste	Annual amount (Mt)	Percentage
Agricultural (crop residue)	14	3
Agricultural (animal waste)	155	36
Agricultural (wood residue)	5	1
Mineral	115	27
Industrial	85	20
Sewage	28	6.5
Municipal and medical	17	4
Dredged spoil	11	2.5

In this section we are concerned with the possibility of using waste as a fuel. Some waste which is combustible is also useful if recycled and the decision to burn or to recycle will depend on the relative economic advantages; for example recycled paper and cardboard can be used for packaging or can even be recycled and used for print.

The potential of solid waste as a fuel is shown by the calorific values on a moisture free basis given in Table 4.7; the calorific values of some fuels are given as a comparison.

Municipal waste is made up of varying proportions of most of the constituents in Table 4.7 and the average GCV will vary also with the amount of moisture and incombustibles present. (It is also interesting to note that the calorific value of municipal waste increases as the standard of living increases, due mainly to

Table 4.7 GCV of some typical waste materials

Constituent	GCV (MJ/kg)
Vegetable	6.7
Dust	9.6
Paper	14.6
Rags	16.3
Wood	17.6
Straw	18.0
Animal waste	18.0
Rubber	34.7
Plastics	37.0
Coal	26.4
Oil	44.0
Natural gas	52.4 (38.6 MJ/m^3)

Note: The figures for coal and oil in the above table are the values used by the Department of Energy when quoting UK Energy Consumption in Mt of Coal Equivalent and Mt of Oil Equivalent

an increase in the proportions of paper and plastics; of course the value will decrease if recycling and reclamation are introduced for combustibles such as paper.) Table 4.8 gives the proximate analysis and GCV of municipal waste as given by J H S Smart for the Greater London Plant at Edmonton in 1986.[4.3]

Table 4.8 Analysis of municipal waste at Edmonton, London

Constituent	% by mass
Fixed carbon	12.5
Volatiles	34.3
Moisture	25.2
Ash	28.0
GCV	10.68 MJ/kg

Hospital waste may have a GCV as high as 19 MJ/kg; the higher figure compared with domestic waste is due in part to the much lower moisture content. The composition of industrial waste varies widely according to the industry from which the waste comes; an average figure is about 16 MJ/kg.

If the total municipal, medical and industrial waste generated each year in the UK as given by the figures in Table 4.6 were converted into energy it would yield about 60 Mt coal equivalent. This compares with a total annual energy consumption for the UK of about 330 Mt coal equivalent. It can be seen that even if only a fraction of this waste were converted into heat and/or power a considerable saving in fossil fuels would be made. Whether this saving in the use of fossil fuel would also lead to a saving in money would depend on the capital and running costs of the plant necessary to prepare and burn the waste compared with the current cost of disposing of the waste. This point will be considered again later.

There are four main ways in which waste can be used as a substitute fuel:

(a) direct incineration;
(b) production of Refuse Derived Fuel (RDF);
(c) pyrolysis;
(d) production of gas through bio-degradation.

Each of these methods will be considered in turn.

Direct incineration

Without pre-sorting

The most direct way of using solid refuse as a fuel is to incinerate it without pre-sorting; this is only feasible with a large plant such as the Greater London refuse-fired power station at Edmonton. A diagrammatic arrangement of the Edmonton plant is given in Figs 4.19 and 20 by permission of the Institution of Mechanical Engineers.

Figure 4.19 Schematic arrangement of Edmonton incineration plant

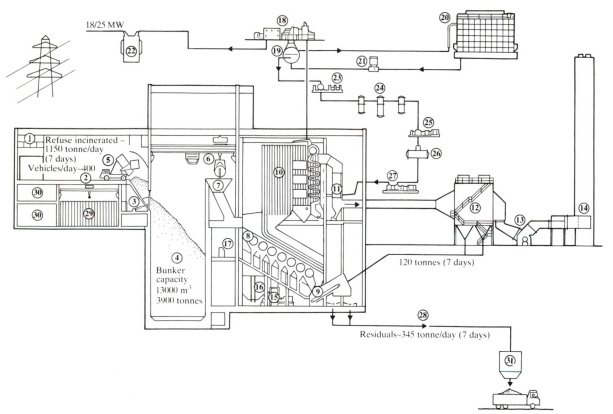

1 Control room	**12** Electrostatic precipitator	**23** Extraction plant
2 Tipping apron	**13** Induced draught fan	**24** Low-pressure heater and drain cooler
3 Trimming door	**14** Chimney	**25** De-aerator lift pump
4 Refuse bunker	**15** Forced draught fan	**26** De-aerator
5 Tipping vehicle	**16** Steam air preheater	**27** Boiler feed pump
6 Grab cranes	**17** Boiler instrument panel	**28** Main residuals conveyor
7 Boiler feed chute	**18** Turbogenerator	**29** Workshop
8 Grate	**19** Condenser	**30** Administration
9 Clinker quenching bath	**20** Induced draught cooling plant	**31** Residuals hopper
10 Boiler	**21** Cooling water pumps	
11 Economizer	**22** Transformer	

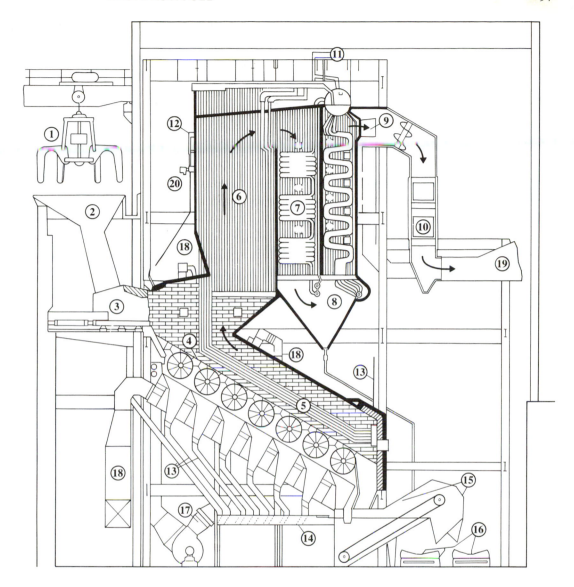

1 Bunker refuse grab	11 Superheater outlet
2 Refuse hopper	12 Downcomer tubes
3 Refuse feeder	13 Dust chutes
4 Incineration grate	14 Dust conveyor
5 Combustion chamber	15 Residual quenching tank
6 Radiation shaft	16 Residual conveyors
7 Superheater	17 Secondary air fan and ducting
8 Boiler bank (entrance)	18 Forced draught fan discharge to grate
9 Economizer gas bypass	19 Flue gas to precipitator and induced draught fan
10 Economizer	20 Sootblowers

Figure 4.20 Steam raising units with Dusseldorf system incineration plant

Apart from loads such as builders' rubble, the plant accepts all refuse without any attempt to remove incombustibles. Provided the refuse fed to the grate has less than 50 % moisture by weight, less than 60 % incombustibles by weight, and more than 25 % combustibles by weight, then the furnace can operate satisfactorily without the addition of a secondary fuel. Combustion takes place with excess air of 60 % to 80 % with the air pre-heated to about 150 °C using steam; the flue gas entering the stack is at about 230 °C. There are five boilers each capable of producing superheated steam at 43 bar, 450 °C which is fed via a ring main to four steam turbines each with a rated output of approximately 10 MW; normally four out of the five boilers are in operation with two or three of the turbines. The annual electricity generated is approximately 175 000 MW h at a station efficiency of about 13 %, using approximately 0.4 Mt refuse each year; the latter figure represents about 13 % of the total refuse collected from Greater London.

The economics of such a plant are complex since the capital and net running costs of this method of disposal of refuse must be compared with the net costs of alternative methods. For example, disposal of the refuse to land-fill sites may also be designed to exploit the production of land-fill gas (see p. 103). Smart[4.3] gives some figures for the economics of the Edmonton plant for the year 1982/83. These figures in terms of annual costs are given in Table 4.9.

Table 4.9 Greater London refuse-fired power station: running costs 1982/1983

Item	Annual cost (£M)
Staff and other fixed costs	2.87
Repairs and maintenance	1.16
Amortization	1.31
Materials and upkeep of mobile plant	0.46
Transport costs and disposal of residual ash	0.79
Overheads	0.60
Total	7.19
Income from sale of electricity	3.80
Net cost of disposal	3.39
Cost of conventional disposal	5.83
Capital cost of plant in 1972	£11.5 M

It can be seen from Table 4.9 that the annual saving over conventional disposal to land-fill sites without exploitation of land-fill gas was £2.44 M in 1982 compared with an initial capital cost of £11.5 M in 1972.

With pre-processing

In smaller plant, and plant using waste to supplement other fossil fuels, it is essential to prepare the waste for combustion. This is done in a variety of ways depending on the condition of the waste on delivery. If it can be assumed that all large objects (e.g. concrete lumps, metal shafts etc.) have been separated out then the waste can be delivered to a pulverizer which will reduce the size to about 150 mm; further screening by magnetic separators can be done to remove any remaining metal particles. In schemes in which the waste is more carefully separated (e.g. in-house waste collection or waste delivered from a municipal authority) then a simple shredder may be all that is required.

There are several types of furnace that can be used, the choice depending on the scale of the operation; it may be that a certain volume of waste must be incinerated each week and some form of heat recovery is desirable (e.g. a hospital); on the other hand a plant may be designed specifically to generate heat and/or power from waste and it may then be economical to import waste to supplement the waste produced in-house.

A boiler like the one shown diagrammatically in Fig. 4.5 can be used with a travelling grate or sprinkler stoker, or with the waste supplied pneumatically; a gas burner could be added to provide supplementary firing; alternatively the boiler could be designed for coal firing with a supplementary waste feed. A diagrammatic arrangement of such a plant is shown in Fig. 4.21; note the use of a filter to remove unburnt particles of waste carried over with the flue gas, particularly important for city locations.

Figure 4.21 Coal-fired boiler with supplementary waste feed

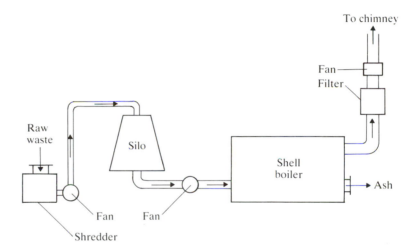

A larger scheme may use rotary furnaces of the starved air type (say 80 % of the stoichiometric air requirement) with combustion being completed in a post combustion chamber; the hot gases are then passed to the boiler. A typical example of such a plant is shown diagrammatically in Fig. 4.22; note that the waste-fired plant produces saturated steam and a separately fired boiler is used

Figure 4.22 Rotary furnace with post-combustion chamber

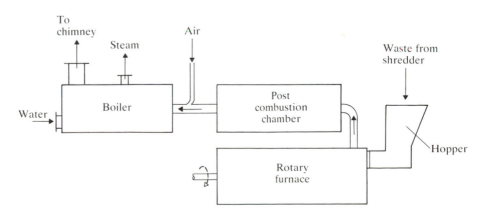

for the superheater; this is done by keeping the gas temperature of the waste fired boiler down to less than about 300 °C (by adding air – known as air attemperation) to avoid corrosion due to the chlorine formed from the plastics in the waste. (Note: the problem of corrosion in the Edmonton plant described earlier is tackled in various ways including adding chemicals to the combustion chamber and coating the superheater tubes with tungsten carbide, or making the tubes of 25/20 Cr/Ni steel.)

A similar system of combustion is that pioneered by the fully-automatic Consumat incinerator (see Fig. 4.23). The waste is ram-fed into the main combustion chamber which is a refractory lined vessel and in which combustion is initiated by gas or oil burners; combustion takes place under starved air, quiescent conditions, so that the products are passed at low velocity to the second chamber where oxidation is completed at a high temperature with the use of a flame which pierces the gas stream; the gas leaving is cooled by attemperating air drawn in as shown. The burners for the main combustion chamber are switched off when the process becomes autothermic; the burner in the second chamber operates for about 50 % of the time being automatically controlled by the gas temperature. The hot gas leaving the incinerator passes to a waste heat boiler for hot water or steam; the boiler can be by-passed, the by-pass operating on steam or hot water demand, with fail-safe in the open position.

Figure 4.23 Consumat incinerator

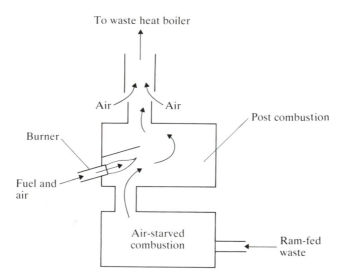

A fluidized-bed boiler (see page 73) is very suitable for waste combustion since waste will usually have a high moisture and ash content (see Table 4.8) similar to brown coal. It is also comparatively easy to burn waste and coal in the same fluidized boiler.

Example 4.4

A new hospital is being planned and it is estimated that it will produce 1400 kg of waste per day, 85 % of which is combustible. Hospital waste when burned is expected to produce ash equivalent to 10 % of its initial weight. It is proposed to use an incinerator with a waste heat boiler as an alternative to an incinerator alone.

Using the data given below estimate the break even point for the energy recovery scheme neglecting depreciation and discounting; assume that the staff and maintenance costs are approximately the same for both cases.

Data

Density of waste, 150 kg/m^3. Density of ash, 350 kg/m^3. Waste heat boiler efficiency, 54 %. Boiler efficiency for a coal-fired plant of similar size, 80 %. GCV of waste, 17 MJ/kg. GCV of coal as an alternative, 28 MJ/kg. Cost of coal, £67 per tonne. Capital cost of incinerator, £46 000. Capital cost of incinerator with waste heat boiler, £95 000. Skip size available, 12.5 m^3. Cost of removal of loaded skip, £60.

Solution

The volume of incombustible waste is

$$1400 \times 0.15 \times 7/150 = 9.8 \text{ m}^3/\text{week}$$

and the volume of ash formed is

$$1400 \times 0.85 \times 0.10 \times 7/350 = 2.38 \text{ m}^3/\text{week}$$

i.e. the volume to be removed each week is 12.18 m^3. Therefore one skip will be required to be emptied once per week at an annual cost of $60 \times 52 = £3120$.

For the energy recovery scheme the useful output of the boiler is equal to

$$1400 \times 0.85 \times 7 \times 17 \times 0.54 = 76\,469 \text{ MJ/week}$$

Therefore coal consumption to give this output is

$$76\,469/(28 \times 0.8) = 3413.8 \text{ kg/week}$$

i.e. annual cost of coal for this output is equal to

$$3413.8 \times 67 \times 52/1000 = £11\,894$$

Hence annual cost of incineration only is

$$£11\,894 + £3120 = £15\,014$$

and annual cost with energy recovery is £3120.

i.e.   break-even point $= (£95\,000 - £46\,000)/(£11\,894) = 4.12$ years

This makes the incinerator with energy recovery marginally more attractive. If on the other hand the alternative to energy recovery were collection of all the refuse (not an option for a hospital with noxious and hazardous waste but a possibility in some industrial applications) then the annual cost of the latter would include the additional cost of collection, i.e. the total volume of refuse is

$$1400 \times 7/150 = 65.3 \text{ m}^3/\text{week}$$

and therefore the skip is to be emptied, say six days a week at an annual cost of

$$6 \times 60 \times 52 = £18\,720.$$

Then the annual cost is

$$£18\,720 + £11\,894 = £30\,614$$

Hence the pay-back period of the energy recovery scheme is given by:

$$\text{pay-back period} = £95\,000/(£30\,614 - £3120) = 3.5 \text{ years}$$

Production of Refuse Derived Fuel (RDF)

The production of pellets or briquettes of combustible material manufactured from waste is an alternative to direct use of waste as a fuel. This is known as Refuse Derived Fuel (RDF) or sometimes as Waste Derived Fuel (WDF). The initial preparation is the same as that described earlier in the section on direct incineration with pre-processing. After shredding, the waste is fed into a pelletizer and pellet cooler before being passed to storage. The pellets can be burned using a variety of methods similar to other solid fuels (e.g. chain grate stoker, ram-fed stoker, underfeed stoker, sprinkler stoker or fluidized bed).

Since the waste is more carefully prepared the combustion is more dependable and RDF can be used as the sole source of fuel or as an addition to coal feed. Several Metropolitan Counties in the UK, including those at Newcastle, Doncaster and Liverpool, have each installed RDF plants; the Liverpool scheme aims to produce 19 500 t/year of RDF with a GCV of 15.5 MJ/kg from 62 000 t/year of pulverized waste; the Doncaster scheme produces 7800 t/year with a claimed GCV of 20 MJ/kg; in both of these cases the ash content of the fuel has been reduced to about 12 % and the moisture content to under 15 %.

Because of the extra capital cost involved in producing RDF it is normally necessary to be able to collect refuse from a large number of customers and to market and sell the resultant fuel to a wide range of customers. Large producers of waste may find it economic to install their own pellet producing facility particularly if they are currently generating power and heat using coal fired boilers. It has been estimated that a minimum of 2000 t/year of combustible waste is necessary to make the installation of a pellet producing facility an economic proposition. Such a plant costing about £400 000 and generating 6000 t/year of RDF would give an annual saving in the coal bill of about £240 000; it can be seen that this is an economic proposition with a pay-back period of less than two years.

For a refuse derived fuel plant to operate economically it is essential that a constant, reliable supply of refuse is available; this appears to be the Achilles heel of such schemes. In the case of Municipal Authority disposal the annual cost per tonne of refuse is higher by this method than by that used at Edmonton (see pp. 96–98); Doncaster quote an annual cost per tonne of refuse disposed using RDF of 1.7 times that using incineration with heat recovery.

Pyrolysis

Pyrolysis is a process in which a material is heated in the absence of oxygen. When refuse is heated at a high temperature (> 500 °C) in the absence of oxygen then hydrogen, carbon monoxide, methane, and carbon dioxide are formed; if the temperature is not too high oil is also formed, and there is always a residual of char.

If the pyrolysis process is used to produce oil then the gas generated can be used in the process itself to provide the energy input. Pyrolytic oil is of poor quality compared to petroleum oils; it is corrosive and has a high ash content.

For gas production the process can be modified to introduce air less than the stoichiometric quantity; the process is then similar to that used for the production of gas from coal. In the future, waste may provide a useful additional feed-stock for the production of substitute natural gas.

Pyrolysis for the production of oil or gas as a fuel is not at present an economic alternative to the other methods considered in this chapter.

Incinerators described previously such as the Consumat are frequently described as using 'partial pyrolysis'; this is because the initial combustion takes place in air-starved quiescent conditions (see Fig. 4.23).

Production of Gas Through Bio-Degradation

An increasingly important source of fuel is bio-mass which can include such diverse sources as agricultural crop waste, forestry waste, animal waste, sewage, municipal waste, and sea-weed. In many underdeveloped countries or in remote rural areas of developed countries the use of agricultural and forestry waste for heat and/or power production is being considered more and more frequently. Many of the methods of preparing the waste are similar to those described earlier for municipal, medical and industrial refuse, and combustion can be by fluidized-bed or in a conventional boiler if the waste is suitably processed.

An interesting example of the use of straw as a fuel is in the heating of historic Woburn Abbey in Bedfordshire. Bales of straw delivered from surrounding farms are shredded into 5–15 cm lengths before passing directly by auger into the ceramic hearth of a shell boiler. Low pressure hot water is produced for heating at a rate of approximately 0.8 MW. The current UK production of straw is about 13.5 Mt and Silsoe College of Agriculture estimate that a realistic figure for straw as a fuel could be as high as 0.6 Mt by the year 2000; this is equivalent to 0.4 Mt coal equivalent using the GCV figures in Table 4.7.

Perhaps the most interesting potential for bio-mass is in the exploitation of *refuse-land fill*. One of the most common methods of disposing of municipal and industrial refuse is by burying it in the ground in so-called land-fill sites. It is estimated that in the UK about 25 Mt of refuse are buried in land-fill sites each year. When organic material such as refuse is buried then methane gas is formed by a natural process. As land-fill sites became more prevalent the formation of methane was seen to be a nuisance and in some cases even a hazard because of the danger of explosion. Exploitation of methane from land-fill sites began to be undertaken in the UK in the 1970s, but its application has been surprisingly slow to develop; there are still many land-fill sites where the methane goes to waste and is a hazard.

The organic matter is initially broken down by acid-forming bacteria to give organic acids; these are in turn broken down by methane-forming bacteria. The gas finally formed contains on average about 40% to 60% methane with the remainder mainly carbon dioxide with small amounts of nitrogen and hydrogen sulphide; the GCV is approximately 15 to 18 MJ/m^3, which compares with 38.6 MJ/m^3 for natural gas.

For the production of land-fill gas a moisture content of greater than 40% is desirable but care must be taken not to waterlog the site by extending below the natural water table, or by allowing the site to flood. A clay pit is ideal to prevent gas leakage with a one metre thickness of clay topsoil; a site of area greater than 10 ha is recommended (i.e. 10^5 m^2 or about 25 acres), with depths of waste at least 10 m and preferably 30 m or more. A typical example of a land-fill site in the UK is at Aveley in Essex. This scheme imports 0.25 Mt of waste each year from East London; the site area is 25 ha and is 40 m deep. The gas which is produced at a rate of 3500 m^3/h consists of 45 % to 50 % methane by volume with a GCV of 18 MJ/m^3.

The gas from a land-fill site is collected by a system of pipes connected to a compressor as shown in Fig. 4.24, then dried and pumped to the boiler where it can be fired using specially adapted burners. Distances of up to 5 miles (about 8 km) are possible between the site and the consumer.

Figure 4.24 Exploitation of land-fill gas

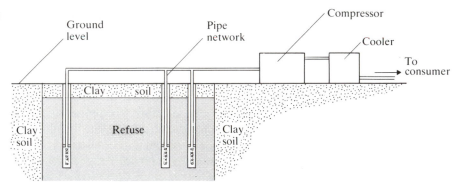

4.4 STEAM AND GAS CYCLES

In the previous sections of this chapter methods of converting chemical energy into thermal energy were considered. This section will look at ways in which thermal energy may be converted to mechanical energy using gas or steam as the working substance.

Steam Cycles

Modern steam plant for electrical power generation uses feed heating and may also use re-heating of the steam at an intermediate pressure. One of the feed heaters is usually an open heater at approximately atmospheric pressure which also acts as a de-aerator to remove the dissolved gases in the feed water. A typical power station plant is shown diagrammatically in Fig. 4.25.

Example 4.5
In a power plant developing 500 MW the steam supply conditions to the turbine are 160 bar and 600 °C. The steam leaving the high pressure turbine at 20 bar and 300 °C is re-heated to 600 °C before expanding through the intermediate pressure turbine to 2 bar, 290 °C and then through the low pressure turbine to the condenser pressure of 0.035 bar where the dryness fraction is 0.96.

There are six stages of feed heating as in Fig. 4.25 at pressures of 90 bar, 45 bar, 20 bar, 8 bar, 2 bar and 0.35 bar. The feed water leaves the open heater at the saturation temperature corresponding to the saturation pressure of the bleed steam; for all the closed feed heaters the temperature of the feed water leaving the water is 5 K below the saturation temperature corresponding to the saturation pressure of the bleed steam. The condensate leaving each heater is

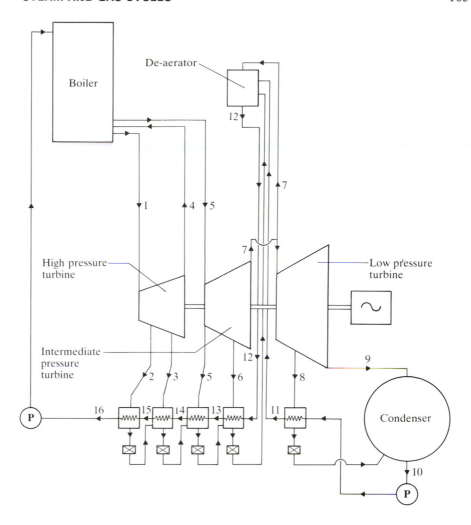

Figure 4.25 Steam plant with six stages of feed-heating

saturated at the saturation pressure of the bleed steam and is flashed through a reducing valve into the preceding heater as shown in Fig. 4.25. Assume that the condition line for each turbine is a straight line on the h–s chart. Neglecting pump work and all losses other than those indicated above, calculate:

(i)   the required steam flow from the boiler;
(ii) the specific steam consumption;
(iii) the cycle efficiency;
(iv) the specific steam consumption and cycle efficiency for a cycle with the same steam conditions but without feed heating.

Solution
The enthalpies of the steam at the various points throughout the cycle can be found from the h–s chart. Referring to the numbered points on Fig. 4.25 we have from the chart or from superheat tables:

$$h_1 = 3573 \text{ kJ/kg} \quad \text{and} \quad h_4 = 3025 \text{ kJ/kg}$$

On the h–s chart a line joining points 1 and 4 is drawn; where this line cuts the pressure lines of 90 bar and 45 bar fixes the points 2 and 3. A sketch of the

Figure 4.26 h–s chart for Example 4.5

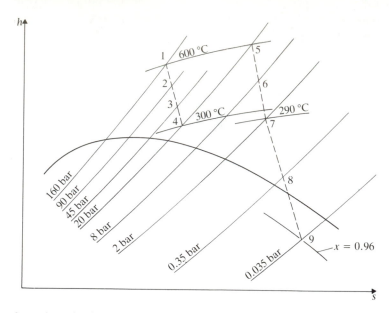

h–s chart is shown in Fig. 4.26. Then from the chart:

$$h_2 = 3400 \text{ kJ/kg} \quad \text{and} \quad h_3 = 3210 \text{ kJ/kg}$$

In a similar way points 5 and 7 are joined by a straight line and where this cuts the 8 bar line point 6 is fixed; point 7 is joined to point 9 and where this line cuts the 0.35 bar line point 8 is fixed. Therefore we have:

$$h_5 = 3690 \text{ kJ/kg}, \quad h_7 = 3052 \text{ kJ/kg}, \quad h_6 = 3410 \text{ kJ/kg},$$
$$h_9 = 2453 \text{ kJ/kg}, \quad h_8 = 2750 \text{ kJ/kg}, \quad h_{10} = 112 \text{ kJ/kg}$$

The enthalpy of water is approximately equal to the saturation value, h_f, at the same temperature. Hence the enthalpies of the feed water leaving the closed feed heaters are found as follows: $h_{11} = h_f$ at 5 K less than the saturation temperature at 0.35 bar

i.e. $\qquad h_{11} = h_f \text{ at } (72.7 - 5)\,°\text{C} = 284 \text{ kJ/kg}$

and $\qquad h_{13} = h_f \text{ at } (170.4 - 5)\,°\text{C} = 699 \text{ kJ/kg}$
$\qquad\qquad h_{14} = h_f \text{ at } (212.4 - 5)\,°\text{C} = 886 \text{ kJ/kg}$
$\qquad\qquad h_{15} = h_f \text{ at } (257.4 - 5)\,°\text{C} = 1098 \text{ kJ/kg}$
$\qquad\qquad h_{16} = h_f \text{ at } (303.3 - 5)\,°\text{C} = 1336 \text{ kJ/kg}$

The enthalpy leaving the open feed heater is saturated at the pressure of 2 bar.

i.e. $\qquad h_{12} = 505 \text{ kJ/kg}$

Let the bleed steam flow rates per unit mass flow rate from the boiler be y_1, y_2, y_3, y_4, y_5, and y_6. Then applying an energy balance on each heater in turn we have (see Fig. 4.27) for the first heater:

$$y_1(3400 - 1364) = 1336 - 1098$$
$$\therefore \quad y_1 = 0.1169$$

For the second heater, referring to Fig. 4.28:

$$3210y_2 + (1364 \times 0.1169) + 886 = 1122(y_2 + 0.1169) + 1098$$
$$\therefore \quad y_2 = 183.7/(3210 - 1122) = 0.0880$$

Figure 4.27 Energy
analysis for heater 1

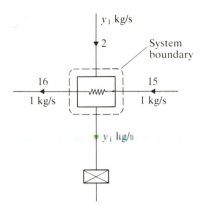

Figure 4.28 Energy
analysis for heater 2

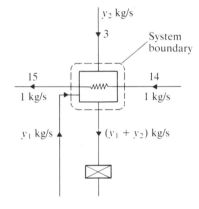

Similarly for the third heater:

$$3690y_3 + 1122(0.1169 + 0.0880) + 699$$
$$= 909(y_3 + 0.1169 + 0.088) + 886$$
$$\therefore \quad y_3 = 143.4/(3690 - 909) = 0.0516$$

For the fourth heater:

$$3410y_4 + 909(0.1169 + 0.0880 + 0.0516) + 505$$
$$= 721(y_4 + 0.1169 + 0.0880 + 0.0516) + 699$$
$$\therefore \quad y_4 = 145.8/(3410 - 721) = 0.0542$$

For the open heater, referring to Fig. 4.29 we have:

$$3052y_5 + 284(1 - 0.1169 - 0.0880 - 0.0516 - 0.0542 - y_5)$$
$$+ 721(0.1169 + 0.0880 + 0.0516 + 0.0542) = 505$$
$$\therefore \quad y_5 = 85.2/(3052 - 284) = 0.0308$$

For the sixth heater:

$$(2750 - 304.5)y_6$$
$$= (284 - 112)(1 - 0.1169 - 0.0880 - 0.0516 - 0.0542 - 0.0308)$$
$$\therefore \quad y_6 = 113.3/2445.5 = 0.0463$$

Figure 4.29 Energy analysis for open heater

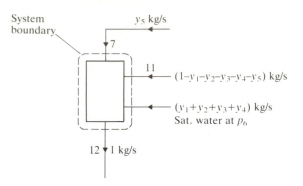

(i)   The power output per unit mass flow rate from the boiler is then given by:

$$(h_1 - h_2) + (h_2 - h_3)(1 - y_1) + (h_3 - h_4)(1 - y_1 - y_2)$$
$$+ (h_5 - h_6)(1 - y_1 - y_2 - y_3)$$
$$+ (h_6 - h_7)(1 - y_1 - y_2 - y_3 - y_4)$$
$$+ (h_7 - h_8)(1 - y_1 - y_2 - y_3 - y_4 - y_5)$$
$$+ (h_8 - h_9)(1 - y_1 - y_2 - y_3 - y_4 - y_5 - y_6)$$

i.e.

$$(3573 - 3400) + (3400 - 3210)(1 - 0.1169)$$
$$+ (3210 - 3025)(1 - 0.1169 - 0.0880)$$
$$+ (3690 - 3410)(1 - 0.1169 - 0.0880 - 0.0516)$$
$$+ (3410 - 3052)(1 - 0.1169 - 0.0880 - 0.0516 - 0.0542)$$
$$+ (3052 - 2750)(1 - 0.1169 - 0.0880 - 0.0516 - 0.0542 - 0.0308)$$
$$+ (2750 - 2453) \times$$
$$(1 - 0.1169 - 0.0880 - 0.0516 - 0.0542 - 0.0308 - 0.0463)$$

Therefore the power output is

$$173.0 + 167.8 + 147.1 + 208.2 + 246.8 + 198.9 + 181.8$$
$$= 1323.6 \text{ kW per kg/s steam flow}$$

Then the required steam flow from the boiler to give a power output of 500 MW is

$$\text{boiler steam flow rate} = 500\,000/1323.6 = 377.8 \text{ kg/s}$$

(ii)   Specific steam consumption $= 3600/1323.6 = 2.72 \text{ kg/kW h}$

(iii)   The heat supplied in the boiler per unit mass flow rate is

$$(h_1 - h_{16}) + (h_5 - h_4)(1 - y_1 - y_2)$$
$$= (3573 - 1336) + (3690 - 3025)(1 - 0.1169 - 0.0880)$$
$$= 2765.7 \text{ kW per kg/s}$$
$$\therefore \quad \text{cycle efficiency} = 1323.6/2765.7 = 0.479 \text{ or } 47.9 \%$$

(iv)   For the case of no bleed steam for feed heating then power output per unit mass flow rate is given by

$$(h_1 - h_4) + (h_5 - h_9) = 3573 - 3025 + 3690 - 2453$$
$$= 1785 \text{ kW per kg/s}$$

and specific steam consumption is 2.017 kg/kW h.

$$\therefore \quad \text{cycle efficiency} = 1785/\{(3573 - 112) + (3690 - 3025)\}$$
$$= 43.3 \text{ or } 43.3 \%$$

It can be seen that the cycle efficiency is increased significantly by feed heating although the power output for a given mass flow rate is reduced. The steam mass flow rate for the power output of 500 MW is now $500\,000/1785 = 280.1$ kg/s which implies that a smaller boiler could be used. When feed heating is used the capital cost is therefore much greater due to the feed heaters, a larger boiler, and more pumps, pipe runs etc. The increase in efficiency yields a reduction in fuel burned which gives an annual cash saving; this must be great enough to offset the other expenditure otherwise feed heating is not an economic proposition.

For the plant above assuming coal burning and taking a boiler efficiency of 90 %, the fuel consumption for the case without feed heating is $500/(0.9 \times 0.433 \times 28.4) = 45.2$ kg/s; with feed heating the fuel consumption is $45.2 \times 0.433/0.479 = 40.9$ kg/s. For coal at £48 per tonne this would give an annual saving of

$$(45.2 - 40.9) \times 3600 \times 24 \times 365 \times 48/100 = £6.5 \text{ M}$$

It is not necessary to do any further calculations to see that a considerable financial advantage is obtained by using feed heating. (Note: the actual cycle efficiency of a power station similar to the one above would be about 42 %; the difference is due to heat losses, mechanical losses, steam used for gland sealing, pumping power etc.)

Gas Cycles

Internal Combustion (IC) engines have been developed to a high standard of efficiency for all forms of transport and scope for cost savings in industry lies mainly in effective maintenance. For the production of mechanical power and the generation of electricity in industry using the gas cycle the diesel engine is most widely used although natural gas-fired engines have become more common due to the availability and price competitiveness of natural gas. Gas turbine engines are not normally used for mechanical drive applications due to their low part-load efficiency and high shaft speed, although their lower capital cost and ease of starting make them suitable as standby generators: the shaft speed is then such that an expensive gear box is not required.

One of the main energy losses in an IC engine is the energy of the exhaust gases. There are two main ways of minimizing this energy loss: one is by recovering as much of the energy as possible within the gas cycle; the second is by using the energy of the exhaust to provide heating or to generate steam. IC engines are increasingly being used in such combined heat and power schemes (see Chapter 8).

Exhaust energy recovery within the cycle in the diesel engine is achieved by passing the exhaust gases through a turbine to compress the air entering the engine thus increasing its density and hence increasing the power output; ideally the air should be cooled after compression since density is inversely proportional to temperature. This process is known as *supercharging*.

In the gas turbine engine, energy recovery can be obtained by passing the exhaust gases through a heat exchanger to heat the air leaving the compressor thus reducing the energy required to be provided by the fuel in the combustion chamber; the heat recovery is more effective if the air is compressed in two stages with inter-cooling between the stages thus reducing the temperature of the air after compression. This is illustrated in Example 4.6 below. It should be noted that such a plant is more expensive and more complex than the simple gas turbine but if a supply of cooling water is readily available then the increased efficiency will more than offset the increased cost. An industrial gas turbine is of a more robust design and can also use a cheaper fuel than its aircraft counterpart.

For further information on IC engines consult the bibliography at the end of the chapter.

Example 4.6

An industrial gas turbine plant is shown diagrammatically in Fig. 4.30. There are two compressors requiring the same shaft power input and driven by a turbine which produces just enough power to drive the compressors and overcome mechanical losses. There is an air intercooler between the compressor stages. The air leaving the second compressor stage passes through a heat exchanger to the combustion chamber and the gases produced enter the turbine which drives the compressors. The gases leaving this turbine pass to a second combustion chamber where they are re-heated before entering the power turbine which drives an electric generator. Using the data given below, calculate:

(i) the electrical power output;

(ii) the overall thermal efficiency of the plant.

Figure 4.30 Gas turbine plant for Example 4.6

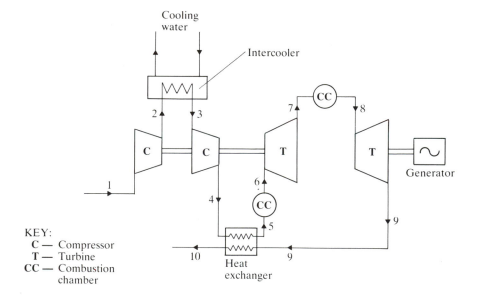

KEY:
  C — Compressor
  T — Turbine
  CC — Combustion
      chamber

Data

Compressors (taken to be identical): pressure ratio, 3.3; isentropic efficiency, 0.82.

Intercooler; air cooled back to ambient temperature.

Combustion chambers (taken to be similar): combustion efficiency, 0.99.

Heat exchanger: effectiveness, 0.9.

Compressor turbine: inlet temperature, 1150 K; isentropic efficiency, 0.87; mechanical efficiency of drive to compressors, 0.98.

Power turbine: inlet temperature, 1150 K; isentropic efficiency, 0.89; mechanical efficiency of drive to generator, 0.98; electrical transmission efficiency, 0.97.

Pressure losses: air side of heat exchanger, 0.3 bar; gas side of heat exchanger 0.05 bar; intercooler, 0.15 bar; each combustion chamber, 0.2 bar.

Ambient conditions: 1.013 bar, 15 °C.

For air take c_p and γ as 1.005 kJ/kg K and 1.4; for the gases take c_p and γ as 1.15 kJ/kg K and 1.333; in each combustion chamber assume that the heat supplied is approximately equal to the exit temperature minus the inlet temperature multiplied by a specific heat at constant pressure of 1.15 kJ/kg K. Take the mass flow rate throughout as 120 kg/s. (Note that this is a good approximation in practice since air is bled off from the compressor for cooling of the turbine discs etc. and this loss compensates for the fuel added where the fuel–air ratio is of the order of 0.01.)

Solution

The cycle is shown on a T–s diagram in Fig. 4.31 with the state point numbers corresponding to those of Fig. 4.30.

Figure 4.31 T–s diagram for Example 4.6

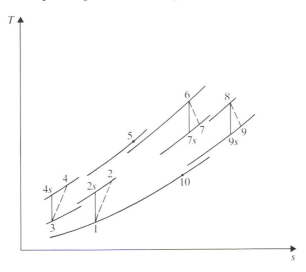

(i) For each compressor

$$T_{2s}/T_1 = T_{4s}/T_3 = (3.3)^{(1.4-1)/1.4} = 1.4065$$

The inlet air temperature to each compressor is the same, therefore

$$T_{2s} = T_{4s} = 1.4065(15 + 273) = 405.07 \text{ K}$$

Also the isentropic efficiencies are equal, hence

$$T_2 = T_4 = 288 + (405.07)/0.82 = 430.77 \text{ K}$$

The specific power required from the expansion in the first turbine is then

$$2 \times c_{pa}(430.77 - 288)/0.98 = 292.82 \text{ kJ/kg}$$

i.e. $$c_{pg}(T_6 - T_7) = 292.82 \text{ kJ/kg}$$
$$\therefore \qquad T_7 = 1150 - (292.82/1.15) = 895.37 \text{ K}$$

To proceed to calculate the power from the second turbine it is first of all necessary to calculate the cycle pressures.

$$p_2 = 3.3 \times 1.013 = 3.343 \text{ bar} \qquad p_3 = 3.343 - 0.15 = 3.193 \text{ bar}$$

Then
$$p_4 = 3.3 \times 3.193 = 10.537 \text{ bar}$$
and
$$p_5 = 10.537 - 0.3 = 10.237 \text{ bar}$$
$$p_6 = 10.237 - 0.2 = 10.037 \text{ bar}$$

Knowing the temperatures at points 6 and 7 and the isentropic efficiency of the turbine we can find the pressure ratio.

i.e.
$$T_{7S} = 1150 - (1150 - 895.37)/0.87 = 857.32 \text{ K}$$
and
$$p_6/p_7 = (1150/857.32)^{(1.333/0.333)} = 3.2404$$
Hence

$$p_7 = 10.037/3.2404 = 3.097 \text{ bar}$$

and
$$p_8 = 3.097 - 0.2 = 2.897 \text{ bar}$$
Taking $p_{10} = p_1 = 1.013$ bar, then $p_9 = 1.013 + 0.05 = 1.063$ bar.
Therefore

$$p_8/p_9 = 2.897/1.063 = 2.726$$

and

$$T_8/T_{9S} = (2.726)^{(1.333 - 1)/1.333} = 1.285$$
$$\therefore \qquad T_{9S} = 1150/1.285 = 895.19 \text{ K}$$

Hence

$$T_9 = 1150 - (1150 - 895.19) \times 0.89 = 923.22 \text{ K}$$

and

$$\text{electrical output} = (1150 - 923.22) \times 0.98 \times 0.97 \times 120 \times 1.15$$
$$= 29\,750 \text{ kW}$$

(ii) To find the heat supplied it is necessary to calculate the temperature at point 5 using the effectiveness of the heat exchanger; Eq. [3.30] gives the effectiveness, E.

i.e.
$$E = 0.9 = (T_5 - T_4)/(T_9 - T_4)$$
$$\therefore \qquad T_5 = 430.77 + 0.9(923.22 - 430.77) = 873.98 \text{ K}$$

(Note that T_{10} is then $535.9 \text{ K} = 262.9\,°\text{C}$.)
Heat supplied is then

$$\{1.15(1150 - 873.98) + 1.15(1150 - 895.37)\} = 610.25 \text{ kJ/kg}$$

i.e. input from fuel supplied is

$$610.25 \times 120/0.99 = 73\,970 \text{ kW}$$

and
$$\text{plant thermal efficiency} = 29\,750/73\,970 = 40.2\,\%$$

A simple cycle with the same overall pressure ratio but without the intercooler, reheater and heat exchanger would have a thermal efficiency of about 29 % and an exhaust gas temperature of about 440 °C (compared with 262.9 °C with inter-cooling, re-heat and heat exchange).

Combined Cycles

The gas turbine engine, despite its low thermal efficiency, is a very versatile unit because of its simplicity which leads to lower capital cost. These advantages are even greater if the energy of the exhaust gas can be used; because of the high air–fuel ratio the exhaust gas has a high proportion of oxygen and hence further combustion can also take place.

In gas pumping stations for natural gas or land-fill gas a gas turbine unit can be used where the fuel burned is the gas itself. Gas turbine units have also been used to produce compressed air; in these units the turbine produces just enough power to drive the compressor. An example of such a use is in a blast furnace where the blast furnace gas is used to fuel the turbine and where the turbine exhaust gases are used to pre-heat the compressed air which is then supplied to the furnace.

One of the most effective ways of using the gas turbine is in combination with a steam plant where the turbine exhaust is used to raise steam as in Example 4.7 below. Other examples of combining IC engines with steam raising plant and other energy devices will be covered in more detail in Chapter 8 on Total Energy.

Example 4.7
The exhaust gases from a simple gas turbine unit are used in a waste-heat boiler to raise steam for a steam power plant as shown in Fig. 4.32. Using the data below and neglecting feed pump work, and all thermal losses, calculate:
(i)   the mass flow rate of gas through the gas turbine unit;
(ii) the mass flow rate of steam generated;
(iii) the overall thermal efficiency for the combined plant.

Figure 4.32 Combined cycle plant of Example 4.7

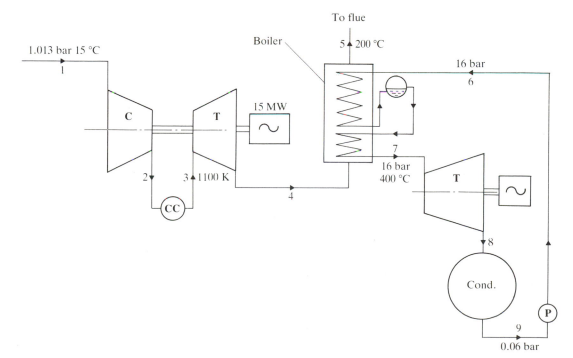

Data

Gas turbine plant: electrical power output, 15 MW; ambient conditions, 1.013 bar and 15 °C; compressor pressure ratio, 10; compressor isentropic efficiency, 0.82; turbine isentropic efficiency, 0.88; combustion efficiency, 0.99; temperature of gases leaving the combustion chamber, 1100 K; temperature of gases leaving waste-heat boiler, 200 °C; pressure loss in combustion chamber, 0.2 bar; pressure loss in waste heat boiler, 0.1 bar; combined mechanical and electrical transmission efficiency, 0.97; for air, c_p and γ, 1.005 kJ/kg K and 1.4; for gases in the turbine, c_p and γ, 1.15 kJ/kg K and 4/3; for the combustion process assume an equivalent c_p of 1.15 kJ/kg K.

Steam turbine plant: boiler pressure 16 bar; steam temperature at superheater exit, 400 °C; condenser pressure, 0.06 bar; steam turbine isentropic efficiency, 0.85; combined mechanical and electrical transmission efficiency, 0.97; neglect all pressure losses and assume that the condensate leaving the condenser is saturated.

Solution

T–s diagrams for the gas turbine unit and for the steam plant are shown in Figs 4.33(a) and 4.33(b) with state point numbers corresponding to those of Fig. 4.32.

Figure 4.33 *T–s* diagrams for Example 4.7

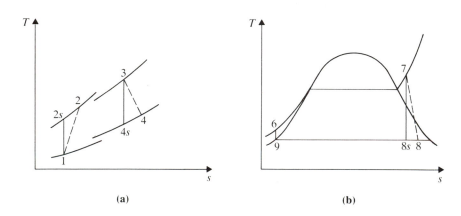

(a)   (b)

(i)   For the gas turbine plant:

$$T_{2S} = 288(10)^{0.286} = 556.41 \text{ K}$$

and $$T_2 = 288 + (556.41 - 288)/0.82 = 615.33 \text{ K}$$

Then

$$p_2 = 10 \times 1.013 = 10.13 \text{ bar} \qquad \text{and} \qquad p_3 = 10.13 - 0.2 = 9.92 \text{ bar}$$

Also

$$p_4 = 1.013 + 0.1 = 1.113 \text{ bar} \qquad \text{and} \qquad p_3/p_4 = 8.922$$

Therefore for the turbine

$$T_{4S} = 1100/(8.922)^{0.25} = 636.5 \text{ K}$$

and

$$T_4 = 1100 - (1100 - 636.5) \times 0.85 = 706 \text{ K}$$

Electrical power output = 15 MW.

i.e.
$$15\,000 = \{1.15(1100 - 706) - 1.005(615.33 - 288)\} \times 0.97 \times \dot{m}$$

$$\therefore \quad \text{mass flow rate, } \dot{m} = 124.58 \text{ kg/s}$$

(ii)   The gases enter the waste heat boiler at 706 K and leave at $200\,°C = 473$ K. Therefore the heat transferred from the gases is

$$1.15 \times 124.58 \times (706 - 473) = 33\,381 \text{ kW}$$

From the h–s chart at 16 bar, $400\,°C$, $h_7 = 3255$ kJ/kg; neglecting pump work, the enthalpy of the feed water is the enthalpy of saturated water at 0.06 bar, i.e. $h_6 = h_9 = 152$ kJ/kg. Hence neglecting thermal losses

$$\text{mass flow rate of steam} = 33\,381/(3255 - 152) = 10.758 \text{ kg/s}$$

(iii)   From the h–s chart, $h_{8s} = 2225$ kJ/kg. Therefore

$$h_8 = 3255 - (3255 - 2225) \times 0.85 = 2379.5 \text{ kJ/kg}$$

Electrical power output is

$$10.758 \times 0.97 \times (3255 - 2379.5) = 9136.1 \text{ kW}$$

i.e. total electrical power output is

$$9136.1 + 15\,000 = 24\,136.1 \text{ kW}$$

The heat supplied by the fuel in the gas turbine plant is

$$124.58\{1.15(1100 - 615.33)\}/0.99 = 70\,645.3 \text{ kW}$$

Hence

$$\text{overall thermal efficiency} = 24\,136.1/70\,645.3 = 34.2\,\%$$

(Note that the thermal efficiency of the gas turbine plant on its own would be $15\,000 \times 0.99/124.58\{1.15(1100 - 615.33)\} = 21.4\,\%$. The thermal efficiency of the steam plant on its own would be, $9136.1/33\,381 = 27.4\,\%$.)

The overall efficiency of the combined plant in the above example is still only 34.2 % which means that 65.8 % of the energy of the fuel is not being used. It should be possible to use part of the energy of the turbine exhaust (which is at $200\,°C$) for space-heating or hot water heating. Similarly the heat rejected in the condenser might be usable although in this case the condensing temperature is $36.2\,°C$ (the saturation temperature at 0.06 bar) which is too low for most types of heat recovery. It would be possible to expand to a higher condenser pressure in order to make the heat rejected in the condenser available at a higher temperature; in that case the power output from the turbine would be reduced. Chapter 8 considers ways in which the primary energy input to a plant may be used to its maximum potential through total energy schemes.

4.5 REFRIGERATION, HEAT PUMPS AND AIR CONDITIONING

The use of coal, oil, gas or electricity to produce thermal energy is essential in all types of industry, not only for space-heating and hot water supply, but for a myriad of processes involving furnaces, ovens, driers, vats, and direct chemical processing. Applications where the requirement is to reduce the temperature below the ambient by refrigerating are also common; obvious examples are in the food industry and for air conditioning.

Industrial refrigeration requirements cover the temperature range down to about $-60\,°C$; below this temperature, for example for gas liquefaction, special techniques are used and this field is known as *cryogenics*.

Refrigeration is a major topic and the references given at the end of this chapter should be consulted for a more detailed coverage. This section will give a concise treatment of the plant requirements and basic mode of operation of the most common refrigeration systems.

Evaporative Cooling

Evaporative cooling in its simplest form is one of the oldest ways of keeping things cool. Water exposed to an air stream will evaporate, provided the air itself is not saturated with water vapour, and on evaporation the enthalpy of vaporization is supplied from the bulk of the water thus reducing the temperature. Cooling of the human body by sweating is the most obvious example of evaporative cooling.

In practice, evaporative cooling is commonly used to reduce the temperature of cooling water in a wide variety of applications using a cooling tower. In the most common design of cooling tower the water to be cooled is sprayed into the top of a tower and allowed to flow down through a fill of material in counter flow to an ambient air stream drawn into the base of the tower. The small percentage of water which is lost by the evaporation is replaced by make-up water which is mixed with the water at exit from the tower.

The basic theory of air/water vapour mixtures is covered in Eastop and McConkey.[4.4]

Example 4.8

In a cooling tower the water to be cooled enters near the top of the tower and flows down through packing in counter flow to an air stream as shown diagrammatically in Fig. 4.34. Air is induced from the base of the tower by a fan mounted on the top of the tower. Using the data given below and neglecting heat losses and pressure losses, calculate:

(i) the required mass flow rate of air induced by the fan for the given cooling duty;
(ii) the mass flow rate of make-up water required.

Data
Water: mass flow rate, 15 kg/s; temperature of water at entry, 27 °C; temperature of water at exit, 21 °C; mean specific heat, 4.18 kJ/kg K.
Air: ambient conditions, 1.013 25 bar, 23 °C dry bulb (DB), 17 °C wet bulb (WB); air at exit, saturated with water vapour at 25 °C; mean specific heat of dry air, 1.005 kJ/kg K; fan air power input, 5 kW.
For air at standard atmospheric pressure it can be assumed that the vapour pressure is given by:

$$\frac{p_S}{[\text{bar}]} = \frac{p_{g\text{WB}}}{[\text{bar}]} - 6.748 \times 10^{-4} \times \frac{(t_{\text{DB}} - t_{\text{WB}})}{[\text{K}]}$$

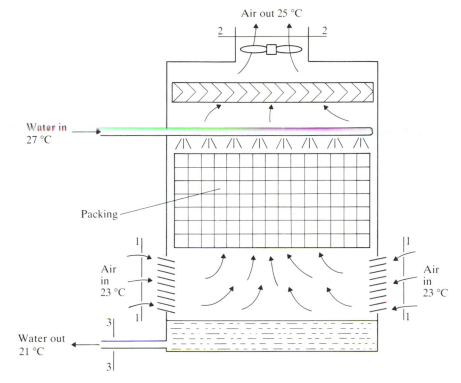

Figure 4.34 Cooling tower

For water vapour at low vapour pressures it is a good approximation to assume that the enthalpy is equal to the saturated value at the same dry bulb temperature.

Solution

(i)   Referring to Fig. 4.34, relative humidity, $\phi_2 = 1$. Hence

$$p_{s2} = p_{g2} \text{ at } 25\,^\circ\text{C} = 0.031\,66 \text{ bar}$$

Then

$$\omega_2 = 0.622 \times 0.031\,66/(1.013\,25 - 0.031\,66) = 0.020\,06$$

For the air at intake

$$p_{s1} = (p_g \text{ at } 17\,^\circ\text{C}) - 6.748 \times 10^{-4}(23 - 17)$$
$$= 0.019\,36 - 0.004\,05 = 0.015\,31 \text{ bar}$$

Therefore

$$\omega_1 = 0.622 \times 0.015\,31/(1.013\,25 - 0.015\,31) = 0.009\,54$$

Let the mass flow rate of dry air be $\dot{m}_a$. Then the mass flow rate of water removed in the air stream is given by

$$\dot{m}_a(\omega_2 - \omega_1) = \dot{m}_a(0.020\,06 - 0.009\,54)$$
$$= 0.010\,52\dot{m}_a$$

Hence the mass flow rate of water at the tower exit is given by

$$\dot{m}_{w3} = (15 - 0.010\,52m_a) \text{ kg/s}$$

The enthalpies of the water vapour in the inlet and exit air streams can be found using the approximation given in the question.

i.e. $h_{s1} = h_g$ at $23\,°C = 2543$ kJ/kg

and $h_{s2} = h_g$ at $25\,°C = 2546.6$ kJ/kg

Then taking the specific heat of dry air as 1.005 kJ/kg K we can write the enthalpy of the air streams at inlet and exit as

$$h_1 = (2543 \times 0.009\,54) + (1.005 \times 23) = 47.38 \text{ kJ/kg of dry air}$$
$$h_2 = (2546.6 \times 0.020\,06) + (1.005 \times 25) = 76.21 \text{ kJ/kg of dry air}$$

Now applying an energy balance to the tower:

$$4.18\{(27 \times 15) - 21(15 - 0.010\,52\dot{m}_a)\} + 5 = \dot{m}_a(76.21 - 47.38)$$
$$\therefore \qquad \dot{m}_a = 13.12 \text{ kg/s}$$

Therefore

$$\text{mass flow rate of air induced by the fan} = 13.12 \times 1.020\,06$$
$$= 13.39 \text{ kg/s}$$

(ii) Since the mass flow rate of water vapour removed by the air stream is given by $0.010\,52\dot{m}_a$ kg/s then

$$\text{mass flow rate of make-up water required} = 0.010\,52 \times 13.12$$
$$= 0.138 \text{ kg/s}$$

It should be noted that the water at exit from the tower in this example is at $21\,°C$ and has been cooled below the temperature of the air entering (i.e. $23\,°C$). This is only possible because of the nature of evaporative cooling in which the limiting temperature for the water at exit is the wet bulb temperature, in this case $17\,°C$. For evaporative cooling it can be shown that the driving force for combined heat and mass transfer is the difference between the enthalpy of the air at the water/air interface and the enthalpy of the bulk of the air at that point. At the exit of the water from the tower in the above example the water/air interface is saturated at $21\,°C$, then calculating ω_g, and taking $h_s = h_g$ at $21\,°C$, we can show that the enthalpy at the interface is 60.83 kJ/kg; the enthalpy of the air at the same section of the tower is $h_1 = 47.38$ kJ/kg as calculated in the example, and hence there is an enthalpy difference driving force of 13.45 kJ/kg.

The smaller the mean enthalpy difference throughout the tower then the larger will be the required surface area for a given cooling duty; as with a heat exchanger a penalty of increased size and hence cost is paid as the driving force is reduced.

(Note: the bacterium known as *legionella pneumophila* tends to be active at temperatures of about $30\,°C$ and hence is a particular threat in cooling tower use; regular and adequate water treatment must be undertaken and the tower designed with drift eliminators to minimize aerosol generation.)

Vapour-Compression Refrigeration

Vapour-compression refrigeration using a reversed heat engine cycle with a wet vapour is the most common refrigeration system in current use; the basic system is shown in Figs 4.35(a) and (b). Many different refrigerants are used or have been used including sulphur dioxide, methyl chloride, ammonia, hydrocarbons such as methane, ethane, propane, and the family of halocarbons. The first two

Figure 4.35 Refrigerant plant and cycle diagram

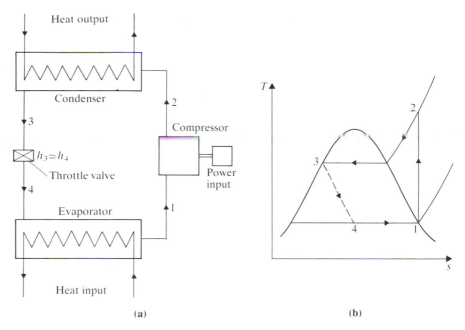

(a)

(b)

named were discontinued because of their toxicity and the hydrocarbons are seldom used because of their flammability; ammonia although toxic is still used because of its relatively low cost.

The halocarbons are compounds containing carbon, fluorine and chlorine in varying proportions; these substances are now known colloquially as CFCs, and are also widely used as propellants in aerosol cans of various types. Despite being the most commonly used refrigerants because of their suitable properties and non-toxicity they are now under threat because of the discovery that, due to the chlorine content, their release into the earth's atmosphere is damaging the ozone layer; under an international agreement effective from January 1989 the use of most CFCs will be phased out and safe alternative fluids developed. (For the same reason the use of insulation materials such as polyurethane foam, extruded expanded polystyrene, phenolic foam, and polyisocyanurate, should be discontinued.)

Research is now underway in the major suppliers of refrigerants to develop acceptable alternatives. The refrigerant HFA 134a is one such alternative which the UK Heating and Ventilating Contractors Association (HVCA) is recommending for countries planning a large, short term expansion of domestic refrigeration (e.g. India and China).

The designation of CFCs is by a two or three digit number; the digit on the right represents the number of fluorine (F) atoms, the middle digit represents one more than the number of hydrogen (H) atoms, and the digit on the left represents one less than the number of carbon (C) atoms and is omitted when there is only one carbon atom. The most commonly used halocarbons are R11 (CCl_3F), R12 (CCl_2F_2), and R22 ($CHClF_2$); used less frequently are R113 (CCl_2FCClF_2) and R114 ($CClF_2CClF_2$); R502 which is an azeotropic mixture of R22 and R115 is sometimes used (an azeotropic mixture acts as a single substance and cannot be separated into its components by distillation). Of these compounds R22 has a very low ozone depletion potential (one-twentieth of that for R11 and R12), and may be exempt from the proposed ban on CFCs.

The commonly accepted designation for refrigerants other than halocarbons is to put the digit 7 in front of the relative molecular mass (e.g. ammonia, NH_3, is 717).

When a typical refrigeration cycle is considered with different refrigerants there is very little variation in coefficient of performance (COP), but the required pressures and the specific volume at compressor suction do vary considerably and these are important factors which determine which refrigerant should be used for a given temperature range, and which type of compressor is most suitable.

Table 4.10 compares different refrigerants for an evaporator temperature of $-15\,°C$, a condensing temperature of $30\,°C$, and assuming dry saturated vapour at compressor inlet and saturated liquid at condenser outlet. The size of the compressor required for a given cooling duty is proportional to the specific volume divided by the theoretical refrigerating effect. As can be seen from the table the compressor volume flow rate for a given cooling load is much greater for R11 because of its high specific volume; the compressor size for ammonia (R717), despite its high specific volume, is similar to that for the halocarbons because of its high enthalpy of evaporation.

Table 4.10 Comparison of some refrigerants

Refrigerant	Evaporator pressure (bar)	Condenser pressure (bar)	Specific volume at compressor suction (m³/kg)	Refrigerating effect (kJ/kg)	Compressor size per unit cooling load (m³/MW)
R11	0.20	1.26	0.762	155.1	4.913
R12	1.83	7.45	0.091	116.4	0.782
R22	2.96	11.92	0.078	162.9	0.479
R502	3.49	13.19	0.050	104.4	0.479
R717	2.35	11.64	0.509	1103.5	0.461

Positive displacement compressors of the reciprocating, rotary vane, or screw type are in common use for refrigeration. Reciprocating compressors are widely used in the range up to 600 kW with the vane type also used in the lower end of the range particularly for domestic refrigerators and air conditioners. Centrifugal compressors are used for larger plant in the range above 300 kW up to 15 MW. The screw type competes with both the reciprocating and centrifugal compressors covering the range 300 kW to 3 MW. It can be seen from Table 4.10 that R11 is more suitable for use with a centrifugal compressor because of its high specific volume.

When reciprocating and screw type compressors are used the lubricating oil is carried over with the refrigerant through the condenser and into the evaporator. Some refrigerants, such as ammonia for example, are not miscible with oil and hence the oil must be drained from the evaporator and returned to the compressor. R12 is miscible with oil but R22 is only partially miscible; R502 was introduced because it has similar properties to R22 but is also miscible with oil.

There are many different designs of evaporator depending on the application and the type of compressor and refrigerant. In a *direct expansion*, or *dry expansion*, shell-and-tube type the refrigerant flows through the tubes leaving with a slight degree of superheat (see Fig. 4.36(a)). In a *flooded evaporator* the

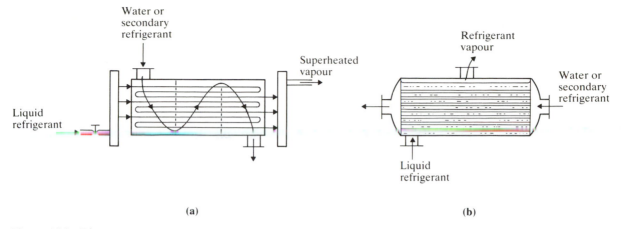

(a)   (b)

Figure 4.36 Direct expansion and flooded evaporators

refrigerant is on the shell side and the liquid covers the tubes with the vapour drawn off the top of the shell (see Fig. 4.36(b)). Heat transfer is improved when liquid refrigerant wets the heat transfer surface and hence a *flooded coil* is sometimes used in which the refrigerant flows to an accumulator where the vapour separates and goes direct to the compressor with the liquid flowing through the tubes within the shell (see Fig. 4.37). In the cases when the refrigerant expands in the tubes any oil is carried to the compressor due to the comparatively high velocity of the refrigerant; in the flooded coil type the oil separates out with the vapour and does not pass through the evaporator tubes, and hence the heat transfer in the evaporator is improved. A flooded shell-and-tube evaporator would normally be used with a centrifugal compressor because of the larger volumes of refrigerant flowing through the evaporator and hence there is no problem with oil separation.

In all the above cases the fluid which is being cooled as the refrigerant evaporates may be water or brine or a secondary refrigerant. Some evaporators of the direct expansion type are air cooled and may be used for air conditioning plant or for cold stores etc.

Condensers can be of three main types: water-cooled, air-cooled or evaporative. Water-cooled condensers can be of the shell-and-tube type or the double pipe type (see Fig. 4.38); the air-cooled type would generally be used with a higher

Figure 4.37 Flooded coil evaporator

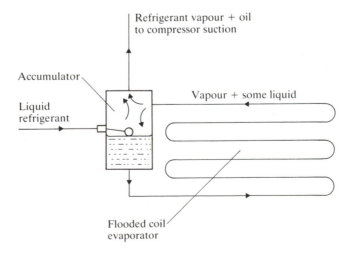

Figure 4.38 Double-
pipe water-cooled
condenser

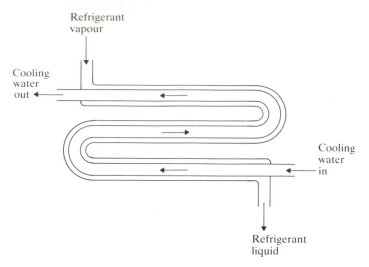

Figure 4.38 Double-
pipe water-cooled
condenser

condensing temperature since air is not available normally at as low a
temperature as water; in the evaporative type the refrigerant is supplied to a
cooling tower where water is sprayed over the coil as shown in Fig. 4.39. The
evaporative type has the advantage of more efficient air cooling combined with
the ability to cool water from some other source.

When a centrifugal compressor is used then the larger volumes and hence
larger pipes make it necessary to site the condenser near the compressor which
leads to the use of a water-cooled condenser with the water being pumped to
a cooling tower for cooling.

Figure 4.39 Sprayed-
coil evaporative
condenser

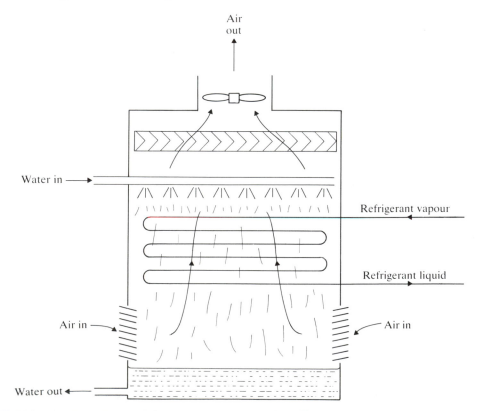

Vapour-Absorption Refrigeration

In an absorption plant the refrigerant passes through a condenser, an expansion device such as a throttle valve, and an evaporator exactly as in the vapour compression system but the compressor is replaced by a generator and absorber. The refrigerant is absorbed into a suitable fluid after evaporation, heat being rejected in the process, and the resulting solution is pumped to the generator; in the generator heat is supplied to drive refrigerant vapour out of solution, the vapour passing to the condenser and the remaining weak solution passing back to the absorber. Since the pressure in the generator and condenser are equal these can be contained in one vessel; similarly the absorber and evaporator can be contained in one vessel. A typical system is shown diagrammatically in Fig. 4.40.

The most commonly used vapour-absorption system uses water as refrigerant with lithium bromide as the absorbing fluid; the only other system commercially used has ammonia as the refrigerant with water as the absorbing fluid.

Vapour-absorption systems have a much lower coefficient of performance than a vapour-compression system but they have the advantage that they can use a supply of waste heat or solar radiation as the energy input. It is possible in small units to eliminate the pump by using gravity circulation due to density differences and hence the system can operate without the need of mechanical or electrical power.

Water chilling plants using a combined vapour-compression and vapour-absorption system have been used; in this case waste heat from the power input for the vapour-compression plant is used in the generator of the absorption unit.

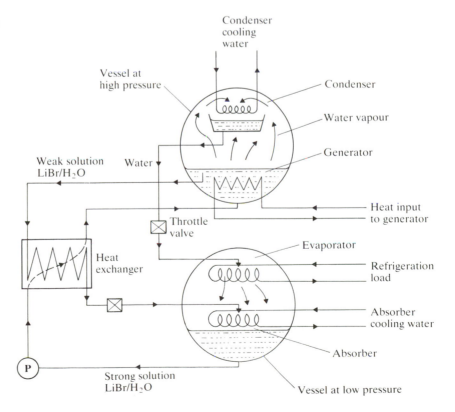

Figure 4.40 Absorption refrigeration plant

Example 4.9

A water chilling plant uses a combined R12 vapour-compression and lithium bromide/water absorption system. The compressor of the vapour-compression unit is driven by a gas engine the exhaust from which is used to provide the energy input to the generator of the absorption unit. The plant is shown diagrammatically in Fig. 4.41.

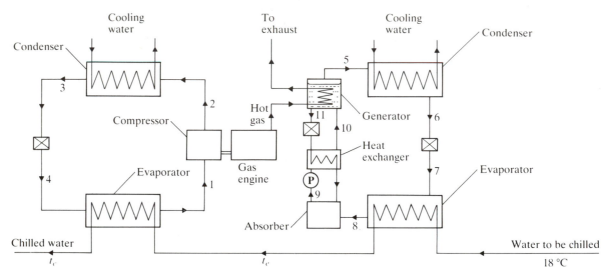

Figure 4.41 Combined vapour-compression and vapour-absorption plant

Using the data given and neglecting pressure losses, heat losses and pump work, calculate:

(i)   the cycle coefficient of performance of the vapour-compression unit;
(ii) the cycle coefficient of performance of the absorption unit;
(iii) the exit temperature of the chilled water;
(iv) the ratio of the chilled water cooling load to the energy input from the gas supply.

Data

Vapour-compression unit: saturation temperature in condenser, $30\,^{\circ}C$; saturation temperature in evaporator, $0\,^{\circ}C$; liquid leaving condenser is saturated; vapour leaving evaporator is dry saturated; assume isentropic compression; power input to refrigerant in the compressor, 50 kW; overall thermal efficiency of gas engine/compressor unit, 25 %.

Absorption unit: temperature of solution in generator, $85\,^{\circ}C$; temperature of solution in absorber, $30\,^{\circ}C$; temperature of solution leaving heat exchanger before entry to the generator, $45\,^{\circ}C$; condenser saturation temperature, $34\,^{\circ}C$; evaporator saturation temperature, $8\,^{\circ}C$; liquid leaving condenser is saturated; vapour leaving evaporator is dry saturated; heat supplied from engine exhaust is 50 % of the energy input from the fuel supplied to the engine; for lithium bromide/water solutions take the values in the table below.

Chilled water: mass flow rate, 10 kg/s; temperature entering evaporator of absorption unit, $18\,^{\circ}C$; mean specific heat, 4.19 kJ/kg K. Take the enthalpy of superheated water vapour as the enthalpy of saturated vapour at the same temperature.

	Concentration (mass LiBr/mass mixture)	Enthalpy (kJ/kg)
at 85 °C and 0.05318 bar	0.64	−77.1
at 30 °C and 0.01072 bar	0.51	−172.1
at 45 °C and 0.05318 bar	0.51	−138.9

Solution

(i) Refer to Fig. 4.41 for the state point numbers of both units; the vapour compression cycle is as shown on the T–s diagram of Fig. 4.35(b). Values of enthalpy for R12 can be calculated using the tables of Rogers and Mayhew,[4.2] or can be read from a pressure–enthalpy chart such as that given in the CIBSE *Guide*.[4.5] It should be noted that the zeroes for enthalpy and entropy are taken at different temperatures depending on the source of the tables or chart. For example the tables for R12 of Rogers and Mayhew[4.2] are based on h_f and s_f equal to zero at −40 °C; in the chart of the CIBSE *Guide*[4.5] the enthalpy and the entropy at the critical point are taken as 1000 kJ/kg and 1 kJ/kg K.

Using the Rogers and Mayhew tables:[4.2]

$$h_1 = h_g \text{ at } 0\,°C = 187.53 \text{ kJ/kg} \qquad \text{and}$$
$$s_1 = s_2 = 0.6966 \text{ kJ/kg K}$$
$$h_2 = 199.62 + \frac{(0.6966 - 0.6853)}{(0.7208 - 0.6853)} \times (210.63 - 199.62)$$
$$= 203.13 \text{ kJ/kg}$$

and
$$h_3 = h_4 = h_f \text{ at } 30\,°C = 64.59 \text{ kJ/kg}$$

Hence for a power input of 50 kW the mass flow of R12 is given by $50/(203.13 - 187.53) = 3.205$ kg/s and the refrigerating effect is, therefore

$$3.205(187.53 - 64.59) = 394 \text{ kW}$$
$$\therefore \qquad \text{coefficient of performance} = 394/50 = 7.88$$

If the engine efficiency is taken into account then the overall coefficient of performance is given by $394/200 = 1.97$.

(ii) The absorption unit is as shown in Fig. 4.41. The pressure in the generator is the same as that in the condenser and is therefore the saturation pressure corresponding to 34 °C, i.e. 0.053 18 bar. Similarly the pressure in the absorber is the same as that in the evaporator and is the saturation pressure at 8 °C, i.e. 0.010 72 bar. The liquid leaving the generator, state point 11, is therefore at 85 °C and 0.053 18 bar and the liquid leaving the absorber, state point 9, is at 30 °C and 0.010 72 bar (neglecting pump work); the concentrations at points 9 and 10 must be equal. Hence for a lithium bromide balance for the generator

$$0.64\dot{m}_{11} = 0.51\dot{m}_{10} \qquad \therefore \qquad \dot{m}_{11} = 0.7969\dot{m}_{10}$$

The overall mass balance for the generator gives

$$\dot{m}_5 = \dot{m}_{10} - \dot{m}_{11}$$

Therefore

$$\dot{m}_5 = \dot{m}_{10} - 0.7969\dot{m}_{10} = 0.2031\dot{m}_{10}$$

i.e.
$$\dot{m}_{10} = 4.923\dot{m}_5 \qquad \text{and} \qquad \dot{m}_{11} = 3.923\dot{m}_5$$

The fuel energy input to the compressor of the vapour compression unit is given by $50/0.25 = 200$ kW, and hence the heat supplied by the engine exhaust gases in the absorption unit generator is given by $200 \times 0.5 = 100$ kW. (Note that in a total energy scheme (see Chapter 8) some of the remaining 50 kW of heat rejected by the engine would also be reclaimed.)

Applying an energy balance to the generator:

$$100 = \dot{m}_5 h_5 + 3.923 \dot{m}_5 h_{11} - 4.923 \dot{m}_5 h_{10}$$

From the information given $h_{11} = -77.1$ kJ/kg and $h_{10} = -138.9$ kJ/kg; h_5 is h_g for steam at $34\,°C$, i.e. 2562.9 kJ/kg. Therefore

$$\dot{m}_5 = 100/2944.24 = 0.033\,97 \text{ kg/s}$$

For the evaporator the refrigerating effect is given by

$$\dot{m}_5(h_8 - h_7)$$

and $h_8 = h_g$ at $8\,°C = 2515.5$ kJ/kg
$h_7 = h_6 = h_f$ at $34\,°C = 142.4$ kJ/kg

Therefore the refrigerating effect is

$$0.033\,97(2515.5 - 142.4) = 80.6 \text{ kW}$$

Hence

coefficient of performance $= 80.6/100 = 0.806$

(iii)   For the water in the absorption unit evaporator:

$$10 \times 4.19(18 - t_e) = 80.6$$

(where t_e is the exit temperature),

i.e. $t_e = 16.08\,°C$

For the water in the vapour compression unit evaporator:

$$10 \times 4.19(16.08 - t_c) = 394$$

(where t_c is the exit temperature of the chilled water)

i.e. $t_c = 6.68\,°C$

The water is therefore cooled from $18\,°C$ to $6.68\,°C$.
(iv)   The chilled water cooling load is $(80.6 + 394) = 474.6$ kW. The energy input from the fuel is at a rate of 200 kW. Hence

overall performance ratio $= 474.6/200 = 2.373$

Other Refrigeration Systems

Vapour compression is by far the most widely used refrigeration system with absorption units of importance if a cheap source of thermal energy is available and/or if mechanical power is not readily available.

The reversed air cycle is used in aircraft air conditioning systems where air bled from the main compressor is conditioned as shown diagrammatically in Figs 4.42(a) and (b).

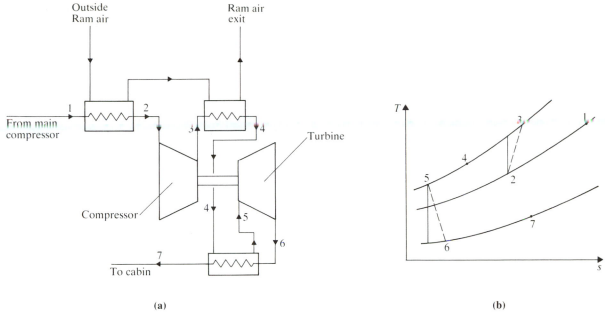

(a)   **(b)**

Figure 4.42 Aircraft cabin air conditioning plant and cycle diagram

Steam-jet water chilling plants are used in the food industry. In this system steam expands in a nozzle to a low pressure where it entrains water vapour from the evaporator; the water vapour on evaporation cools the bulk of the water remaining, in the same way as in a cooling tower. An evaporation of less than 1 % of the water in the evaporator can yield a 5 K reduction in the water temperature. A typical unit is shown diagrammatically in Figs 4.43(a) and (b).

Thermoelectric refrigeration making use of the Peltier effect in semiconductors is limited to very small scale applications at present.

Figure 4.43 Steam-jet water chilling plant and cycle diagram

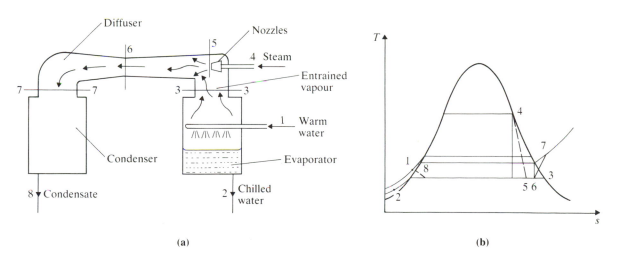

(a)   **(b)**

The Heat Pump

A heat pump is a refrigeration system in which the emphasis is placed on the heat rejected in the condenser rather than the heat supplied in the evaporator.

For an ideal refrigeration system operating between a condenser temperature of T_C and an evaporator temperature of T_E the maximum coefficient of performance (COP) is given by the Carnot cycle.

i.e. $$COP_{ref} = T_E/(T_C - T_E)$$
and $$COP_{HP} = T_C/(T_C - T_E)$$

It can be seen that in both cases the COP increases as the temperature difference between the condenser and evaporator decreases. For example, for a heat pump required to supply heat at 80 °C, say, the maximum COP for a low temperature source at 10 °C would be $(80 + 273)/(80 - 10) = 5.04$; the maximum COP would be increased to 6.42 if the source were increased to 25 °C. A real cycle has a lower efficiency than the Carnot cycle; also temperature differences between the working fluid and the source, and between the sink and the working fluid (say about 10 K) are needed to provide the heat transfer driving force; hence a practical heat pump has a much lower efficiency than the maximum possible efficiency between the temperatures of the source and sink. A $T–s$ diagram for a heat pump with typical temperature differences is shown in Fig. 4.44.

Figure 4.44 $T–s$ diagram for a heat pump

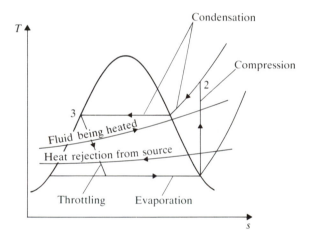

Heat pumps are categorized by the fluid which is used as the source and sink. Four categories are in common use: air-to-air, air-to-water, water-to-air, and water-to-water; an experimental application currently under development uses a U-tube buried in soil, giving two further possible categories of earth-to-air or earth-to-water. For example for a heat pump in the air-to-water category the refrigerant in the evaporator would receive heat from ambient air or hot waste air, and the refrigerant in the condenser would reject heat to water which could be used for heating purposes.

The UK market for the various categories in 1985 and the projected figures of the Building Services Research and Information Association (BSRIA)[4.6] are given in Table 4.11 by permission of BSRIA. The 1987 total of £33.7M was divided into Commercial 72 %, Industrial 9 %, Public 7 %, and Domestic 12 %, the latter being mainly for swimming pool heating. The air-to-air category takes the largest share of the market due to the large commercial use of this type which in 1987 was £19.25M or 57.1 % of the entire market.

A hierarchy of use drawn up by BSRIA[4.7] is as follows: (a) dehumidification for swimming pools; (b) combined heating and air conditioning for large office

Table 4.11 UK market share of heat pump categories for 1987 and 1990

Category	Market 1987 £M	Market 1987 %	Market 1990 £M	Market 1990 %
Air-to-air	25.0	74.2	30.5	74.0
Air-to-water	3.6	10.7	3.5	8.5
Water-to-air	4.6	13.7	6.8	16.5
Water-to-water	0.5	1.5	0.4	1.0
Total	33.7		41.2	

blocks, superstores, hypermarkets, high-street shops, banks, and building societies; (c) heat recovery systems; (d) heating only. It can be seen from the market statistics that the greatest use by far is in the commercial sector due to the various uses listed under categories (b) and (c) above.

It will be noted that heat pumps used for heating only are put at the foot of the hierarchical list and yet on first consideration they appear to be an attractive proposition. Compared with other forms of heating there is an immediate advantage (e.g. a gas boiler efficiency may be about 0.8 whereas the COP of a heat pump in practice may be about 3). The major disadvantages are in the high capital cost (four or five times the cost of an equivalent gas boiler) and in the lack of flexibility in use.

The most commonly used heat pump for heating only is the air-to-water type using the ambient air as the source. If the heating effect for such a system is plotted against the ambient air temperature (see Fig. 4.45), it is found that the heating effect is reduced as the ambient air temperature falls. This happens because, as the evaporator temperature and hence pressure is reduced, the specific volume of the refrigerant at compressor inlet is increased, and therefore the mass flow rate for a given compressor swept volume is reduced.

Figure 4.45 Heat pump balance point

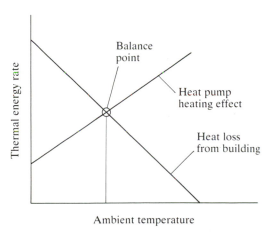

It can be seen from Fig. 4.45 that there is a balance point where the heat output from the heat pump equals the heat loss from the building. The balance point is recommended to be in the range 2 °C to 3 °C[4.7]; for ambient temperatures below this value supplementary heating is required. Supplementary heating (from a boiler say) can be supplied as an alternative to the heat pump below the balance point; this is known as a *bivalent alternative* system.

Alternatively, the heat pump output below the balance point can be supplemented by a secondary source; this is known as a *bivalent supplementary* system, or *bivalent parallel*. The selection of the correct balance point depends on the type of bivalent system selected, the capital costs of the heat pump and the secondary source of heating, and the fuel tariffs.

A heating-only heat pump will operate more efficiently without the need for a bivalent system if the temperature of the source is constant and independent of the ambient temperature. Good alternatives to ambient air as a source temperature are a lake or a river or the sea, all of which have a much more constant winter temperature. An even better alternative which has recently found favour is in the use of ground water as a source: a soak-way or store is built under the earth to provide a reservoir of ground water which is then pumped to the evaporator and returned by gravity to the soak-way.

A typical recent design for an office block uses an underground store of 30 000 m^3 of rock in a permeable plastic membrane as a source for a 350 kW electric heat pump. In another scheme for heating a school in Essex using ground water for an electric heat pump, measurements showed only a small variation in the ground water temperature from a maximum of 13.5 °C in October to a minimum temperature of 9.5 °C in April.

The use of hot waste water as a source (e.g. from a laundry) is an attractive proposition which allows a higher temperature for the heat supply without reducing the COP.

The BSRIA survey[4.6] shows that in the UK in 1987 heat pumps of a greater size than 50 kW represented only 9 % of the market with by far the largest use being in the 2 to 5 kW range, (55 %). Other European Countries have developed the use of large scale heat pumps linked to District Heating schemes (see Chapter 8). For example in Sweden there are several heat pump installations in excess of 100 MW; a 175 MW plant near Stockholm uses a sewage treatment plant as its source, while another 100 MW installation at Ropsten is built on a 100 m × 25 m barge and uses sea water as source; in both these cases the heat supplied is used for district heating.

Heat pump compressors are usually driven by electric motors although gas engine drive is becoming more common with energy recovery from the cooling water and engine exhaust; the overall system COP for an IC engine-driven heat pump is lower than for an electric heat pump because of the low thermal efficiency of the engine. (Of course the electricity generated has been obtained from a cycle with a similarly low thermal efficiency.) The choice must be made on economic grounds taking into account electricity tariffs, fuel costs, and capital costs.

Absorption-type heat pumps are not at present in common use although they may become important in the future, particularly if a more suitable fluid pair can be found as an alternative to LiBr/water or water/NH$_3$.

Example 4.10
A small nursing home is to be heated by a low-thermal-mass underfloor hot water heating system. Two ways of providing the heat input are being considered.
(a) An electric heat pump with a low temperature source provided by ground water; the ground water is pumped to the evaporator and returned by gravity to a rubble-filled soak-way.
(b) A gas-fired condensing boiler.

The required design heating load is 100 kW and the under-floor system has an average water temperature of 35 °C. The annual hours of use of the heating system at the average conditions may be taken as 4000. Using the data below, neglecting heat losses and pumping power, make an initial estimate of the time after which the heat pump system will start to make a net saving neglecting all annual costs other than the fuel costs.

Data
Vapour compression heat pump: refrigerant, R12; approximate minimum temperature difference for heat transfer in both the condenser and evaporator, 10 K; overall mechanical and electrical efficiency of motor/compressor unit, 86 %; average temperature of ground water, 10 °C; dry saturated vapour at entry to compressor; no undercooling in the condenser; assume isentropic compression; capital cost of the complete system, £21 000; cost of electricity, 3.7 p/kW. Condensing boiler: efficiency, 90 %; capital cost of the complete system, £5500; cost of gas, 1.2 p/kW.

Solution
The vapour-compression cycle is shown on a T–s diagram in Fig. 4.46. From Tables[4.2]:

at 0 °C, 3.086 bar, $h_1 = 187.53$ kJ/kg and $s_1 = s_2 = 0.6966$ kJ/kg
at 45 °C, 10.84 bar, $h_3 = 79.71$ kJ/kg

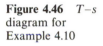

Figure 4.46 T–s diagram for Example 4.10

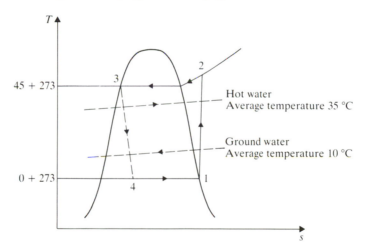

Also, interpolating

$$h_2 = 204.87 + (216.47 - 204.87)\frac{(0.6966 - 0.6811)}{(0.7175 - 0.6811)}$$
$$= 209.93 \text{ kJ/kg}$$

and
$$\text{COP}_{\text{HP}} = \frac{h_2 - h_3}{h_2 - h_1} = \frac{209.93 - 79.71}{209.93 - 187.53} = 5.81$$

Then allowing for the mechanical and electrical efficiency of 86 % we have an electrical power input of

$$100/(5.81 \times 0.86) = 20 \text{ kW}$$

The running cost of the heat pump is therefore $20 \times 3.7/100 = £0.74$ per hour.

For the gas boiler the running cost for a heat supply of 100 kW with a boiler efficiency of 90 % is given by $100 \times 1.2/(0.9 \times 100) = £1.333$ per hour.

The expressions for total cost for a given time for the two systems are then:

$$£(21\,000 + 0.74y) \qquad \text{for the heat pump}$$

and,

$$£(5500 + 1.333y) \qquad \text{for the condensing boiler}$$

where y is the number of hours of running.

By equating these expressions the break-even time can be found

i.e. $$21\,000 + 0.74y = 5500 + 1.333y$$
$$\therefore \qquad y = 26\,124 \text{ hours}$$

The running hours per year at the average design conditions is given as 4000, therefore

$$\text{break-even time} = 26\,124/4000 = 6.53 \text{ years}$$

This length of time would be considered too long to make the heat pump a feasible proposition particularly since pumping power and heat losses have been neglected; there would also be greater annual maintenance costs for the heat pump.

Air Conditioning

One of the main uses of refrigeration is in the air conditioning of buildings. Cooling coils supplied with chilled water are used to provide the required cooling and de-humidification load. Specialist texts such as Jones[4.8] should be consulted for a full description of air conditioning systems; a brief summary and a simple treatment of single-zone systems for summer and winter is given in Eastop and McConkey.[4.4]

Problems are most easily solved using a psychrometric chart similar to the CIBSE chart given in a reduced scale version as Fig. 4.47 with the permission of CIBSE. The following example illustrates the energy requirements for a simple air conditioning plant.

Example 4.11
An air conditioning plant is designed to maintain a room at 20 °C, percentage saturation 50 %, with an air supply to the room of 1.8 kg/s at 14 °C, percentage saturation 60 %. The design outside air conditions are 27 °C, percentage saturation 70 %. The plant consists of a mixing chamber for re-circulated and fresh air, a cooling coil supplied with chilled water, heating coil, and supply fan. The ratio of re-circulated air to fresh air is 3; the cooling coil has an apparatus dew point of 5 °C, and the refrigeration unit supplying the chilled water has an overall coefficient of performance 2. Neglecting all losses and fan and pump work, calculate:
(i)   the total air conditioning load for the room;
(ii) the required total energy input;
(iii) the required energy input if the energy to the heating coil is supplied from the refrigeration plant condenser cooling water.

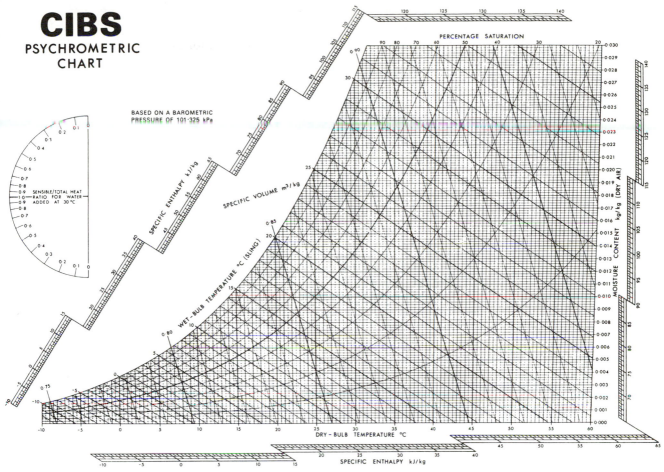

CIBS
PSYCHROMETRIC
CHART

BASED ON A BAROMETRIC
PRESSURE OF 101·325 kPa

Figure 4.47 CIBSE
psychrometric chart.
(Reproduced by
courtesy of CIBSE.)

Solution

The plant is shown diagrammatically in Fig. 4.48(a) with the corresponding
state points on a sketch of the psychrometric chart in Fig. 4.48(b). The apparatus
dew point, A, represents the temperature of the cooling coil surface to which

Figure 4.48 Summer
air conditioning

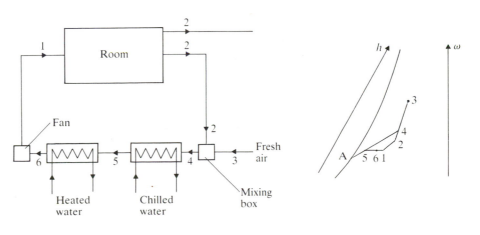

(a)   (b)

the air would be cooled in an ideal heat exchange process; the air leaving the coil is at state point 5 which is on a horizontal line through point 1; the heating coil then heats the air at constant specific humidity to the required condition for entry to the room. Points 6 and 1 coincide if the fan work and duct heat gains are neglected.

(i)   From the psychrometric chart:

$$h_1 = 29.3 \text{ kJ/kg} \qquad h_2 = 38.8 \text{ kJ/kg}$$

i.e.   total load on room $= 1.8(38.8 - 29.3) = 17.1$ kW

(ii)   Since there are three parts re-circulated air to one part fresh air the state of the air after mixing at point 4 can be found by dividing the line joining points 3 and 2 in the ratio of 3 to 1.

i.e.   line 2–4 = (line 2–3)/4

Then from the chart, $h_4 = 46.1$ kJ/kg. Point 4 can now be joined to point A which is at a temperature of $5\,^\circ$C (given); where this line cuts the horizontal line through point 1 gives the point 5. From the chart, $h_5 = 22.7$ kJ/kg. Then the cooling load from the chilled water is

$$1.8(46.1 - 22.7) = 42.1 \text{ kW}$$

and the energy input to the compressor of the refrigeration unit is

$$42.1/2 = 21.05 \text{ kW}$$

Also the energy input to the heating coil is

$$1.8(29.3 - 22.7) = 11.88 \text{ kW}$$

i.e.   total energy input required $= 21.05 + 11.88 = 32.93$ kW

(iii)   The heat rejected to the condenser cooling water is given by the refrigeration effect plus the power input to the compressor, i.e. the heat rejected to cooling water is

$$42.1 + 21.05 = 63.15 \text{ kW}$$

There is therefore adequate energy available to provide hot water for the heating coil, provided of course that the condenser evaporation temperature is sufficiently high; the hot water to the coil must be a sufficient margin above the air outlet temperature of $14\,^\circ$C to allow for heat transfer. The required energy input is therefore only the power input to the refrigeration unit compressor,

i.e.   total energy input $= 21.05$ kW

In some cases the water vapour gains in a space are not great and the exact control of humidity is not always as important as the control of temperature. The heater can then be dispensed with and the final state of the air supplied to the room is controlled by mixing the air from the cooling coil with some of the air re-circulated from the room. This is shown in Figs 4.49(a) and (b); for a mixing process the state point 6′ must lie on a line joining points 2 and 5′. Note that the state point 4′ is now nearer to point 3 on the line 2–3 since some of the re-circulated air by-passes the first mixing box; the point 5′ can be fixed by assuming that the apparatus dew point is unchanged and that the coil

Figure 4.49 Simple air conditioning plant

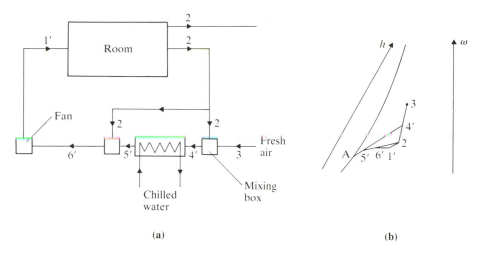

(a)   (b)

contact factor, $(h_4 - h_5)/(h_4 - h_A)$ remains the same, and hence is equal to $(h_{4'} - h_{5'})/(h_{4'} - h_A)$.

An alternative way of dispensing with the heating coil is to mix the air leaving the cooling coil with some by-passed air at state 4; this method will also control the supply temperature but not the humidity.

In complex multi-zone buildings different heating and moisture loads exist in different zones and requirements for temperature and humidity may vary from zone to zone. If humidity control is not required then a simple dual duct system can be used with cooled air supplied in one duct and the mixed fresh and re-circulated air in a second duct; mixing of the two streams then takes place in terminals at the room giving control of each room separately.

Another method is to deliver separate supplies of chilled and heated water to individual units in each room; air is drawn from outside by a fan in the unit, mixed with room air which is cooled or heated as required depending on the season. A similar system supplies a small quantity of primary conditioned air to each room where it enters the unit through a nozzle thus inducing a large quantity of room air across a cooling coil which then mixes with the primary air before being supplied to the room.

In recent years the Variable Air Volume (VAV) method has come into prominence; the conditioned air is supplied at constant temperature and control is effected by varying the volume flow rate of the supply to each room.

Any complex plant must be able to be controlled to satisfy all zones in summer, winter and the mid-season periods. A proper economic assessment is difficult since the cost of the basic system plus controls must be offset against the running costs including pump and fan costs, the latter depending on the size and length of pipes and ducting as well as on the velocity of the fluid. From pure energy considerations a system must attempt to supply the energy requirements for a given space as efficiently as possible. Systems which do not require complete control of humidity are more energy efficient since it is wasteful on energy in summer to cool air down to de-humidify it and then re-heat it to the required condition for the room; similarly in winter it is wasteful to supply more energy than the energy loss from the room when this may be necessary in order to humidify the air to the required condition.

For further information on air conditioning systems see Jones[4.8] and the CIBSE *Guide*.[4.9]

4.6 ELECTRICAL CONVERSION

By far the most common way of producing electricity is by converting thermal energy to mechanical energy to electrical energy. Since the electrical energy may then be converted back to thermal energy, or to mechanical energy using an electric motor, it can be seen that the tortuous conversion path incurs large losses. When comparing a mechanical drive using an electric motor with a direct drive from say an IC engine it must be remembered that the electricity has been produced from a heat engine cycle with an efficiency very little different from that of the IC engine. This is of course reflected in the comparative prices to the consumer of the electricity and the fuel for the IC engine. In theory direct conversion from chemical energy to electrical energy or from thermal energy to electrical energy would be more efficient but in practice such devices have even greater inefficiencies than the complex conversion and reconversion path in common use.

Chemical energy may be stored for later use in a *battery* but the supply is limited to low voltages and the energy stored is limited by size. The *fuel cell* is similar in principle to a battery, with an electrolyte, a cathode and an anode, but unlike a battery, fuel reactants are continuously supplied to the anode (e.g. hydrogen) and the cathode (e.g. oxygen). To date fuel cells have only been used in special relatively short term applications such as in space flights.

Thermoelectric conversion is based on the Seebeck effect; when junctions of dissimilar metals are maintained at different temperatures then a voltage is produced between the junctions. Peltier found that if a direct current flows in a circuit of two dissimilar metals then one junction will be heated and the other cooled. Hence thermoelectric devices can be used either to produce electricity or to convert electricity to achieve a refrigerating effect. Both systems are bound by the limitations of the Second Law and hence the maximum thermal efficiency is that of a Carnot cycle between the same temperature limits; at the current state of research practical devices have thermal efficiencies no higher than 5 to 10 % and the materials used (e.g. Lead telluride) are expensive.

Other devices such as thermionic converters are even more expensive operating at high temperatures with low efficiencies. All of these systems require a major breakthrough in the materials used before they can become commercially viable for electricity generation.

Electromagnetic energy can be converted to electricity using the photovoltaic cell or *solar cell*. In the 19th century it was discovered that when light is directed on an electrode a voltage is produced, but it was not until the development of semiconductor materials that practical use was made of this effect. Solar cells have been widely used in space and for applications such as water pumping in underdeveloped countries but their low efficiency and the relatively expensive materials required rule them out for large scale commercial use.

Generation of Electricity

Conversion of mechanical energy to electrical energy is effected when an electric conductor cuts a magnetic field thus producing a voltage gradient in the conductor. The magnetic field is created by putting coils carrying direct current (dc) around steel cores to form North and South poles; these coils are known as the field coils (see Section 3.3 for a summary of the basic theory).

Electric generators operate with a rotating shaft with either, (a) the field coils on the fixed annulus or *stator*, with the coils in which the current is produced wound on a rotating element called the *rotor*; or, (b) the field coils on the rotor and the coils producing the current on the stator (see Figs 4.50(a) and (b)). In either case the current is produced due to the relative movement of the conductor and the magnetic field; since the relative motion is rotational then the current produced will vary from zero to a maximum and back to zero in each cycle of rotation. The current produced is therefore basically alternating current (ac). (Note: the element which contains the conductors in which electricity is generated is known as the *armature*; from the above it will be seen that either the rotor or the stator can be the armature of the machine.)

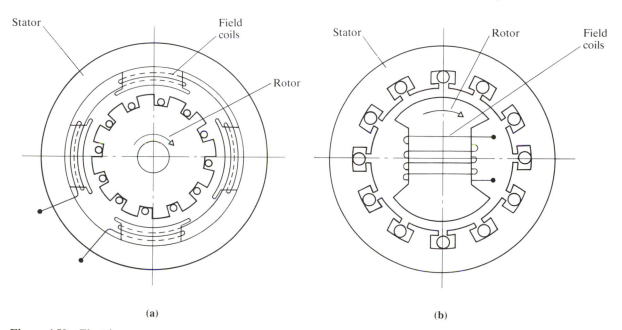

(a)

(b)

Figure 4.50 Electric generator with field coils on stator or rotor

The Direct Current Generator

In a direct current (dc) generator the current reversal in the armature, which is the rotor, is prevented by *commutation*; this process is accomplished by connecting the coils to commutators which are in contact with stationary carbon 'brushes' which short circuit small segments of the commutator as it rotates. The output current is obtained from conductors connected to the brushes.

The field coils in a dc generator may be separately-excited by supplying a current from a source of supply other than that from the generator itself (e.g. a battery, or a rectified ac supply), or they may be self-excited by a current from the generator (for smaller machines permanent magnets may be used). In the case of self-excitation the field coils may be connected in series with the armature windings (series-wound), or connected across the armature terminals (shunt-wound), or in a combination of series and shunt windings (compound-wound). A self-excited machine may be started using the residual magnetic flux present

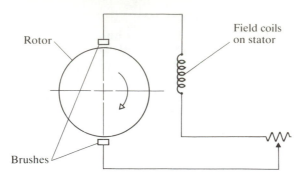

Figure 4.51 Series-wound dc generator

in the machine. A typical series-wound dc generator is shown diagrammatically in Fig. 4.51.

The Alternating Current Generator

In an ac generator (formerly called an alternator) the field coils are wound on the rotor and the coils producing the current wound on the stator; brushes on slip rings are used to bring the exciting current to the field coils on the rotor. By putting the field coils on the rotor the brushes need take only the comparatively small exciting current rather than the generated current, thus reducing the losses considerably.

The exciting current for the ac generator can be supplied from a separate dc supply but it is more usual, particularly in larger machines, for a small dc generator, known as the *exciter*, to be driven from the main shaft. In the very large power station generators a pilot exciter is also used; the pilot exciter can be started from its residual magnetic flux and it then supplies the exciting current to the main exciter which in turn supplies the exciting current for the rotor of the generator (see Fig. 4.52). In some large machines a small ac generator driven by the main shaft acts as exciter; this small generator, known as a *brushless exciter*, has its field coils on the stator and generates ac current in the rotor windings; a solid state rectifier rotates with the shaft, converting the ac output from the small generator into dc which is then supplied to the rotating field coils of the main generator without the need of brushes; a pilot exciter provides the necessary dc current to excite the field coils on the stator of the small ac generator.

Figure 4.52 Typical power station ac generator

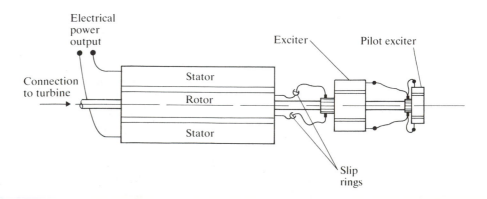

Figure 4.53 Salient
pole and slot-type pole
ac generators

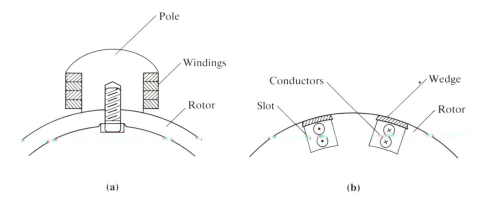

(a) (b)

The rotor of an ac generator can be of the *salient pole* construction which is in use for machines driven at relatively low speed, for example by an IC engine, or water turbine (see Fig. 4.53(a)). As discussed in Section 3.3 the frequency of the ac current generated is given by the product of the shaft rotational speed and the number of pairs of poles (Eq. 3.39). In the UK and in Europe generally the frequency of the supply is fixed at 50 Hz and hence the shaft speed for one pole pair is $50 \times 60 = 3000$ rev/min; for two pole pairs it is 1500 rev/min etc. (In other countries a different frequency is chosen; for example in the USA the frequency is 60 Hz.) Applying Eq. 3.39 it can be seen that an IC engine running at 375 rev/min would require 8 pole pairs to give the 50 Hz frequency. This type of construction requires a relatively large diameter rotor of short axial length. The large power station generators are driven by steam turbines, or sometimes gas turbines, rotating at 3000 rev/min with one pole pair. To reduce the centripetal forces and increase the mechanical strength the rotor is made from a solid steel forging with longitudinal slots containing copper conductors held in position by phosphor bronze wedges (see Fig. 4.53(b)).

In Section 3.3 it was pointed out that a magnetic field is created round the secondary windings when a current is induced in them and that for three phase generation the field created in the stator rotates at the same speed as the rotor. The normal design of an ac generator described above is therefore known as a *synchronous generator*. Another type of generator used in small combined heat and power systems is described at the end of this section.

Electricity Supply

In the USA at the end of the 19th century a debate raged on whether to standardize the national electricity supply as direct current or as alternating current; the two main proponents were Tesla (who favoured ac, and had invented the induction motor in 1888), and Eddison (who had by far the greater reputation as an inventor but who for once in his life was wrong). The same argument took place in the UK in the 1890s with Sebastian De Ferranti supporting Tesla, recognizing that the polyphase induction motor would be the industrial tool of the future. The main factor which caused the decision to come down in favour of ac was the ease with which the voltage of ac can be increased and decreased using a transformer. This allows electricity to be generated at a relatively high voltage, transformed up to a very high voltage for transmission across long distances, and then transformed down to a low voltage at the point of use. Large quantities of power can be transmitted in this way with relatively small currents and hence with low losses.

There are disadvantages in the transmission of ac, particularly when it is required to provide a link between different countries; the importing and exporting of power is becoming increasingly part of the international scene. Long transmission lines of high power develop a considerable reactive current; also when different countries are involved there is a need to synchronize the systems which is impossible if different frequencies are used. With the advent of solid state devices and the development of the thyristor as a rectifier the use of dc links is becoming more widespread. For example England and France are linked under the Channel by a dc link: at each end there are rectifiers which convert high voltage ac to high voltage dc, and also at each end there are inverters which convert high voltage dc to high voltage ac; either country can therefore import or export electricity as required. The use of such dc links is still at an early stage and the main internal distribution system of all countries is ac at present.

In the UK electricity is generated at about 25 kV at the stator of the generator, and is distributed at either 400 kV, 275 kV , or 132 kV. (In the USA the values used are 500 kV and 345 kV.) The 400 kV lines are used for distributing electricity over large distances, the 275 kV lines supply the major cities, and the 132 kV lines distribute the electricity within a local area. This system is known as the National Grid.

Transformers in substations reduce the voltage down to 33 kV for a given locality, some of which may be supplied to large industrial consumers; further substations within the area reduce the voltage to 11 kV for light industry and large hospitals, say; finally the voltage is transformed down to 415 V for the general community with lighting and power sockets supplied with 240 V using a single phase.

The different types of connection of a three-phase supply are outlined briefly in Section 3.3. Since a truly balanced system can never exist due to different customers taking different loads, a neutral conductor is always provided which is earthed at the distribution transformer. The line voltage of the three-phase supply is 415 V and therefore in the UK for the single phase supply the voltage across one line and the neutral conductor is $415/1.732 = 240$ V. A separate earth wire is also provided to customers which is at a true earth potential.

Synchronous Motors

Electric machines should properly be called electromechanical energy converters since a basic machine will operate either as a generator of electricity using mechanical power or as a motor supplying mechanical power from an electrical supply.

A synchronous generator as described above will operate as a synchronous motor if ac current is supplied to the stator windings. In that case the rotor will be driven at the synchronous speed (i.e. f/p), but otherwise the machine is identical with a generator. The number of pole pairs, p, on the rotor will determine the rotational speed for the normal UK frequency of 50 Hz; for example, a motor with six pole pairs will rotate at $50 \times 60/6 = 500$ rev/min.

The main disadvantage of synchronous motors is that they are not self starting. The three-phase current supplied to the stator creates a rotating magnetic field but until the rotor is excited and is revolving no effective torque is applied to the shaft. If the rotor is turned by an auxiliary drive then the two rotating magnetic fields will start to interact and the rotor will be dragged into synchronization.

For any fixed value of excitation there is a maximum load for any synchronous motor; if this load is exceeded the machine will stop. For a certain load, varying the excitation changes the power factor for the same active power from the supply. This is a particular advantage since by supplying an exciting current in excess of the synchronous value a leading power factor can be obtained; the machine then draws active power from the mains and at the same time returns reactive power to the mains hence acting as a large capacitor. A single synchronous motor can therefore be used to correct the power factor of a large inductive load (for example a number of induction motors) and is then called a *synchronous capacitor*.

Because of the need for a dc supply via brushes to the rotating field coils (possibly with a small shaft-mounted dc generator to generate the exciting current) a synchronous motor is expensive and would normally be used only when a large amount of mechanical power was required, or as an effective large capacitance for power factor correction. Modern synchronous motors have a brushless ac exciter, (i.e. an ac generator with field windings on the stator), mounted on the shaft with rotating rectifiers to provide the dc field current to the rotor windings of the motor without the need for slip rings.

Induction Motors

By far the most common type of motor used in industry is the *three-phase induction motor*. The stator of an induction motor is similar to that of a synchronous motor but in the most common type of induction motor the rotor conductors consist of a number of copper or aluminium bars in slots on the rotor and joined at the ends to circular rings. This type is known as the *cage rotor*, or *squirrel-cage*, induction motor.

When a three-phase current is supplied to the stator windings of an induction motor the rotating magnetic field induces an emf in the rotor. The current in the rotor conductors produces its own magnetic field which distorts the flux in the air gap thus causing a torque on the rotor which tends to drag it round in the direction of the stator's rotating field. The current in the rotor conductors, and hence the magnetic field, depends on the relative motion between the rotor and the rotating magnetic field which is rotating at synchronous speed. As the rotor speeds up the relative speed between it and the rotating magnetic field gets less and hence the torque decreases. Conversely as the rotor speed falls the relative speed increases and hence the torque increases. In the steady running condition the rotor reaches a speed below the synchronous speed such that the torque is balanced by the load on the motor and the losses in the machine itself. The difference between the synchronous speed and the rotor speed is called the *slip*, s, and is expressed as a fraction or a percentage of the synchronous speed,

i.e. $$s = (N_s - N)/N_s \qquad [4.9]$$

(where N_s = synchronous speed; N = rotor rotational speed). At full load the percentage slip varies from about 2 % for large machines to about 6 % for smaller machines; the shaft speed therefore is essentially constant as the load varies.

The synchronous speed of a three-phase induction motor depends on the stator windings. These can be arranged to give one pole pair or several pole pairs and since the speed of the rotating flux is given by Eq. [3.39] (i.e. f/p), the synchronous speed can be designed to be 3000 rev/min (two poles), 1500 rev/min (four poles), 1000 rev/min (six poles), etc.; (this represents motor shaft speeds of 2880 rev/min, 1440 rev/min, and 960 rev/min for a slip of 0.04 for example).

Induction motors with a three-phase *wound rotor* are also in use; in this type the rotor current is taken via slip rings to a variable external resistance (see Fig. 4.54). The operating principle is exactly the same as for a cage rotor machine, although its capital cost is considerably more.

Figure 4.54 Three-phase wound-rotor induction motor

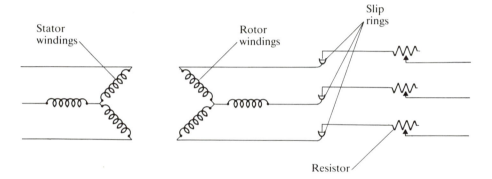

The torque-speed characteristics of a three-phase induction motor are as shown in Fig. 4.55(a); it can be seen that the torque falls to zero at the synchronous speed, N_s, but there is a starting torque available at standstill. The shape of the curve depends on the ratio of the resistance to the impedance of the rotor conductors; as this ratio increases for a given machine the shape of the curve changes as shown in Fig. 4.55(b). By connecting variable resistances in the external circuit of a wound rotor it is therefore possible to vary the speed of the motor by varying the slip; this system also provides a means of increasing

Figure 4.55 Torque against speed for a three-phase induction motor

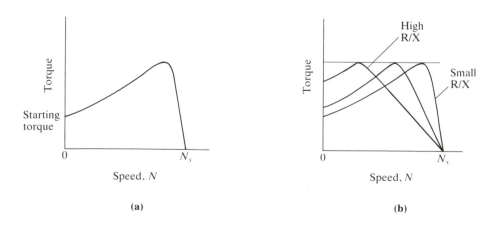

the starting torque. The wound rotor with external resistances is limited in speed control to a range of about 1.5/1.

Since the motor speed depends on the synchronous speed and since this can be changed by increasing the number of poles in the stator windings (i.e. $N_s = f/p$), then one method of speed changing is a system of switching so that the stator windings are altered to give a different number of pole pairs. This method gives a number of discrete steps in speed with very little loss of efficiency and at a small increase in capital cost.

The speed of an induction motor can also be varied by altering the frequency of the supply and hence the synchronous speed. Modern solid state devices provide a means of frequency control which was not previously available. The widest range of control can be obtained using a solid state rectifier to change the normal ac supply to dc; the dc supply thus produced is then converted to ac using a solid state inverter which gives the possibility of a wide range of ac frequencies. Using this method of control the cheaper cage rotor can be used although the total capital cost of the solid state controller plus motor must be compared with the capital cost of a motor of the wound rotor type depending on the application required.

Both the efficiency and the power factor of an induction motor fall off as the load is reduced below the design value; this is illustrated in Figs 4.56(a) and (b). It is therefore important not to run such motors at less than about half load. For the same reason it is important not to install a motor which is oversized for the required duty. The input power to an induction motor is reduced by the I^2R loss and the iron loss in the stator; the power at the rotor is further reduced by the I^2R loss and iron loss in the rotor; there are then further friction and windage losses giving finally the useful shaft power. As the load is increased the input current increases and hence the stator losses increase as the square of the current. It can also be shown that the rotor I^2R loss is equal to the fractional slip times the input power to the rotor (the current induced in the rotor depends on the relative speed between the rotating magnetic flux of the stator and the rotor conductors, i.e. the slip).

Figure 4.56 Efficiency and power factor variation with load for a three-phase induction motor

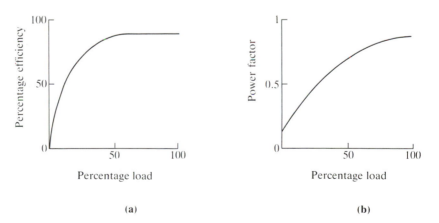

(a)   (b)

Example 4.12

A three-phase star-connected induction motor runs from a 415 V, 50 Hz supply and at full load the electrical power input is 100 kW. The stator is wound to

give twelve poles and the slip at full load is 4 %. The stator loss at full load is 6 kW when the current per phase is 170 A. The friction and windage losses are 4 kW. When the load on the motor is reduced so that the mechanical shaft power is one quarter of that at full load the slip is then 1 % and the stator current is 65 A. Assuming that the friction and windage losses are unchanged as the load changes, but that the stator loss is reduced to 1.5 kW, calculate:
(i)   the motor speed at full load and part load;
(ii) the efficiency and power factor at full load;
(iii) the efficiency and power factor at part load.

Solution
(i)   Using Eq. [3.39], for a twelve pole machine the synchronous speed is given by

$$N_s = 50 \times 60/6 = 500 \text{ rev/min}$$

Then when the slip is 4 %, at full load we have, from Eq. [4.9]

$$\text{rotor speed} = N_s - sN_s = 0.96 \times 500 = 480 \text{ rev/min}$$

When the slip is 1 % at part load

$$\text{rotor speed} = 0.99 \times 500 = 495 \text{ rev/min}$$

(ii)   At full load the power input to the rotor is given by $(100 - 6) = 94$ kW. The rotor electrical loss can be taken as this value times the fractional slip, i.e. $94 \times 0.04 = 3.76$ kW. Hence the mechanical power at the shaft is given by

$$\text{power output} = 100 - 6 - 3.76 - 4 = 86.24 \text{ kW}$$

Therefore motor efficiency is

$$86.24/100 = 86.24 \%$$

The power factor is given by Eq. [3.44]

i.e.   power factor $= 100/VI = 100\,000/(1.732 \times 170 \times 415) = 0.82$

(iii)   At part load the stator loss is given as 1.5 kW, the slip is 1 %; the friction and windage losses are unchanged at 4 kW. The mechanical power at part load is $86.24/4 = 21.56$ kW. Therefore, the power from the rotor is

$$21.56 + 4 = 25.56 \text{ kW}$$

Then the input power to the rotor is

$$25.56/0.99 = 25.82 \text{ kW}$$

Hence the power input to the motor is

$$25.82 + 1.5 = 27.32 \text{ kW}$$

and efficiency is given by

$$\eta = 21.56/27.32 = 78.92 \%$$

Also

$$\text{power factor} = 27\,320/(1.732 \times 415 \times 65) = 0.59$$

It can be seen from this example that the speed of the motor changes by a small percentage, $(495 - 480)/480 = 3.1\%$, as the load changes from 86.24 kW to 21.56 kW. The efficiency drops from 86.24 % to 78.92 % and the power factor goes down from 0.82 to 0.59. If the motor runs for long periods at low load then because of the low power factor the system has a large reactive current; this can be reduced by connecting a capacitor in parallel with the motor (see Example 3.10).

Single-phase induction motors are widely used in the smaller power range. Since there is no rotating magnetic flux in the stator as in the three-phase case, there is therefore no starting torque. The torque speed relationship is as shown in Fig. 4.57. The normal method of starting such a machine is by having a second, or auxiliary, winding on the stator which makes it effectively a two-phase machine for starting; the current in the auxiliary winding leads the current in the main winding with the same voltage across both windings. The auxiliary winding is not designed to remain in circuit after the rotor is up to speed and is normally cut out by a centrifugal device operating at about 70 % of the running speed of the motor. Improved starting characteristics can be obtained by putting a capacitor in series with the auxiliary winding; such motors are called capacitor-start. Another way of solving the starting problem is by incorporating a copper ring on a spur of each of the stator poles; this creates a separate flux which is sufficient to create a starting torque; this type is known as a *shaded-pole* motor. Motors which have separate main and auxiliary windings to give phase differences for starting are called *split-phase motors*.

Figure 4.57 Torque against speed for a single-phase induction motor

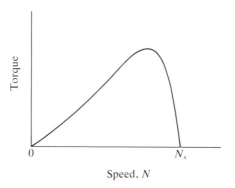

The Chartered Institution of Building Services Engineers give a useful summary[4.10] of characteristics of motors used in the building services industry outlining starting methods, speed control, and applications.

Commutator Motors

At the beginning of the description on generators it was pointed out that electromechanical machines may have their field coils on the rotor (as is the usual design for ac machines) or on the stator (as is always the case for dc machines). The placing of field windings on the rotor for ac machines was done in order to avoid the brushes on the rotor having to conduct the large current of the supply rather than the field current. It is possible to supply ac through brushes to a commutator on the rotor which is then surrounded by field coils on the stator also supplied with ac. This is then the ac equivalent of a dc motor (see Fig. 4.51 which is a series-wound dc generator or motor), but the ac supply creates further emfs which make commutation more difficult leading to a circulating current through the brushes. For this reason this type of motor has a high capital cost and incurs more maintenance. *Three-phase commutator motors* can provide speed control up to ratios of about 4/1, and may therefore be cheaper overall when the capital cost, although high, is compared with other machines with more expensive methods of speed control.

When the brushes are raised from the commutator the machine runs as an induction motor; when an ac voltage is fed through the brushes the slip can be increased thus lowering the rotor rotational speed; by reversing the supply voltage to the brushes the motor can be made to run at supersynchronous speed.

Single-phase commutator motors are used in high numbers mainly in the fractional kW range; the stator coils can be in series with the rotor and for this type the motor will run on either ac or dc. The other common type is the *repulsion motor* in which the brushes are shorted to provide a closed path for the rotor current which is therefore induced from the stator (see Figs 4.58(a) and (b)).

Figure 4.58
Commutator motors

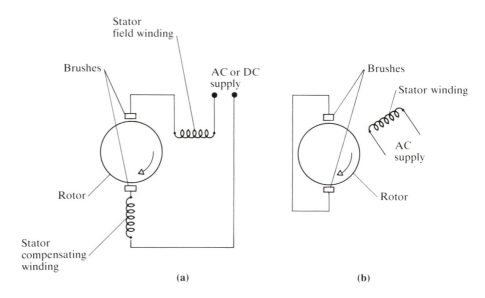

Motors running on dc are very versatile and speed can be controlled easily over a wide range. The power range is large, from tiny motors with permanent magnets running off small batteries up to machines of thousands of kW. The cost on the other hand is high because of the brush/commutator system and the need to convert the normal ac supply to dc in all cases other than the very small battery-driven motors.

Private Generation

The generation of power by private consumers has been a feature for many years, not only as a means of energy saving by using waste heat etc., but also to allow a plant, or hospital for example, to become independent of the mains supply in the event of failure of that supply. The latter contingency is covered by installing a *standby generator* which is driven typically by a diesel engine and is not connected to the grid. Where a plant generates its own electricity for use within the plant this is known as *standalone-operation*. When a generator is connected to the grid this is known as *parallel-operation*; electricity can be exported to the grid or taken from it in this system.

The 1983 Energy Act first introduced the possibility of a private consumer generating electricity and selling it to the grid. Since that date many companies have made use of small combined heat and power units, usually using diesel or gas engines, to generate electricity which can be supplied to the grid when generated in excess of requirements. Generators used in such systems can be synchronous generators but a cheaper alternative is the *induction generator* (also called the *asynchronous generator*).

The induction generator is simply an induction motor with the rotor shaft driven above synchronous speed; the rotor can be of the simple cage type. The characteristic of the motor previously given in Fig. 4.55(a) can be extended as shown in Fig. 4.59. The torque becomes negative at speeds above the synchronous value indicating that there is an input of power to the shaft. The stator windings are normally connected to the three-phase mains supply which supplies the reactive power necessary for excitation; the stator connections thus receive reactive power from the mains and deliver active power to the mains or to an electrical load. This machine is called a *mains-excited induction generator*. Since a mains supply is required such machines are not suitable for standalone systems or standby-generation.

Figure 4.59 Torque against speed for an induction motor or induction generator

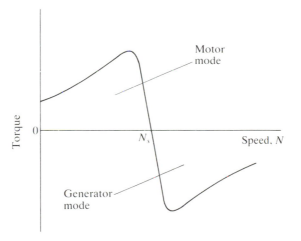

It is possible for an induction generator to receive its reactive power from a separate source such as a capacitor bank or a synchronous machine; this source then defines the frequency of the stator rotating flux and hence the synchronous speed above which the rotor shaft must be driven. A system using a synchronous machine to provide the reactive power is shown in Fig. 4.60(a), and one with a capacitor bank in Fig. 4.60(b).

When a synchronous machine is used (Fig. 4.60(a)), it is first of all run up to speed as a generator using a separate drive; the mechanical drive is then disconnected and the two machines will settle into an equilibrium state with the synchronous machine supplying the required reactive power to the inductive

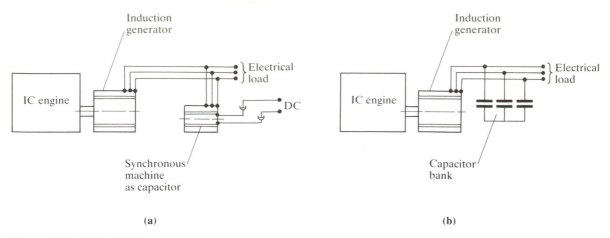

Figure 4.60 Self-excited induction generators

generator. The synchronous machine can be considerably smaller in size than the induction machine.

When a capacitor bank is used (Fig. 4.60(b)), the stator of the induction machine is initially connected to the mains to provide the excitation; when the machine is running the mains supply can be switched off leaving the bank of capacitors to provide the reactive power.

The machine in Fig. 4.60 is known as a *self-excited induction generator*. Both methods of self-excitation can be used on standalone operation, but the use of a capacitor bank is not suitable for standby generation since a supply from the grid is needed on startup. The type with a synchronous machine can be used as a standby generator if the synchronous machine is driven mechanically to start it up, with a battery supplying the dc excitation current to the rotor.

Power Factor Correction

As we have already seen, a low power factor implies a large reactive component of the current which leads to larger transmission lines, transformers etc. and greater losses. In Section 3.3 power triangles were introduced and Example 3.10 illustrates how the introduction of a capacitor in parallel with an induction motor can reduce the supply current by a large percentage with the electrical power unchanged.

Banks of fluorescent light tubes are highly inductive and the load taken by the transformers of arc welders has an uncorrected power factor in the range 0.3 to 0.5, but by far the largest inductive load in industry is from induction motors. In the section above on Induction Motors it was shown that the power factor decreases with reduction in load (see Fig. 4.56(b)). It is important not to install a motor which is too large for the duty required or to run a motor at part load for long time periods. Small motors have much lower power factors than large motors; also slow speed machines are less efficient and have a lower power factor due to the greater leakage and other losses because of the large number of stator poles. For the range above 50 kW for two-pole up to 12-pole machines the power factor varies from about 0.9 down to 0.8; below 5 kW the power factor range for the same number of poles is in the range 0.85 down to 0.7.

There are basically three ways of correcting power factor: (a) connecting in a capacitor or bank of capacitors; (b) using a synchronous motor for part of the load and running it at a leading power factor; (c) using a synchronous motor on no load as a synchronous capacitor. Alternative (a) is the cheapest in terms of both capital and maintenance costs. The synchronous motor is more expensive initially since it requires a dc exciting current supplied via slip rings; the exciting current is normally supplied by a separate dc generator on the motor shaft. The use of a synchronous capacitor for power factor correction is economical only on a large scale.

The most effective way of correcting power factor is by connecting capacitors at individual items of plant, or groups of items, as shown in Fig. 4.61(a). In the alternative method of Fig. 4.61(b) the bank of capacitors is connected across the feed lines coming from the in-plant transformer; in this system the feeder lines to the various loads carry the reactive current for each of these loads and hence the internal losses are much higher. It would appear therefore that it is always better to use individual capacitors at each item of load or at least for groups of similar items. This is not necessarily the case since the cost of a capacitor falls with size and hence the decision must be based on a proper economic appraisal setting the capital cost against the saving in running costs.

It should be noted that for an existing plant the introduction of power factor correction will not only save money but will also increase the potential capacity of the plant; this is illustrated in the following example.

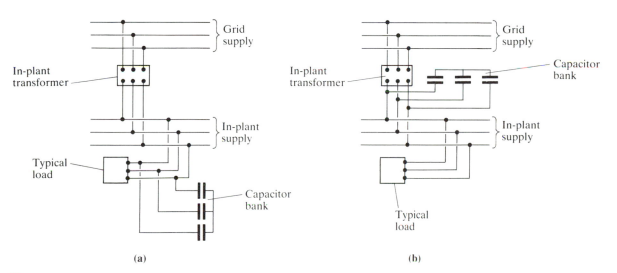

(a)   (b)

Figure 4.61 Two methods for correcting power factor

Example 4.13

An electrical load of 100 kW is provided at a power factor of 0.8. Calculate:
(i) the kVAr capacity of the capacitor bank required to increase the power factor to 0.9;
(ii) the increased load that can be installed at a power factor of 0.8 for the same kVA capacity of the original system.

Solution

(i) Figure 4.62 shows the power triangle for the original system and with the power factor corrected. Initially the kVA line is at an angle of 36.87° to the horizontal (i.e. cos 36.87° = 0.8). After correction the kVA line is at an angle of

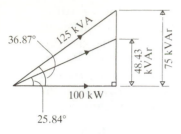

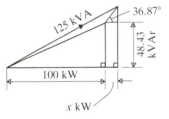

Figure 4.62 Power triangle for existing system of Example 4.10

Figure 4.63 Power triangle for new system of Example 4.10

$25.84°$ to the horizontal (i.e. $\cos 25.84° = 0.9$). The reactive power is therefore reduced from $100 \tan 36.87° = 75.00$ kVAr, to $100 \tan 25.84° = 48.43$ kVAr. Hence

$$\text{rating of capacitor bank required} = (75.00 - 48.43)$$
$$= 26.57 \text{ kVAr}$$

(ii) Let the added load be x kW then the new total load is $(100 + x)$ kW. The original kVA capacity is $100/0.8 = 125$ kVA and this is to remain the same with the new load. Figure 4.63 shows the new system. The reactive power for the added load is given by $x \tan \phi = 0.75x$, where $\phi = 36.87°$ (i.e. $\cos 36.87° = 0.8$ and $\tan 36.87° = 0.75$) i.e. the new total reactive power is

$$48.43 + 0.75x$$

Also, the new reactive power is given by

$$\sqrt{\{125^2 - (100 + x)^2\}}$$
$$\therefore \quad (48.43 + 0.75x)^2 = 125^2 - (100 + x)^2$$

i.e. $$1.5625x^2 + 272.65x - 3279.54 = 0$$
$$\therefore \quad x = 11.29 \text{ kW}$$

Therefore just over 10 % increase in power output can be introduced for the same feeder lines and feeder line losses with the addition of the capacitor bank. Hence

$$\text{new overall power factor} = 111.29/125 = 0.89$$

Problems

4.1 An anthracite, rank 101, has the following composition by mass: carbon 84.7 %; hydrogen 2.6 %; nitrogen 1.0 %; sulphur 1.1 %; oxygen 1.6 %; ash 5.0 %; moisture 4.0 %.

     Construct a graph of carbon dioxide and oxygen in the dry flue gas against the percentage excess air supplied over that required for stoichiometric combustion.

     (at 10 %, CO_2 17.77 %, O_2 1.94 %; at 80 %, CO_2 10.80 %, O_2 9.41 %)

4.2 Derive expressions for the percentages of carbon dioxide and oxygen in the dry flue gas in terms of the percentage excess air by volume, P, for a boiler using natural gas as fuel. Use the volumetric analysis given in Table 4.3 (p. 76).

     ($1066.75/(P + 89.94)$ % CO_2; $21.005P/(P + 89.94)$ % O_2)

4.3 A coal-fired steam boiler has a blow-down system in which a flash vessel is used with the flash steam passing to the feed tank. The boiler is in continuous use 24 hours per day all year. Using the data below calculate:
(i) the percentage flue gas loss;
(ii) the percentage blow-down loss;
(iii) the boiler efficiency based on GCV;
(iv) the annual cash saving due to returning the flash blow-down steam to the feed tank.

Data

Coal: composition by mass as fired; 71.6 % C, 4.3 % H, 1.6 % N, 1.7 % S, 3.8 % O, 8.0 % ash, 9.0 % moisture; GCV, 29.55 MJ/kg; mass flow rate, 600 kg/h; cost, £65/tonne.

Steam: condition, dry saturated at 10 bar; flow rate, 5425 kg/h; feed water temperature at boiler inlet, 35 °C; temperature of make-up water, 15 °C; blow down flow rate, 8 %; pressure in flash vessel, 1.1 bar; cost of chemical treatment of feed, 0.02 p/kg.

Air: excess air supplied, 25 %; temperature at entry, 20 °C.

Flue gas: temperature at boiler exit, 277 °C; pressure at boiler exit, 1.03 bar. Heat loss from casing, 100 kW.

Specific heats at constant pressure at 550 K may be taken as follows: carbon dioxide, 1.046 kJ/kg K; oxygen, 0.988 kJ/kg K; nitrogen, 1.065 kJ/kg K; sulphur dioxide, 0.824 kJ/kg K.

Take the specific heat of air at 20 °C as 1.0046 kJ/kg K.

(15.99 %; 1.51 %; 80.5 %; £982.7)

4.4 (a) For the boiler of Problem 4.3 calculate the percentage dry flue gas loss and the percentage water vapour loss using the approximations given by Eqs 4.7 and 4.8.

(b) Using the approximate formulae of Eqs 4.7 and 4.8 calculate the boiler efficiency for flue gas temperatures of 277 °C, 250 °C, 225 °C, and 200 °C. Assume that the fuel consumption changes but that all other data are unchanged.

(11.1 %; 5.7 %; 79.9 %; 81.2 %; 82.3 %; 83.4 %)

4.5 A natural gas-fired condensing boiler for hot water heating has an output of 2 MW when operating with an exit gas temperature of 52 °C. The air inlet conditions are 20 °C, 1.013 bar with 25 % excess air supplied over that required for stoichiometric combustion. The condensate collected at a temperature of 46 °C is found to be 40 % of the maximum possible rate of production from the gas as supplied. The boiler has a casing loss of 1.4 % and is in use at full load for 2500 h per year. Using the additional data below, calculate:

(i)   the boiler efficiency;

(ii) the mass flow of condensate collected per unit time;

(iii) the break-even period for the condensing boiler compared with a conventional gas boiler with an efficiency of 82 % and a capital cost 2/3 that of the condensing boiler.

Data

For natural gas take GCV as 38.6 MJ/m³, use the analysis by volume given in Table 4.3 (p. 76), and take the price as 37 p/therm. The specific heats at constant pressure for the constituents of the dry flue gas may be taken from Rogers and Mayhew,[4.2] or other similar, tables. The capital cost of the condensing boiler is £36 000.

(91 %; 13.2 kg/h; 1.58 years)

4.6 A city centre department store has an average hot water requirement for all purposes of 11 500 GJ each year; the load is constant over the year with an average period of use of 60 h per week. It is suggested that this requirement could be met by burning the waste produced in-house which has a high proportion of cardboard, paper etc. with an average GCV of 16 000 MJ/tonne; the waste produced averages 13 tonne/week. Make preliminary calculations

using the additional data given to find:
(i) the percentage of the hot water load which could be taken by the waste heat boiler;
(ii) the rating of the boiler required assuming an average load factor of 0.75;
(iii) the annual cash saving assuming that the alternative method of heating is natural gas with a boiler efficiency of 80 %;
(iv) the simple payback period for the incineration system.

Data
Waste heat boiler efficiency, 70 %; capital cost of complete incineration system, £135 000; cost of gas, 37 p/therm; cost of refuse disposal, £25 per tonne.

(65.8 %; 900 kW; £50 089; 2.7 years)

4.7 In a power plant steam is supplied to the high pressure turbine at 40 bar, 500 °C and exhausts from the low pressure turbine to a condenser at a pressure of 0.03 bar; the overall isentropic efficiency of the expansion is 0.8.

The plant has four feed heaters at bleed pressures of 18 bar, 6 bar, 1.7 bar, and 0.3 bar; the heater at 1.7 bar is an open heater used also as a de-aerator, and all the other heaters are closed heaters. The condensate from each closed heater is flashed into the previous heater with the condensate from the last heater being flashed into the condenser.

The temperature of the condensate leaving each closed heater and the temperature of the condensate leaving the condenser is the saturation value corresponding to the pressure. The feed water leaving the open heater is at the saturation temperature corresponding to the bleed steam pressure, and the feed water leaving each closed heater is 10 K below the saturation temperature at the bleed pressure.

Neglecting pump work and all thermal losses, calculate:
(i) the cycle efficiency;
(ii) the cycle efficiency for the same plant without feed heating assuming the overall isentropic efficiency is unchanged;
(iii) the mass flow rate of steam from the boiler for a power output of 30 MW.

Assume that the process line for the expansion is a straight line on the $h-s$ chart.

(35.16 %; 32.18 %; 32.72 kg/s)

4.8 A gas turbine plant consists of two compressors, a turbine, and an electric generator on a common shaft. Air from ambient is drawn into the first compressor and after compression passes through an intercooler before entering the second compressor. The air after compression passes through a heat exchanger in counter-flow to the exhaust gases from the turbine before entering the combustion chamber. The gases from the combustion chamber expand in the turbine then pass through the heat exchanger to the chimney. Using the data below, neglecting thermal losses, and assuming that the mass flow rate is constant throughout, calculate the overall thermal efficiency.

Data
Ambient conditions, 10 °C and 1.013 bar; pressure ratio for each compressor, 2.5; isentropic efficiency of each compressor, 84 %; temperature of air leaving the intercooler, 20 °C; effectiveness of heat exchanger, 0.75; combustion efficiency, 98 %; temperature of gases entering turbine, 800 °C; isentropic efficiency of turbine, 87 %; c_p and γ for air, 1.005 kJ/kg K and 1.4; c_p and γ for the turbine

gases, 1.15 kJ/kg K and 1.333; for the combustion process take an equivalent c_p of 1.15 kJ/kg K; mechanical transmission efficiency, 99 %; efficiency of electric generator, 88 %.

Pressure loss as a percentage of the inlet pressure to each component: intercooler, 1 %; air side of heat exchanger, 3 %; combustion chamber, 2 %; gas side of heat exchanger including chimney, 4 %.

(32.9 %)

4.9 A plant provides power, process heat and hot water heating using a combination of gas and steam turbines with natural gas as the fuel for the turbine and incorporating a waste-incineration system.

The waste is shredded and passed to a rotary furnace the gases from which are attempered with air to reduce their temperature before passing to a steam-raising boiler. The boiler produces dry saturated steam which is passed to a separate superheater. The heat input to the superheater is from the exhaust gases from the gas turbine unit; on exit from the superheater the gases pass through a water heater and thence to the chimney.

The superheated steam produced is expanded in a steam turbine to a back pressure suitable for the process steam requirement; the condensate from the process is recovered and passed through a water heater before being pumped back to the boiler.

Using the data given, neglecting pressure drops and thermal losses in the first instance, and assuming continuous firing for six days every week for 50 weeks per year, sketch the plant lay-out and calculate:
(i)   the required steam flow rate to the steam turbine;
(ii) the required air mass flow rate for the gas turbine neglecting the mass of fuel added;
(iii) the heat available for the process;
(iv) the overall efficiency of the waste-incineration furnace/boiler system;
(v)   the hot water heating available from the process steam heat exchanger and from the gas turbine gases heat exchanger;
(vi) the overall efficiency defined as the ratio of the total useful energy output to the total energy input from the waste and gas supplied;
(vii) the simple pay-back period for the scheme, assuming that at present the power is bought from the grid at 4.2 p/kW h and the heat is obtained from gas-fired boilers with an efficiency of 80 %.

Data

Refuse: amount supplied, 2000 tonne/week; GCV, 15 MJ/kg; cost of disposal if not incinerated, £18/tonne.

Steam system: supplied from boiler to superheater at 30 bar, dry saturated; supplied to turbine at 30 bar, 450 °C; power output, 5 MW; combined electrical and mechanical transmission efficiency, 85 %; isentropic efficiency of turbine, 85 %; process steam pressure, 6 bar; condition of process steam leaving the process, saturated water at 6 bar; condition of process condensate leaving water heater and pumped to the boiler, 40 °C at 6 bar.

Gas turbine unit: air temperature at compressor intake, 15 °C; compressor pressure ratio, 10; isentropic efficiency of compressor, 80 %; isentropic efficiency of turbine, 85 %; power output from unit, 5 MW; combined mechanical and electrical transmission efficiency, 85 %; gas temperature at exit from turbine, 550 °C; gas temperature at exit from superheater, 380 °C; gas temperature at exit from water heater, 120 °C; c_p and γ for air, 1.005 kJ/kg K and 1.4; c_p and γ for gases in turbine, 1.15 kJ/kg K and 1.333; equivalent c_p for combustion chamber, 1.15 kJ/kg K; combustion efficiency, 0.98; cost of natural gas, 1.2 p/kW h.

Capital cost of complete new plant, £7 M; additional annual maintenance and wage costs, £100 000.

(15.91 kg/s; 43.95 kg/s; 36.65 MW; 72.5%; 8 MW, 13.14 MW; 75.9%; 10.1 months)

4.10   An industrial process requires 6 kg/s of cooling water which leaves the process heat exchanger at 46 °C. It is proposed to use an induced draught cooling tower for the cooling water to restore it to the required inlet temperature of 25 °C. Using the data below and the psychrometric chart, neglecting heat losses and assuming that the make-up water is added external to the tower, calculate:

(i)   the volumetric flow rate of air at fan outlet;
(ii) the mass flow rate of make-up water required.

Data

Atmospheric air, 15 °C and percentage saturation 50%; power input to fan, 4 kW; air leaving tower, 25 °C saturated; pressure throughout tower, 1.01325 bar; temperature of make-up water, 10 °C.

(9.78 m³/s; 598 kg/h)

4.11   A vapour-compression refrigeration plant using R22 refrigerant has evaporator and condenser pressures of 4.214 bar and 10.439 bar and an electrically-driven centrifugal compressor with an isentropic efficiency of 0.82 and a combined mechanical and electrical transmission efficiency of 0.88. Assuming that the vapour entering the compressor is dry saturated, and the condensate leaving the condenser is a saturated liquid, neglecting heat losses, calculate using the data for R22 as given below:

R22

Temperature (K)	Pressure (bar)	Specific volume (m³/kg)	Enthalpy Liquid (kJ/kg)	Enthalpy Vapour (kJ/kg)	Entropy Liquid (kJ/kg K)	Entropy Vapour (kJ/kg K)
268	4.214	0.0553	194.2	403.5	0.979	1.759
298	10.439	0.0226	230.3	413.3	1.105	1.718

(i)   the electrical power required for a 100 kW cooling load;
(ii) the volumetric flow rate of refrigerant at compressor inlet.

Superheated vapour at 10.439 bar, 313 K: $h = 425.7$ kJ/kg, $s = 1.759$ kJ/kg K.

(17.76 kW; 0.0319 m³/s)

4.12   A water-chilling plant uses a lithium bromide/water vapour-absorption refrigeration system. The absorber and the evaporator are housed in one vessel at a pressure of 0.008 72 bar and the generator and condenser are housed in a second vessel at 0.091 bar. The heat supplied to the generator is provided by steam which is produced in a gas-fired boiler. Using the additional data below and neglecting all thermal losses and pumping power, calculate for a cooling load of 1 kW:

(i)   the heat supplied to the generator;
(ii) the coefficient of performance, and the running cost of the plant;
(iii) the heat rejected from the absorber;
(iv) the running cost of an equivalent vapour-compression plant with a coefficient of performance of 3, and with the compressor driven by a gas engine.

Data
Solution leaving generator and entering heat exchanger, 98 °C; solution leaving absorber and entering heat exchanger, 40 °C; solution leaving heat exchanger and entering generator, 78 °C; condensate leaving condenser, saturated at 0.091 bar; vapour leaving evaporator, dry saturated at 0.008 72 bar; boiler efficiency, 85 %; overall thermal efficiency of gas engine, 30 %; cost of gas, 1.2 p/kW h.

Lithium bromide solution

Temperature (°C)	Pressure (bar)	Concentration	Enthalpy (kJ/kg)
40	0.00872	0.60	−160
78	0.09100	0.60	−90
98	0.09100	0.64	−55

(1.374 kW; 0.728, 1.94 p/h; 1.344 kW; 1.333 p/h)

4.13 A heat pump using refrigerant R11 is used to heat a building. Some of the air from the building is exhausted and some is recirculated mixing with fresh air from outside, then passing across the condenser coils of the heat pump before being supplied to the building. The coils of the evaporator are in contact with the outside air. A centrifugal compressor is used with dry saturated vapour at entry; there is no undercooling in the condenser. Using the data given and neglecting heat losses, calculate:
(i)   the mass flow rate of refrigerant;
(ii) the mass flow rate of air to the building;
(iii) the percentage reduction in running costs for exactly the same air temperatures and mass flow rates of air and refrigerant but with the evaporator coils placed in the air duct which exhausts to the atmosphere.

Data
Outside air, saturated at 0 °C; air supplied to room, 20 °C; air leaving room, 18 °C and percentage saturation 50 %; ratio of fresh air to air supplied to room, 0.2; electrical power input to compressor, 30 kW; mechanical and electrical transmission efficiency of compressor, 87 %; isentropic efficiency of compression process, 80 %; minimum temperature difference for heat transfer in the evaporator and condenser coils, 5 K; air pressure throughout system, 1.013 25 bar.

R11

Temperature (°C)	Pressure (bar)	Enthalpy Liquid (kJ/kg)	Enthalpy Vapour (kJ/kg)	Entropy Liquid (kJ/kg K)	Entropy Vapour (kJ/kg K)	15 K superheat Entropy (kJ/kg K)	15 K superheat Enthalpy (kJ/kg)
−5	0.323	29.7	219.8	0.119	0.830	—	—
13	0.681	45.2	229.5	0.175	0.818	—	—
25	1.056	55.6	235.5	0.210	0.813	0.837	242.9

Air at 1.01325 bar

Temperature (°C)	Percentage saturation (%)	Enthalpy (kJ/kg dry air)	Specific humidity
0	100	9.48	0.00379
18	50	34.56	0.00649
20	40	35.09	0.00590
20	42	35.84	0.00620

(0.997 kg/s; 33.42 kg/s; 17.68 %)

4.14 (a) In summer a room is to be maintained at 20 °C, percentage saturation 50 % when the outside air is at 26 °C, percentage saturation 50 %. The mass flow rate of dry air supplied to the room is 2 kg/s at 15 °C. Fresh air is mixed with re-circulated air in the ratio three parts re-circulated air to one part fresh air. The air after mixing is passed over a cooling coil with an apparatus dew point of 4 °C and a coil contact factor of 0.9. On leaving the cooling coil the air passes over a heating coil and is then delivered to the room by the fan; the fan and ducting cause a 2 K rise in temperature of the air between the heating coil and the room. Using the psychrometric chart, find:
(i) the cooling coil load;
(ii) the heating coil load.
(b) Since there is no need to control the humidity in the room it is decided to remove the heating coil and use an arrangement as in Fig. 4.49(a) to supply the same mass flow rate of air at the same supply temperature as before. The design conditions for the room are assumed to be unchanged. The mixing ratio for the first mixing box is now one part fresh air to one part re-circulated air. Assuming that the cooling coil has the same apparatus dew point and coil contact factor, and that there is a 2 K temperature rise from the fan and ducting as before, calculate the percentage saturation of the air supplied to the room.

(46.4 kW; 14.6 kW; 60 %)

4.15 A three-phase delta-connected 10-pole cage-rotor induction motor runs off a 415 V, 50 Hz supply. At full load the shaft rotational speed is 584 rev/min with a mechanical power output of 36 kW. The stator loss at full load is 3.8 kW and the friction and windage losses are 0.6 kW; the phase current at full load is 44 A. Calculate:
(i) the percentage slip at full load;
(ii) the overall efficiency at full load;
(iii) the power factor at full load.

(2.67 %; 86.95 %; 0.756)

4.16 A factory has a total electrical load of 500 kW at an overall power factor of 0.75. It is required to increase the load by adding a load of 50 kW with a power factor of 0.75. In order to maintain the same kVA load from the grid a bank of capacitors is connected in parallel with the power supply to the factory load. Calculate:
(i) the new overall power factor;
(ii) the required kVAr rating of the capacitor bank.

(0.825; 108.3 kVAr)

References

4.1 CIBSE 1986 *Guide to Current Practice* **C5**

4.2 Rogers G F C, Mayhew Y R 1989 *Thermodynamic and Transport Properties of Fluids* 4th edn Basil Blackwell

4.3 Smart J H S 1986 The Greater London Council's Refuse-fired Power Station. *Proc. Inst. Mech. Engrs* **200** A4

4.4 Eastop T D, McConkey A 1986 *Applied Thermodynamics for Engineering Technologists* 4th edn Longman

4.5 CIBSE 1986 *Guide to Current Practice* **B14**: 14–17

4.6 Product Profile *Heat Pumps* BSRIA Statistics Bulletin **13** no 4, Dec 1988

4.7 Kew J 1985 *Heat Pumps for Building Services* BSRIA TN 8

4.8 Jones W P 1985 *Air Conditioning Engineering* Edward Arnold

4.9 CIBSE 1986 *Guide to Current Practice* **B3**

4.10 CIBSE 1986 *Guide to Current Practice* **B** Table B.10.5

Bibliography

Armstead H C H 1983 *Geothermal Energy* 2nd edn E & F N Spon

Armstead H C H 1985 The under-valuation of hydro-power potential: a statistical pitfall. Chartered Mechanical Engineer March 1985

Bell P C (ed) 1971 *Industrial Fuels* Macmillan

Bennet D J, Thomson J R 1989 *Elements of Nuclear Power* 3rd edn Longman

Brown D, Hamilton E P 1984 *Electromechanical Energy Conversion* Macmillan

Burley A J, Edmunds W M 1979 *Catalogue of Geothermal Data for the UK* Dept. of Energy

CIBSE 1988 *Design Guidance for Heat Pump Systems* TM15

Cohen J, Rogers G F C, Saravanamuttoo H I H 1987 *Gas Turbine Theory* 3rd edn Longman 1987

Culp (jr) A W 1979 *Principles of Energy Conversion* McGraw-Hill

Dryden I G C (ed) 1975 *The Efficient Use of Energy* IPC

Dunn P D 1986 *Renewable Energies: Sources, Conversion and Application* Peter Peregrinus Ltd

Flood M 1983 *Solar Prospects: the Potential for Renewable Energy* Wildhouse Wood

Flood M 1985 The troughs and crests of wave energy. Chartered Mechanical Engineer June 1985

Fuel Efficiency Booklet No 18 *Boiler Blowdown* Energy Efficiency Office 1983

Fuel Efficiency Booklet No 14 *Economic Use of Oil-fired Boiler Plant* Energy Efficiency Office 1984

Fuel Efficiency Booklet No 15 *Economic Use of Gas-fired Boiler Plant* Energy Efficiency Office 1984

Fuel Efficiency Booklet No 17 *Economic Use of Coal-fired Boiler Plant* Energy Efficiency Office 1984

Fuel Efficiency Booklet No 9 *The Economic Use of Electricity* Energy Efficiency Office 1986

Fuel Efficiency Booklet No 11 *The Economic Use of Refrigeration Plant* Energy Efficiency Office 1986

Goodall P M (ed) 1980 *The Efficient Use of Steam* IPC

Hartnett J P 1976 *Alternative Energy Sources* Academic Press
Hughes E 1977 *Electrical Technology* 5th edn Longman
Laithwaite E R, Freris L L 1980 *Electric Energy*: *its Generation, Transmission and Use* McGraw-Hill
Martin D 1985 *Heat Pumps for Heating in Buildings* Energy Technology Series 5 Energy Efficiency Office
Meinel A B, Meinel M P 1977 *Applied Solar Energy*: *an Introduction* Addison Wesley
Mustoe J 1984 *An Atlas of Renewable Energy Resources* John Wiley
Nowacki P (ed) 1980 *Lignite Technology* Noyes Data Corp., USA
O'Callaghan P (ed) 1979 *Energy for Industry* Pergamon Press
Pita E G 1984 *Refrigeration Principles and Systems* John Wiley
Reay D A 1977 *Industrial Energy Conversion* Pergamon Press
Reay D A, MacMichael 1987 *Heat Pumps* 2nd edn Pergamon
Say M G 1987 *Alternating Current Machines* 5th edn Longman
Scivier J B, Baker A C J 1987 Turbines for tidal power. *Chartered Mechanical Engineer* Feb. 1987
Sorensen B 1979 *Renewable Energy* Academic Press
Stacy R E, Thain I A 1984 Twenty-five years of geothermal power generation. *Chartered Mechanical Engineer* Sept. 1984
Stoecker W F, Jones J W 1982 *Refrigeration and Air Conditioning* McGraw-Hill
Twidell J W, Weir A D 1986 *Renewable Energy Resources* E & FN Spon
Walker D C 1985 *Factors Influencing Boiler Efficiency* BSRIA TN 7
Warme D F 1983 *Wind Power Equipment* E & FN Spon
White L P, Plaskett L G 1982 *Biomass as Fuel* Academic Press

5 ENERGY RECOVERY

A typical system producing useful heat and work from an energy input rejects energy in the form of waste gas and/or liquid, and transfers heat to the atmosphere due to mechanical friction, electrical losses, and heat transfer from surfaces. To minimize the energy input for a given useful output of work and heat the plant must be carefully designed to reduce the energy wastage to a minimum. Energy recovery is one of the ways of minimizing the thermal losses and various energy recovery methods are described and discussed in this chapter.

Later chapters consider the whole system, using for example pinch technology as in Chapter 6, or the total energy approach dealt with in Chapter 8. This chapter will restrict itself to energy recovery techniques and equipment, using the theory given in Chapters 3 and 4.

Energy recovery will not necessarily lead to a saving of money; a proper evaluation using the methods of Chapter 2 is always necessary. For example, in the case of an industrial gas turbine some of the energy of the exhaust gases can be recovered by heating the air between the compressor and the combustion chamber (see Example 4.6). However, the pressure drop on each side of the heat exchanger reduces the pressure ratio across the turbine and hence the work output. Also, the capital cost of the heat exchanger must be recouped from any saving in fuel cost in a reasonable time period. Another simple example of energy recovery is the use of an economizer and/or air pre-heater for a boiler (see Section 4.2). Again the fuel saving must be set against the increased capital cost; also in this case more expensive materials may be necessary in the chimney to avoid possible corrosion if the gas temperature is reduced below the acid dew point.

5.1 INSULATION

One of the most obvious and basic ways of recovering energy is to insulate surfaces which are rejecting heat to the atmosphere. The annual fuel cost is reduced as the thickness of the insulation is increased but the capital cost of the insulation increases with the thickness and hence the financial saving must be off-set against the capital cost. For a typical write-off period there will normally be an economic thickness of insulation for a particular case. The following example illustrates this point.

Example 5.1

A steel pipe carries wet steam from a gas-fired boiler through a small workshop to a process plant. It is proposed to insulate the pipe using a glass fibre insulation with an aluminium alloy casing. Using the data below making suitable assumptions, determine:

(i) the most economic thickness of insulation;
(ii) the simple pay-back period for this thickness.

Data

Pipe outside diameter, 60.3 mm; heat transfer coefficient for outside surface of insulation, 10 W/m² K; thermal conductivity of insulation, 0.07 W/m K; steam temperature, 200 °C; temperature of air in workshop, 15 °C; boiler efficiency, 80 %; price of gas, 0.3 p/MJ; operation of plant for 3000 h per annum; cost of insulation as given in the table below.

Thickness of insulation/mm	19	25	32	38	50	60
Cost per metre length/p	476	531	632	763	1007	1280

Assume that the write-off period for the insulation is five years.

Solution

(i) It will be assumed that the thermal resistances of the steam film on the inside of the pipe and of the pipe wall itself are both negligible, and that the heat transfer coefficient for the outside surface is constant for all diameters and temperatures. The thermal resistance, R, of the insulation is given by Eq. [3.14]:

$$R = \frac{\ln\{(60.3 + 2x)/60.3\}}{2\pi \times 0.07} = \frac{\ln(1 + 0.033\,17x)}{0.44} \text{ m K/W}$$

(where $x =$ thickness of insulation in mm).

The thermal resistance, R, of the fluid film on the outside surface is given by Eq. [3.17]:

$$R = 1/\{10\pi(60.3 + 2x)10^{-3}\} = 1/\{1.894(1 + 0.033\,17x)\} \text{ m K/W}$$

Then using Eq. [3.15], the heat transfer per unit length, Q, is given by:

$$Q = \frac{(200 - 15)}{\{\ln(1 + 0.033\,71x)/0.44\} + \{1/1.894(1 + 0.033\,17x)\}} \text{ W/m}$$

The annual operating time is given as 3000 hours and the cost of the gas to produce this loss is 0.3 p/MJ, therefore the annual cost of the heat loss is given by:

$$Q \times 3000 \times 3600 \times 0.3 \times 10^{-6}/0.8 = 4.05Q \text{ p/m}$$

(for a boiler efficiency of 80 %).

The table below shows the values of heat loss, Q, annual fuel cost and insulation capital cost for the range of insulation thickness given.

If it is assumed that the capital cost is written off every five years then

Thickness (mm)	Capital cost (p/m)	Heat loss (W/m)	Annual fuel cost (p/m)
0	0	350.5	1419.1
19	476	129.0	522.5
25	531	111.4	451.2
32	632	97.4	394.5
38	763	88.6	358.8
50	1007	76.4	309.4
60	1280	69.4	281.1

ignoring depreciation and inflation the total annual cost can be obtained by adding the annual fuel cost to one-fifth of the capital cost to give the table below.

Thickness/mm	0	19	25	32	38	50	60
Total annual cost/(p/m)	1419	618	557	521	511.4	510.8	537

The figures are plotted on Fig. 5.1 It can be seen that the economic thickness of insulation is 50 mm.

Figure 5.1 Insulation cost against thickness for Example 5.1

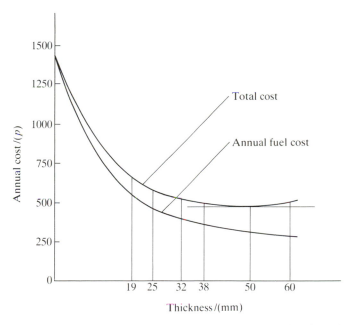

(ii)   The simple pay back period is found by dividing the capital cost by the annual saving in fuel cost by insulating.

i.e.   payback period $= 1007/(1419.1 - 309.4)$
                                    $= 0.907$ years $= 10.9$ months

It is clear that the greater the temperature the greater the heat loss to ambient and one of the greatest sources of heat loss in industry is from high temperature devices such as furnaces, ovens, kilns, crucibles, and liquid metal ladles. For example, heat treatment furnaces can have interior temperatures of 1100 °C, kilns may reach 1250 °C, forge furnaces 1350 °C, and induction furnaces producing molten steel 1650 °C. Not only is the heat loss by conduction through the insulation wall greater in such cases but the radiation loss from the outside surface is also greater due to the high surface temperature.

Furnaces, kilns etc. which are not continuously fired require energy to heat them up to the required temperature and hence the ideal material for their construction should have the following characteristics: ability to withstand high temperatures and temperature cycles without cracking, low thermal conductivity, high thermal diffusivity. In transient conduction the rate of heating or cooling of a body depends on the thermal diffusivity, $\alpha = k/\rho c$, where k is the thermal conductivity, ρ is the density, and c is the specific heat. It can be seen therefore that for a material to have a low thermal conductivity, k, and a high diffusivity, α, it must have a low value of the product ρc, i.e. a low thermal mass per unit volume. If a furnace, kiln etc. has a low thermal mass it is more responsive to changes in working conditions and hence the temperature is easier to control.

Materials have been developed, sometimes called Low Thermal Mass (LTM) refractories, which are of a ceramic fibre. Considerable improvements in efficiency and in response times can be achieved by relining fire-brick furnaces and kilns with a ceramic fibre lining. Fuel savings of between 10 % and 30 % have been achieved.

Example 5.2

A kiln has two complete operating cycles per week and runs for 48 weeks per year; the existing gas consumption is 150 therm/cycle. On examination it is decided that the inside of the kiln brickwork is suitable to be relined with ceramic fibre at a total cost of £1400. It has been decided that the improvement will be made only if the simple pay-back period is less than two years. Assuming the price of gas is 32 p/therm calculate the percentage saving in gas required to achieve the required pay-back period.

Solution
The initial annual gas consumption is $150 \times 2 \times 48 = 14\,400$ therm.
i.e. the annual fuel cost is

$$14\,400 \times 32/100 = £4608$$

The annual fuel saving due to the relining for a pay-back period of two years is given by

$$£1400/2 = £700$$

i.e.   percentage annual fuel saving required $= 700 \times 100/4608 = 15.2\,\%$

This order of reduction is well within that expected for ceramic fibre relining.

The doors of high temperature containers should be kept air-tight to avoid a convective heat loss due to ambient air leaking into the interior replacing hot air which leaks out. Continuous process furnaces and ovens in which material is introduced via conveyors present additional problems of insulation. In such cases a hot air curtain might be feasible; also the pre-heating of conveyors by waste heat would eliminate the loss of energy from the furnace in heating the conveyor as it entered.

Considerable savings have been made in the steel industry by providing insulation to cover the liquid metal surface in ladles during pouring.

Many processes in industry involve the heating of liquid in open tanks and considerable savings in heat loss can be made using Allplas balls. These are hollow polypropylene balls which float on the liquid surface and can be used up to temperatures of about 120 °C. As well as acting as an insulating layer the balls also reduce considerably the evaporation from the tank; a double layer is even more effective. For example, for water at 80 °C a double layer of balls has been shown to reduce the evaporation rate by 90%. Evaporated fumes from tanks must be extracted by fans giving an additional source of heat loss; by reducing the evaporation the extraction rate can be reduced considerably thus reducing further the energy loss.

Example 5.3

A factory has five uninsulated steel tanks used for dip cleaning a product; each tank is 2 m long by 1 m high by 1.5 m wide. The detergent solution in the tanks is heated to a temperature of 65 °C by steam in tubes immersed in the liquid; the steam is provided by a boiler with an efficiency, including steam distribution losses, of 60 %. Using the additional data below, assuming that the heat loss through the floor of the tanks is negligible and neglecting the thermal resistance of the steel tank walls, calculate:

(i)   the annual cash saving if the tank walls are insulated using a 25 mm thick slab of insulating material;

(ii) the annual cash saving if a double layer of Allplas balls is applied to the liquid surfaces;

(iii) the simple pay-back period if the measure in (i) and (ii) are both implemented.

Data

Ambient temperature, 15 °C; heat transfer coefficient for all tank surfaces, 10 W/m² K (assumed to be the same with or without insulation); heat transfer coefficient, liquid surface to ambient air, 10 W/m² K; heat transfer coefficient from liquid to inside of tank wall, 500 W/m² K; heat loss by evaporation from liquid surface, 8000 W/m²; thermal conductivity of insulating material used to lag tanks, 0.035 W/m K; heat loss by evaporation from tanks with Allplas balls, 1500 W/m²; additional thermal resistance due to Allplas balls, 0.2 m² K/W; cost of fuel used in boiler, 0.3 p/MJ; insulation cost, £12/m²; cost of Allplas balls, £60/m²; effective operating time at the given rate of heat input, 4800 hours per annum.

Solution

(i)   For an uninsulated tank the thermal resistance is made up of that of the two fluid films on the inside and outside surfaces since the resistance of the steel wall is negligible. Therefore using Eq. [3.17] the thermal resistance, R, of the tank wall is given by

$$R = (1/500) + (1/10) = 0.102 \text{ m}^2 \text{ K/W}$$

Then using Eq. [3.15], heat loss, Q, through the walls is given by

$$Q = \frac{(65 - 15)\{(2 \times 1 \times 2) + (2 \times 1 \times 1.5)\}}{0.102 \times 1000} = 3.43 \text{ kW}$$

With the tank insulated the thermal resistance of the insulation will be added to the initial value of resistance since the other resistances are unchanged. Using Eq. [3.13] thermal resistance, R, of the insulation is given by

$$R = 0.025/0.035 = 0.714 \text{ m}^2 \text{ K/W}$$

Therefore, the total thermal resistance, R_T, of the walls is given by

$$R_T = 0.714 + 0.102 = 0.816 \text{ m}^2 \text{ K/W}$$

Then, heat transfer, Q, through the walls is given by

$$Q = (65 - 15) \times 7/(0.816 \times 1000) = 0.43 \text{ kW}$$

and reduction in heat loss is

$$3.43 - 0.43 = 3 \text{ kW}$$

This represents a fuel energy input per tank of $3/0.6 = 5$ kW, for the given overall efficiency of boiler plus steam distribution. Hence

$$\text{annual saving by insulating} = 5 \times 5 \times 4800 \times 3600/1000$$
$$= 432\,000 \text{ MJ}$$

and $\text{annual cash saving} = 0.3 \times 432\,000/100 = £1296$

(ii)   The heat loss from the liquid surface is due to the evaporation plus convection from the surface. By evaporation

$$Q = 8000 \times (2 \times 1.5)/1000 = 24 \text{ kW}$$

and by convection

$$Q = 10 \times (2 \times 1.5) \times (65 - 15)/1000 = 1.5 \text{ kW}$$

i.e. the total heat loss from the liquid surface is given by

$$Q_T = 24 + 1.5 = 25.5 \text{ kW}$$

After the introduction of Allplas balls the heat loss by evaporation becomes 1500 W/m^2, i.e. $1500 \times 3/1000 = 4.5$ kW. Also the thermal resistance of the surface is increased to $0.1 + 0.2 = 0.3$ m^2 K/W, and hence the heat loss from the liquid surface is now given by $(65 - 15) \times 3/(0.3 \times 1000) = 0.5$ kW.

i.e. $Q_T = 4.5 + 0.5 = 5 \text{ kW}$

Hence the saving in energy is $5 \times (25.5 - 5) = 102.5$ kW.

i.e. $\text{annual cash saving} = 102.5 \times 4800 \times 3600 \times 0.3/(1000 \times 100)$
$$= £5313.6$$

(iii) $\text{Total cash saving for both measures} = £(1296 + 5313.6)$
$$= £6609.6$$
$$\text{Insulation cost} = £(5 \times 12 \times 7) = £420$$
$$\text{Cost of Allplas balls} = £(5 \times 60 \times 3) = £900$$

i.e. $\text{Total capital cost} = £1320$

Therefore

$$\text{pay-back period} = 1320/6609.6 = 2.4 \text{ months}$$

Improvements to the insulation of the fabric of a building can make considerable cash savings in the heating bill. This is considered in detail in Chapter 7.

5.2 RECUPERATIVE HEAT EXCHANGERS

The heat exchanger most commonly used for heat recovery is the recuperative type (see Section 3.2 for the basic theory). A recuperative heat exchanger is one in which the two fluids are separated at all times by a solid barrier; there are a large number of different kinds of heat exchanger within this definition and some of the more common types are shown diagrammatically in Figs 5.2 to 5.9.

Figure 5.2 Waste-heat water-tube boiler

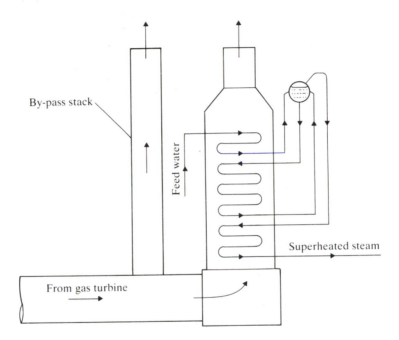

Figure 5.3 Shell boiler using waste gas

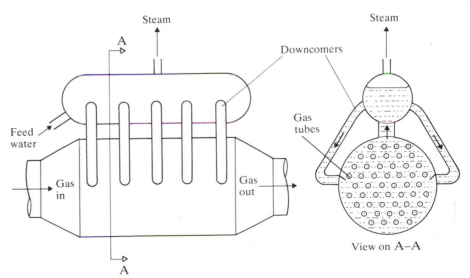

The water-tube boiler shown in Fig. 5.2 is the type used with waste gas from an industrial gas turbine. Typical temperatures quoted for a BBC Brown Boveri 1200 MW power station are as follows: for the gases, 500 °C to 110 °C; for the steam, superheated at 50 bar, 450 °C from feed at 90 °C. The use of a by-pass enables the gas turbine plant to continue running if the steam plant is shut down. Similar water-tube boilers are used in the chemical industry with high temperature gases up to 1200 °C. Two further examples of waste heat boilers are shown in Figs 5.3 and 5.4; these are both fire-tube boilers in which only saturated steam is produced. Boilers of this type would be used to raise steam for process work or space heating since saturated steam is not suitable for expansion in a turbine because of the erosive effect of the water droplets on the turbine blades; alternatively a separate superheater could be added after the boiler if power production is required.

In high temperature furnaces the energy of the furnace gases can be used to pre-heat the combustion air as shown in Fig. 5.5; a typical air pre-heat temperature would be about 450 °C. As an alternative to air pre-heating, or as well as air pre-heating, the gases can be used to raise steam in a heat exchanger such as those of Figs 5.3 and 5.4. A more direct use of the energy of furnace gases is exemplified by the recuperative burner mentioned earlier (see Fig. 4.11).

Figure 5.4 High-temperature waste-heat boiler

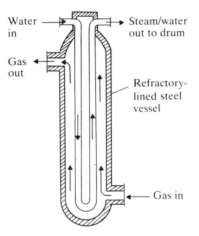

The conventional shell-and-tube heat exchanger can be used for gas-to-liquid, or liquid-to-liquid heat transfer, as well as for steam raising. An example of a two-pass shell-and-tube heat exchanger is given in Fig. 5.6; a typical use would be as shown with waste hot water from, for example, a washing process, being used to heat fresh water.

The plate-fin heat exchanger shown in Fig. 5.7 is widely used in industry up to high temperatures for air-to-air, air-to-liquid, and liquid-to-liquid heat transfer. It is constructed from a series of corrugated plates separated by spacer bars at the edges and sandwiched together. The corrugations form flow passages and the plates are arranged such that flow in successive plates is at right angles to the previous plate giving a cross-flow arrangement as shown in Fig. 5.7; multi-pass or even multi-stream flow is possible by suitable arrangements of the plate corrugations and inlet nozzles.

For gas-to-gas heat recovery at relatively low temperatures a plate-fin block can be fitted at an angle into a split duct to give the convenient heat recovery

Figure 5.5 Furnace gas air pre-heater

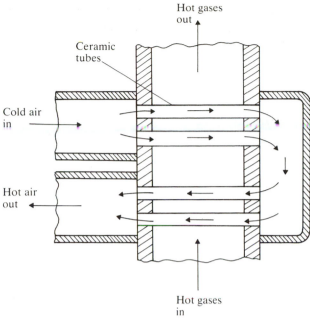

Hot gases out

Ceramic tubes

Cold air in

Hot air out

Hot gases in

Figure 5.6 Two-pass shell-and-tube heat exchanger

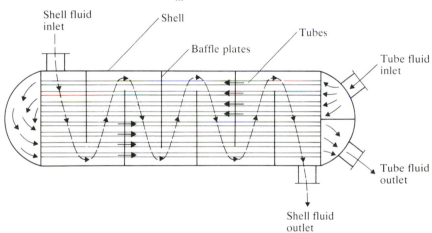

Shell fluid inlet

Shell

Baffle plates

Tubes

Tube fluid inlet

Tube fluid outlet

Shell fluid outlet

Figure 5.7 Plate-fin heat exchanger

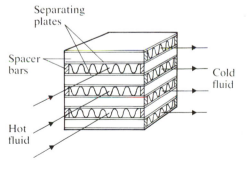

Separating plates

Spacer bars

Cold fluid

Hot fluid

Figure 5.8 Gas-to-gas heat recovery with a plate-fin heat exchanger

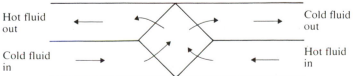

Hot fluid out

Cold fluid out

Cold fluid in

Hot fluid in

arrangement shown in Fig. 5.8. A different type of compact heat exchanger used mainly for liquid-to-liquid heat transfer is shown in Fig. 5.9. Thin corrugated plates of the type shown in Fig. 5.9(a) are bolted together to form a compact unit with the fluids flowing in adjacent passages. This type of heat exchanger is used widely in the food industry.

Design Factors

The basic equations for any recuperative heat exchanger are as given in Section 3.2, Eq. [3.27] to [3.29] reproduced below.

$$Q = \dot{m}_H c_H (t_{H1} - t_{H2}) = \dot{m}_C c_C (t_{C1} - t_{C2}) = U A_o (LMTD) K$$

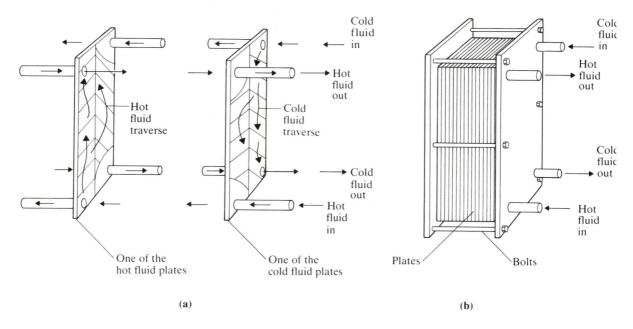

(a)   (b)

Figure 5.9 Liquid–liquid plate heat exchanger

where

$$LMTD = \frac{\Delta t_1 - \Delta t_2}{\ln(\Delta t_1 / \Delta t_2)}$$

and K depends on the type of flow.

$$\frac{1}{UA_o} = \frac{1}{h_i A_i} + R_w + \frac{1}{h_o A_o} + F_i + F_o$$

where the symbols are defined in Section 3.2.

(Note that in the first equation above it is assumed that all the heat rejected by the hot fluid is transferred to the cold fluid. Since the outside casing of the heat exchanger is always at a higher temperature than the ambient then there will be a heat loss to atmosphere but in most cases it can be shown that this heat loss is negligible compared with the heat transferred between the two fluids.)

For a given heat exchanger duty, Q, it can be seen from the equation that the heat transfer area required for given temperature differences depends on the overall heat transfer coefficient, U. The heat transfer coefficients for the two fluids determine the thermal resistances of the fluid film on each surface. Assuming that the fouling factors are outside the designer's influence and that the thermal resistance of the wall is negligible then the thermal resistances of the two fluid films will determine the overall heat transfer. Frequently one of the fluid film coefficients will be much smaller than the other and hence this fluid film controls the heat transfer; for example, when air is heated by condensing steam the heat transfer coefficient on the air side is of the order of 10 to 500 W/m² K whereas the heat transfer coefficient for the wet steam is of the order of 10 000 W/m² K.

In cases where one fluid controls the heat transfer it may be economic to increase the surface area for that fluid by using fins for example. Some typical examples of *extended surfaces*, or *secondary surfaces*, are shown diagrammatically in Figs 5.10(a), (b) and (c). Due to the conduction of heat along the fin there is a temperature gradient from the base (i.e. where it joins the surface) to the tip. Therefore the fin surface is at a mean temperature between the primary, or base, surface and the bulk of the fluid. For this reason the added surface due to the fin is not totally effective. A fin efficiency, η_F, can be defined as the ratio

Figure 5.10 Extended surfaces

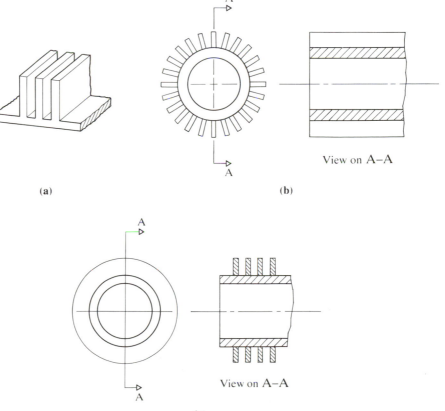

of the heat transferred from the fin surface to that which would be transferred if the whole of the fin surface were at the temperature of the base surface.

Assuming that the heat transfer coefficient is the same for the base surface and the fin surface we can write

$$Q = h\Delta t(A_b + \eta_F A_F) \qquad [5.1]$$

where Δt is the temperature difference between the base surface and the fluid; A_b is the area of the unfinned base surface; A_F is the total surface area of the fins. A thermal resistance for the finned surface, $1/h(A_b + \eta_F A_F)$ can then be used in the equation for $1/UA_o$ (see above and Eq. [3.29]).

Heat transfer coefficients for finned surfaces depend on the fin geometry, fin spacing, fin height etc.; fins should be spaced so that the boundary layers on adjacent fins do not merge. Manufacturers' catalogues should be referred to for details of heat transfer coefficients from finned surfaces.

An extended surface can be considered when one of the fluid films controls the heat transfer and when the heat transfer coefficient on that surface is small; natural convection into still air is one example where fins are very effective. Fins should also have a high thermal conductivity and hence would not normally be used on non-metallic surfaces. For design purposes a good guide is to assume that fins should not be used unless $hb/k < 0.5$, where b is the fin thickness.

Example 5.4

A flat surface as shown in Fig. 5.10(a) has a base temperature of 90 °C when the air mean bulk temperature is 20 °C. Air is blown across the surface and the mean heat transfer coefficient is 30 W/m² K. The fins are made of an aluminium alloy; the fin thickness is 1.6 mm, the fin height is 19 mm, and the fin pitch is 13.5 mm. Calculate the heat loss per m² of primary surface with and without the fins assuming that the same mean heat transfer coefficient applies in each case. Neglect the heat loss from the fin tips and take a fin efficiency of 71 %.

Solution
The heat loss, Q, per m² of primary surface with no fins is given by

$$Q = 30(90 - 20) = 2100 \text{ W}$$

Referring to Fig. 5.11 the number of fins on a 1 m length is given by $1/0.135 = 74$, and hence the relevant areas are

$$A_b = 74 \times (0.0135 - 0.0016) \times 1 = 0.881 \text{ m}^2$$

and $\qquad A_F = 2 \times 74 \times 0.019 \times 1 = 2.812 \text{ m}^2$

Figure 5.11 Finned surface for Example 5.4

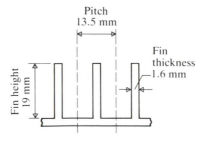

Pitch
13.5 mm

Fin thickness
1.6 mm

Fin height
19 mm

Therefore using Eq. [5.1]

$$\text{heat loss from finned surface} = 30(90 - 20)(0.881 + 0.71 \times 2.812)$$
$$= 6043 \text{ W}$$

Comparing this with the value for the unfinned surface of 2100 W shows an increase of nearly 200 %.

The effectiveness, E, of a heat exchanger (see Eq. [3.30]), is given by

$$E = Q / \{(\dot{m}c)_{min}(t_{Hmax} - t_{Cmin})\}$$

It can be shown that E is a function of the ratio of the thermal capacities of the two fluids, R, and the Number of Transfer Units, NTU, where

$$NTU = (UA)_o / (\dot{m}c)_{min}$$

Graphs of E against NTU for values of R for different arrangements of flow and number of passes are given by Kays and London;[5.1] an example for a shell-and-tube heat exchanger with two shell passes and four, eight, twelve etc. tube passes is shown in Fig. 5.12.

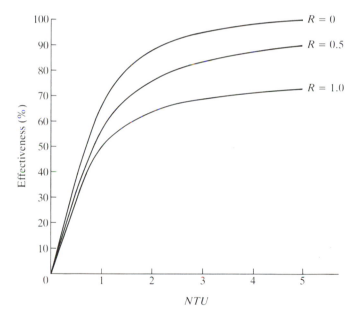

Figure 5.12
Effectiveness against NTU for a shell-and-tube heat exchanger with 2 shell passes and 4, 8, or 12 tube passes

The greater the effectiveness of a heat exchanger used for energy recovery then the greater the fuel saving. For given mass flow rates and specific heats of the two fluids the value of E depends on the NTU and hence on the product $(UA)_o$. Thus for a given value of U the NTU is proportional to A_o. It can then be seen from Fig. 5.12 that increasing the heat transfer area increases the effectiveness and hence the saving in fuel.

The capital cost of the heat exchanger increases as the area increases and Fig. 5.12 shows that at high values of effectiveness a large increase in area produces only a small increase in effectiveness. (For example at $R = 0.5$ for the heat exchanger of Fig. 5.12 a change of effectiveness from 0.7 to 0.8 requires an increase in area of about 56 %, whereas an increase in effectiveness from 0.8 to 0.9 requires an increase in area of about 108 %.)

The NTU, and hence the effectiveness, can be increased for a fixed value of the area by increasing the value of the overall heat transfer coefficient, U; the value of U can be increased by increasing the heat transfer coefficient for one or both of the individual fluids. A typical equation for turbulent heat transfer in a tube is as follows (see Eq. [3.18]):

$$Nu = 0.023 \ Re^{0.8} Pr^{0.4}$$

For flow in a tube of inside diameter, d_i, the Reynolds number can be written as

$$Re = \rho C d_i / \mu = 4 \dot{m}_t / \pi d_i \mu \qquad [5.2]$$

(where $\dot{m}_t$ is the mass flow rate per tube, $\rho C \pi d_i^2 / 4$).

Hence the value of Nu varies with the diameter to the power -0.8 and with the mass flow rate per tube to the power of 0.8 (neglecting changes in property values). Since $Nu = h d_i / k$, it follows that the heat transfer coefficient is given by

$$h = (\text{constant}) \dot{m}_t^{0.8} / d_i^{1.8}$$

(with properties assumed constant). Therefore the heat transfer coefficient can be increased by reducing the tube diameter, and/or increasing the mass flow rate per tube. Now the total mass flow rate is given by

$$\dot{m} = n \dot{m}_t$$

(where n is the number of tubes per pass). Therefore for a constant total mass flow rate the number of tubes per pass must be reduced correspondingly if the mass flow rate per tube is increased. Conversely if the mass flow rate per tube is maintained constant then the number of tubes must remain the same to keep the total mass flow rate constant.

Also, the heat transfer area is given by

$$A_o = n p \pi d_o L \qquad [5.3]$$

(where p is the number of tube passes; d_o is the outside diameter of the tubes; L is the length of one tube pass). Therefore to maintain the same total heat transfer area for a reduced tube diameter in a given type of heat exchanger (say p fixed), it is necessary to increase the length of the tubes per pass, L, and/or the number of tubes per pass.

As stated above, increasing the number of tubes will require the mass flow rate per tube to be reduced for the same total mass flow rate, hence the heat transfer rate may be reduced. The design process is therefore an iterative process in order to arrive at the optimum arrangement of tube diameter, tube length, and number of tubes.

The discussion above has considered flow through the tubes but the same arguments apply to flow on the shell side. Also, altering the inside diameter of a tube to increase the heat transfer coefficient for flow through the tube will alter the heat transfer on the shell side.

A full economic analysis also requires consideration of the pumping power for both fluids. Pressure losses in fluid flow due to friction, turbulence, and fittings such as valves, bends etc. are proportional to the square of the flow velocity. The higher the fluid velocity and the more turbulent the flow the higher is the heat transfer coefficient but the greater the pumping power.

Example 5.5

(a)   A shell-and-tube heat exchanger is used to recover energy from engine oil and consists of two shell passes for water and four tube passes for the engine oil as shown diagrammatically in Fig. 5.13. The effectiveness plotted against NTU for values of thermal capacity ratio, R, in Fig. 5.12 applies to this heat exchanger. For a flow of oil of 2.3 kg/s entering at a temperature of 150 °C, and a flow of water of 2.4 kg/s entering at 40 °C, use the data given to calculate:
(i)   the total number of tubes required;
(ii) the length of the tubes;
(iii) the exit temperatures of the water and the oil;
(iv) the fuel cost saving per year if water heating is currently provided by a gas boiler of efficiency 0.8.

Figure 5.13   Shell-and-tube heat exchanger with 2 shell passes and 4 tube passes for Example 5.5

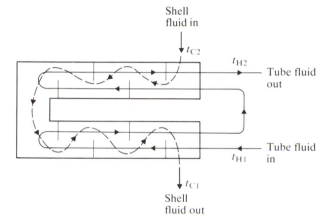

Assume that the velocity of the oil in the tubes is about 0.8 m/s and that the required effectiveness is about 0.7. Neglect heat losses.
(b)   It is suggested that the effectiveness of the heat exchanger, and hence the fuel savings, could be increased by using an eight pass heat exchanger of the same design with all other data unchanged. Calculate:
(v)   the new effectiveness;
(vi) the new fuel cost saving per year.

Data
Tube inside diameter, 5 mm; tube outside diameter, 7 mm; mean specific heat of oil, 2.19 kJ/kg K; mean specific heat of water, 4.19 kJ/kg K; mean density of oil, 840 kg/m^3; overall heat transfer coefficient, 400 W/m^2 K; annual usage, 4000 h; cost of gas for water heating, 1.2 p/kW h.

Solution
(a)   (i)   For the oil the volume flow is given by

$$\dot{V} = 2.3/840 = 0.002\,74 \text{ m}^3/\text{s}$$

Therefore for a fluid velocity of about 0.8 m/s the required cross-sectional area is given by $0.002\,74/0.8 = 0.003\,42$ m^2. For one tube, the cross-sectional area is

$$\pi \times 0.005^2/4 = 0.000\,0196 \text{ m}^2$$

i.e. the number of tubes per pass is $0.003\,42/0.000\,0196 = 174.6$ say 175. Therefore

$$\text{total number of tubes} = 4 \times 175 = 700$$

(ii)   From Eq. [3.32]

$$R = (\dot{m}c)_{min}/(\dot{m}c)_{max} = 2.3 \times 2.19/(2.4 \times 4.19) = 0.5$$

From the graph of Fig. 5.12 at a value of effectiveness of 0.7 and a value of R of 0.5 we read, $NTU = 1.65$. Then from Eq. [3.31]

$$NTU = UA_o/(\dot{m}c)_{min} = 1.65$$

i.e. $A_o = 1.65 \times 2.3 \times 2.19 \times 10^3/400 = 20.78 \text{ m}^2$

Therefore from Eq. [5.3] we have

length of tubes per pass $= 20.78/(4 \times 175 \times \pi \times 0.007) = 1.35 \text{ m}$

(iii)   From Eq. [3.30]:

$$E = (t_{H1} - t_{H2})/(t_{H1} - t_{C2})$$

i.e. $t_{H2} = 150 - 0.7(150 - 40) = 73 \,^{\circ}C$

Then from Eq. [3.28],

$$(\dot{m}c)_H(t_{H1} - t_{H2}) = (\dot{m}c)_C(t_{C1} - t_{C2})$$

i.e.   exit temperature of water, $t_{C2} = 40 + 0.5(150 - 73) = 78.5 \,^{\circ}C$

(iv)   The heat transferred to the water is given by:

$$Q = 2.4 \times 4.19(78.5 - 40) = 387.2 \text{ kW}$$

This is the load that would be required to be delivered by an equivalent gas boiler.

i.e.   fuel cost saving per year $= 387.2 \times 4000 \times 1.2/(0.8 \times 100)$
$$= £23\,230$$

(b)   (v)   Doubling the number of passes with the same number of tubes per pass and the same tube diameter gives a doubling of the surface area for heat transfer. With the same mass flow rate of oil and the same velocity in the tubes then the heat transfer coefficient is unchanged; the heat transfer coefficient for the water on the shell-side will be altered slightly since the two shell-passes now traverse more tubes but since the heat transfer coefficient for the oil is likely to be the controlling factor this change will be ignored. Figure 5.12 is applicable for an eight-pass heat exchanger.

$$NTU = UA_o/(\dot{m}c)_{min} = 1.65 \times 2 = 3.30$$

Then from Fig. 5.12

$$E = 0.854$$

(vi)   The exit temperature of oil is given by

$$t_{H2} = 150 - 0.854(150 - 40) = 56.5 \,^{\circ}C$$

and the exit temperature of water is given by

$$t_{C2} = 40 + 0.5(150 - 56.5) = 86.8 \,^{\circ}C$$

Therefore

$$Q = 2.4 \times 4.19(86.8 - 40) = 470.6 \text{ kW}$$

Then

$$\text{fuel cost saving per year} = \frac{470.6}{387.2} \times £23\,230 = £28\,235$$

and increased annual fuel saving $= (£28\,235 - £23\,230) = £5005$ which is a percentage increase of $5005 \times 100/23\,230 = 21.6\,\%$. To achieve this annual saving the heat transfer area has been doubled; this implies an increased capital cost (say about 70 %), and since the pumping cost for the oil is also approximately doubled the decision to double the number of tube passes is almost certainly not economic.

5.3 RUN-AROUND COIL SYSTEMS

A run-around coil heat recovery system is the name given to a linking of two recuperative heat exchangers by a third fluid which exchanges heat with each fluid in turn as shown diagrammatically in Fig. 5.14.

Figure 5.14 Run-around coil system of heat recovery

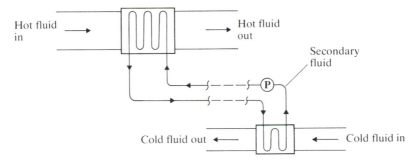

A run-around coil would be used in cases where the two fluids which are required to exchange heat are too far apart to use a conventional direct recuperative heat exchanger. It is also desirable to use such an indirect system if there is a risk of cross-contamination between the two primary fluids (e.g. when a particularly corrosive fluid is involved, or when there is a risk of bacterial contamination as in a hospital).

A particular advantage of the run-around coil is that a suitable heat transfer liquid can be chosen with a low freezing point and good heat transfer characteristics. Also, the capital cost is low since the individual heat exchangers are standard items chosen to suit the fluid. One disadvantage is in the low effectiveness which is reduced even more if effective insulation is not used for the indirect liquid pipe-line.

Typical applications are: recovery of energy from the air leaving a room or building to pre-heat the air entering; recovery of energy from the exhaust of an oven or other process heater to pre-heat the entering air; recovery of energy from a drying process; recovery of energy from a corrosive gas for water heating.

Design Factors

In the following discussion it is assumed that in all cases the heat transfer processes approximate to counter-flow; this is the most efficient form of heat exchanger. Emerson[5.2,5.3] gives a simple analysis of run-around coil systems including some discussion of costs.

One of the most common examples of a run-around coil system is heat recovery from a fluid at one stage in a process to the same fluid at a different stage, see for example the system shown in Fig. 5.15(a) with temperature variations as given in Fig. 5.15(b). In this case the thermal capacity $(\dot{m}c)$ of the cold fluid is the same as that of the hot fluid (neglecting change of specific heat due to temperature). It follows from Eq. 3.28 that

$$Q/(\dot{m}c)_S = Q/(\dot{m}c)_C = Q/(\dot{m}c)_H = (t_{H1} - t_{H2}) = (t_{C1} - t_{C2})$$
$$= (t_{S1} - t_{S2})$$

i.e. $$(\dot{m}c)_S = (\dot{m}c)_H = (\dot{m}c)_C$$

and $$(t_{H1} - t_{S1}) = (t_{H2} - t_{S2}) \qquad (t_{S1} - t_{C1}) = (t_{S2} - t_{C2})$$

The three temperature lines are therefore straight and parallel as shown in Fig. 5.15(b).

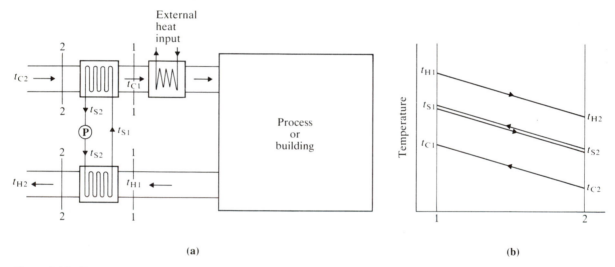

(a)   (b)

Figure 5.15 Run-around coil heat recovery between fluids with the same thermal capacity

The two heat exchangers are identical since they involve the same two fluids exchanging heat with the same fluid temperature differences. (The fluid properties will vary slightly since each heat exchanger operates at a different temperature range but this is assumed to be negligible.)

i.e. $$(UA)_H = (UA)_C = Q/(t_{H1} - t_{S1}) = Q/(t_{S1} - t_{C1})$$

Therefore the temperature line for the secondary fluid must be midway between the temperature line for the hot and cold fluids.

i.e. $$t_{S1} = (t_{H1} + t_{C1})/2 \qquad \text{and} \qquad t_{S2} = (t_{H2} + t_{C2})/2$$

Also

$$Q = (UA)_H(t_{H1} - t_{S1}) = (UA)_C(t_{S1} - t_{C1}) = (UA)_o(t_{H1} - t_{C1})$$

i.e. $$(UA)_H = (UA)_C = 2(UA)_o$$

When the inlet temperatures of the hot and cold fluids are known a convenient expression for the heat recovery is derived as follows:

$$Q = (UA)_H(t_{H1} - t_{C1})/2$$

and since $\quad t_{C1} = t_{C2} + Q/(\dot{m}c)_C$

then $\quad Q = \dfrac{(UA)_H(t_{H1} - t_{C2})}{2 + \{(UA)_H/(\dot{m}c)_C\}}$ [5.4]

Example 5.6

A run-around coil heat recovery system similar to that of Fig. 5.15(a) is used for a room in which the presence of bacteria rules out any possibility of air re-circulation or a direct recuperative heat exchanger. Air enters the room at 24 °C and leaves at 20 °C; the average outside air temperature during the annual period of use is 5 °C. Assuming that the mass flow rate of air is 2 kg/s, mean specific heat 1.005 kJ/kg K, that $(UA)_H = (UA)_C = 4$ kW/K, and that the specific heat of the secondary fluid is 2.5 kJ/kg K, calculate:

(i) the required mass flow rate of secondary fluid;
(ii) the temperature of the air leaving the run-around coil;
(iii) the percentage energy saving by using the run-around coil.

Solution
(i) $\qquad (\dot{m}c)_S = (\dot{m}c)_H = (\dot{m}c)_C = 2 \times 1.005 = 2.01$ kW/K

$\qquad \therefore \qquad$ mass flow rate of secondary fluid $= 2.01/2.5 = 0.804$ kg/s

(ii) Using Eq. [5.4]

$$Q = 4(20 - 5)/\{2 + (4/2.01)\} = 15.04 \text{ kW}$$

and $\qquad t_{C1} = t_{C2} + Q/(\dot{m}c)_C$

i.e. $\qquad$ temperature of air leaving coil $= 5 + (15.04/2.01) = 12.5$ °C.
(iii) Without the heat recovery the air must be heated from 5 °C to 24 °C hence the percentage energy saving is given by:

$$\text{percentage saving} = (12.5 - 5) \times 100/(24 - 5) = 39.4 \%$$

When the thermal capacities of the hot and cold fluids are not equal, as for example for the case shown in Fig. 5.16, then finding the required thermal

Figure 5.16 Run-around coil heat recovery between fluids of different thermal capacity

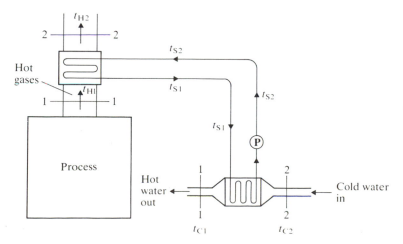

Figure 5.17
Temperature variation
when thermal capacity
of secondary fluid is
intermediate to that of
both hot and cold fluids

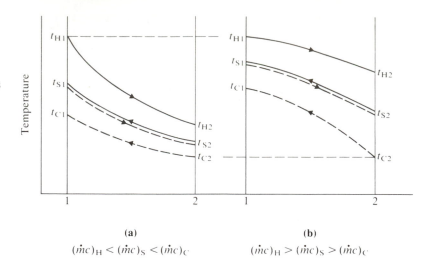

(a)

$(\dot{m}c)_H < (\dot{m}c)_S < (\dot{m}c)_C$

(b)

$(\dot{m}c)_H > (\dot{m}c)_S > (\dot{m}c)_C$

capacity, and hence flow rate, of the secondary fluid is more difficult. The temperature variations can be either as in Figs 5.17(a) and (b), or as shown in Figs 5.18(a) and (b). In Fig. 5.17(a) the thermal capacity of the hot fluid is less than that of the secondary fluid which is in turn less than that of the cold fluid; in Fig. 5.17(b) the thermal capacity of the hot fluid is greater than that of the secondary fluid which is in turn greater than that of the cold fluid. In Fig. 5.18(a) the thermal capacity of both the hot and cold fluids is less than that of the secondary fluid, whereas in Fig. 5.18(b) the thermal capacity of both fluids is greater than that of the secondary fluid.

Figure 5.18
Temperature variation
when thermal capacity
of secondary fluid is
greater or less than that
of both hot and cold
fluids

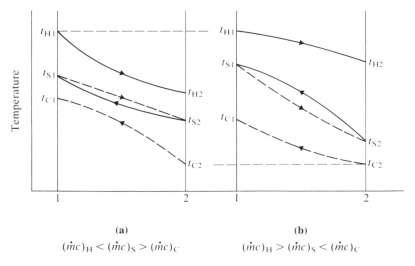

(a)

$(\dot{m}c)_H < (\dot{m}c)_S > (\dot{m}c)_C$

(b)

$(\dot{m}c)_H > (\dot{m}c)_S < (\dot{m}c)_C$

Consider the case in which the run-around coil is replaced by an infinite number of identical coils as shown in Fig. 5.19(a); referring to Fig. 5.19(b), for each small coil, 1, 2, 3 to n

$$\Delta t_{H-S} + \Delta t_{S-C} = \Delta t_{H-C}$$

Hence the temperature variation of the secondary fluid when exchanging heat with the hot fluid is exactly the reverse of the temperature variation when

Figure 5.19 Idealized run-around coil system

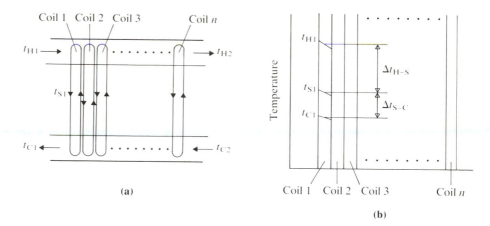

(a)

(b)

exchanging heat with the cold fluid; the temperature variations must therefore conform to that of either Fig. 5.17(a) or (b), and not to that of Fig. 5.18. It follows that for the optimum case:

$$(LMTD)_o = (LMTD)_{H-S} + (LMTD)_{S-C}$$

and therefore

$$1/(UA)_o = 1/(UA)_H + 1/(UA)_C \qquad [5.5]$$

Also, at any cross-section,

$$\frac{t_S - t_C}{t_H - t_S} = \text{const.} = \frac{(UA)_H}{(UA)_C} = Y, \text{ say}$$

$$\therefore \qquad Y = \frac{t_{S1} - t_{C1}}{t_{H1} - t_{S1}} = \frac{t_{S2} - t_{C2}}{t_{H2} - t_{S2}}$$

therefore $\qquad t_{S1} = (t_{C1} + Y t_{H1})/(1 + Y)$
and $\qquad t_{S2} = (t_{C2} + Y t_{H2})/(1 + Y)$

$$\therefore \qquad (t_{S1} - t_{S2}) = \frac{(t_{C1} - t_{C2}) + Y(t_{H1} - t_{H2})}{(1 + Y)}$$

i.e. $\qquad 1/(\dot{m}c)_S = \frac{1/(\dot{m}c)_C + Y/(\dot{m}c)_H}{(1 + Y)}$

$$\therefore \qquad \dot{m}_S = \frac{(\dot{m}c)_H (\dot{m}c)_C \{(UA)_H + (UA)_C\}}{c_S \{(\dot{m}c)_H (UA)_C + (\dot{m}c)_C (UA)_H\}} \qquad [5.6]$$

It is shown by Emerson[5.2] that when the mass flow rate of secondary fluid is less than the correct value the heat recovery falls rapidly away; when the mass flow rate is larger than the correct value the heat recovery still falls away but not so markedly. From Eq. [5.6] it can be seen that the correct mass flow rate of secondary fluid is independent of the temperatures of the fluids; it depends entirely on the thermal capacities of the fluids and on the heat transfer characteristics of the two heat exchangers.

For this case the derivation of an expression for heat recovery is much more complex but if the expressions for effectiveness are used (see Eq. [3.30] and [3.33]) then problems can be solved more easily. In Eq. [3.33] for a counter-flow heat exchanger the values of NTU and R are required.

i.e.

$$E = \frac{Q}{(\dot{m}c)_{min}(t_{H1} - t_{C2})} = \frac{1 - e^{-NTU(1-R)}}{1 - Re^{-NTU(1-R)}}$$

Then for this case, using Eq. [5.5]

$$NTU = (UA)_o/(\dot{m}c)_{min} = \frac{(UA)_H(UA)_C}{(\dot{m}c)_{min}\{(UA)_H + (UA)_C\}}$$

Hence when (UA) for both heat exchangers is known, and when $(\dot{m}c)$ for both hot and cold fluids is known, the NTU for the overall process can be found and then the effectiveness, E. Knowing the inlet temperature of both hot and cold fluids the heat recovery, Q can then be found.

Example 5.7
A corrosive gas at a flow rate of 30 kg/s from a process at 300 °C is to be used to heat 20 kg/s of water entering at 10 °C using a run-around coil as shown previously in Fig. 5.16. Calculate, using the data given:
(i)   the mass flow rate of secondary fluid required;
(ii) the effectiveness of the overall heat transfer;
(iii) the exit temperature of the water;
(iv) the temperatures of the secondary fluid.

Data
Mean specific heat of gases, 1.2 kJ/kg K; mean specific heat of water, 4.2 kJ/kg K; mean specific heat of secondary fluid, 3.8 kJ/kg K; (UA) for the gas to secondary fluid heat exchanger, 40 kW/K; (UA) for the secondary fluid to water heat exchanger, 200 kW/K.

Solution
(i)   For the hot fluid

$$(\dot{m}c)_H = 30 \times 1.2 = 36 \text{ kW/K}$$

and for the cold fluid

$$(\dot{m}c)_C = 20 \times 4.2 = 84 \text{ kW/K}$$

Then using Eq. [5.6]

$$\text{mass flow of secondary fluid} = \frac{36 \times 84(40 + 200)}{3.8\{(84 \times 40) + (36 \times 200)\}}$$
$$= 18.09 \text{ kg/s}$$

(ii) From Eq. [5.5]

$$1/(UA)_o = (1/40) + (1/200) = 0.03$$

$$\therefore \quad (UA)_o = 33.333 \text{ kW/K}$$

(Note that the overall heat transfer process is less effective than either of the processes to and from the secondary fluid; this is always the case for a run-around coil system as can be seen from Eq. [5.5].) Hence

$$NTU = (UA)_o/(\dot{m}c)_{min} = 33.333/36 = 0.926$$

Then using Eq. [3.33]

$$\text{effectiveness} = \frac{1 - e^{-0.926 \times 0.571}}{1 - 0.429e^{-0.926 \times 0.571}} = 0.55$$

(where $R = (\dot{m}c)_{min}/(\dot{m}c)_{max} = 36/84 = 0.429$).

(iii)   From Eq. [3.30]

$$E = \frac{300 - t_{H2}}{300 - 10} = 0.55$$

$$\therefore \quad t_{H2} = 140.5\,°C$$

Then

$$\text{water exit temperature} = t_{C1} + R(t_{H1} - t_{H2})$$
$$= 10 + 0.429(300 - 140.5)$$
$$= 78.4\,°C$$

(iv)   The secondary fluid temperatures can be calculated as follows:

$$\frac{t_{H1} - t_{S1}}{t_{S1} - t_{C1}} = \frac{(UA)_C}{(UA)_H} = \frac{200}{40} = 5$$

$$\therefore \quad t_{S1} = \frac{300 + (5 \times 7.84)}{6} = 115.3\,°C$$

and

$$\frac{t_{H2} - t_{S2}}{t_{S2} - t_{C2}} = 5$$

$$\therefore \quad t_{S2} = \frac{140.5 + (5 \times 10)}{6} = 31.8\,°C$$

The complete temperature changes are as shown diagrammatically in Fig. 5.20.

Figure 5.20
Temperature changes
for Example 5.7

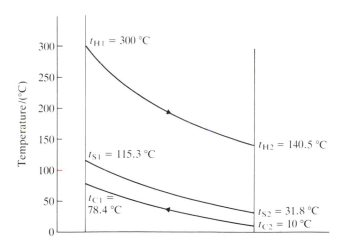

5.4 REGENERATIVE HEAT EXCHANGERS

In a regenerative heat exchanger (sometimes called a capacitance heat exchanger)
the hot and cold fluids pass alternately across a matrix of material; the matrix

is heated up by the hot fluid then cooled down by the cold fluid so that the process is cyclic. There are two main types.

(a)   The first type is stationary with two matrixes as shown in Figs 5.21(a) and (b). In Fig. 5.21(a) matrix B is hot and heats up the cold fluid while matrix A is heated by the hot fluid; in Fig. 5.21(b) the cold fluid is now heated by matrix A while the hot fluid re-heats matrix B; the valves are then switched over and the cycle commences again as in Fig. 5.21(a). By having two matrixes the process of heat transfer between the two fluids is continuous with the inlet and exit temperatures of the fluids remaining constant with time.

Figure 5.21   Stationary regenerative heat exchanger

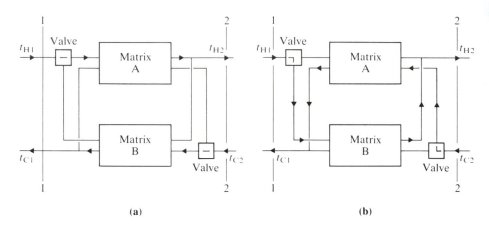

This type has been in use for more than 50 years in, for example, steel plants, where hot furnace gases are used to heat the incoming air. A more recent use is in the heat recovery from buildings as shown in Figs 5.22(a) and (b).

(b)   The second type is the *rotary regenerator* in which a matrix of material is mounted on a wheel which is rotated slowly through the hot and cold fluid streams as shown in Fig. 5.23. This type has been in use for over 50 years as a flue gas/air pre-heater for boilers, known as the Ljungstrom rotary regenerator after its Danish inventor. A more recent use in heat recovery is in air-to-air applications for buildings when it has become known as the *thermal wheel*.

The advantages of the regenerative heat exchanger are in the large surface area to volume ratio of the heat exchanger, the relatively low cost per unit surface area, and the self-cleaning feature due to the periodic flow reversal; also, it is possible to reclaim the enthalpy of vaporization of moisture in the hot fluid in a thermal wheel by using a non-metalic matrix material with a desiccant

Figure 5.22   Double-accumulator regenerative heat exchanger

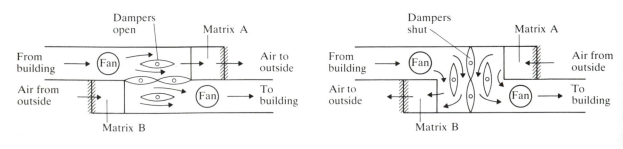

Figure 5.23 Rotary regenerator or thermal wheel

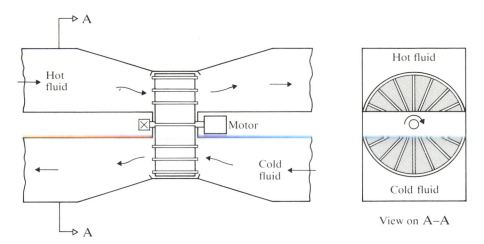

View on A–A

coating. The major disadvantage is in the possibility of cross contamination; in the rotary regenerator this can be reduced considerably by ensuring that the clean fluid is at a higher pressure and by using a purge device. Other disadvantages are the cleaning difficulties, and the susceptibility to corrosion.

It can be seen that this type of heat exchanger is suitable only for two fluids of similar properties; also, in the stationary type the fluid pressures must be the same. The rotary type can be used with similar fluids at different pressures (as for example in the gas turbine unit), but the sealing of the wheel becomes much more difficult.

Design Factors

Because of the periodic nature of the heat transfer processes analysis is more difficult than that for steady state recuperative heat exchangers. A simple treatment is given in Bayley *et al*[5.4] and in Kays and London[5.1] a set of figures and graphs is given for the effectiveness.

The effectiveness, E, is defined as previously (see Eq. [3.30]); NTU is equal to $(UA)/(\dot{m}c)_{min}$ as before with (UA) now defined by

$$1/(UA) = 1/(hA)_{\text{H}} + 1/(hA)_{\text{C}} \qquad [5.7]$$

where (hA) is the product of the heat transfer coefficient between fluid and matrix and the surface area of the matrix; in most practical cases, $(hA)_{\text{H}} = (hA)_{\text{C}}$. (Note that the value of (UA) is equivalent to that for a recuperative heat exchanger in which the thermal resistance of the separating wall is negligible and fouling factors are neglected.)

Let the mass of the matrix be M, and the specific heat of the matrix material be c_{M} (for the stationary type shown in Fig. 5.21 or Fig. 5.22, M is the mass of the matrixes added together). An equivalent thermal capacity rate for the matrix can then be derived either for the stationary type or for the rotary type as follows: for the stationary type

$$(\dot{m}c)_{\text{M}} = Mc_{\text{M}}/\tau \qquad [5.8]$$

(where τ is the total cycle time), and for the rotary type

$$(\dot{m}c)_{\text{M}} = NMc_{\text{M}} \qquad [5.9]$$

(where N is the revolutions of the wheel per unit time).

Kays and London[5.1] give tables and graphs of effectiveness against NTU for various values of the ratio $(\dot{m}c)_{M}/(\dot{m}c)_{min}$; each table or graph is for a different value of $(\dot{m}c)_{min}/(\dot{m}c)_{max}$ in the range 0.5 to 1.0

The same authors give a useful empirical expression which shows the effect of matrix rotational speed (or of the length of time period in the stationary type), on the effectiveness.

i.e.
$$E = E_{c}\left\{1 - \frac{1}{9[(\dot{m}c)_{M}/(\dot{m}c)_{min}]^{1.93}}\right\} \qquad [5.10]$$

where the term E_{c} is the effectiveness of an equivalent counter-flow heat exchanger with NTU defined using (UA) from Eq. 5.7 above.

i.e.
$$E_{c} = (1 - e^{-NTU(1-R)})/(1 - Re^{-NTU(1-R)})$$

or
$$E_{c} = NTU/(1 + NTU) \qquad \text{when } R = 1$$

Figure 5.24 shows the ratio E/E_{c} plotted against the ratio $(\dot{m}c)_{M}/(\dot{m}c)_{min}$; it can be seen that above a value of $(\dot{m}c)_{M}/(\dot{m}c)_{min}$ of about 2 the effectiveness of the heat exchanger is approximately equal to the effectiveness of an equivalent cross-flow heat exchanger.

Figure 5.24 Influence of matrix rotational speed on effectiveness (see Eq. 5.10)

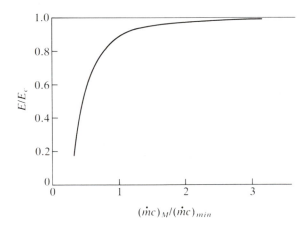

$(\dot{m}c)_{M}/(\dot{m}c)_{min}$

Example 5.8

A rotary regenerator is used to recover energy from a gas stream leaving a furnace at 300 °C at a mass flow rate of 10 kg/s. Heat is transferred to a mass flow rate of air of 10 kg/s entering at 10 °C. The wheel has a diameter 1.5 m, giving an approximate face area of 1.6 m², and a width of 0.22 m; the matrix has a surface area to volume ratio of 3000 m²/m³ and a mass of 150 kg; the rotational speed of the wheel is 10 rev/min. The heat transfer coefficient for both fluid streams is 30 W/m² K and the mean specific heats at constant pressure for the gas and air are 1.15 kJ/kg K and 1.005 kJ/kg K; the specific heat of matrix material is 0.8 kJ/kg K. Calculate:

(i)   the effectiveness of the heat exchanger;
(ii) the rate of heat recovery and the temperature of the air at exit;
(iii) the air temperature at exit if the rotational speed of the wheel is increased to 20 rev/min;
(iv) the air temperature at exit if the rotational speed is reduced to 5 rev/min.

Solution

(i)   Since the heat transfer coefficient and the area are the same for both hot and cold fluids then from Eq. [5.7] we have:

$$1/U = 2/h \qquad \therefore \quad U = 30/2 = 15 \text{ W/m}^2 \text{ K}$$

The heat transfer area is given by

$$A = (\text{volume}) \times (\text{surface area}/\text{volume}) = (1.6 \times 0.22) \times 3000$$
$$= 1056 \text{ m}^2$$
$$NTU = UA/(\dot{m}c)_{min} = 15 \times 1056/(10 \times 1.005 \times 10^3) = 1.576$$

Also $R = (10 \times 1.005)/(10 \times 1.15) = 0.874$

Then using Eq. [3.33] for the effectiveness of an equivalent counter-flow heat exchanger we have

$$E_c = \frac{(1 - e^{-1.576 \times 0.126})}{(1 - 0.874 e^{-1.576 \times 0.126})} = 0.636$$

From Eq. [5.9]

$$(\dot{m}c)_M = 10 \times 150 \times 0.8/60 = 20 \text{ kW/K}$$
$$(\dot{m}c)_M/(\dot{m}c)_{min} = 20/(10 \times 1.005) = 1.99$$

Then substituting in Eq. [5.10]

$$\text{Effectiveness} = 0.636 \left\{ 1 - \frac{1}{9(1.99)^{1.93}} \right\} = 0.636 \times 0.971 = 0.617$$

(ii)   Then using Eq. [3.30], and taking the same notation for temperature as used throughout this chapter (see for example Fig. 5.21)

$$E = 0.617 = Q/\{(\dot{m}c)_{min}(t_{H1} - t_{C2})\}$$

i.e. $Q = 0.617 \times 10 \times 1.005(300 - 10) = 1799 \text{ kW}$
and $Q = 10 \times 1.005(t_{C1} - 10) = 1799 \text{ kW}$

Therefore

exit temperature of air, $t_{C1} = 10 + (1799/10 \times 1.005) = 189 \,°\text{C}$

(iii)   When the rotational speed is increased to 20 rev/min then, assuming that all other data are unchanged

$$(\dot{m}c)_M/(\dot{m}c)_{min} = \frac{20}{10} \times 1.99 = 3.98$$

Then

$$E = 0.636 \left\{ 1 - \frac{1}{9(3.98)^{1.93}} \right\} = 0.631$$

Therefore

exit temperature of air $= 10 + \{0.631(300 - 10)\} = 193 \,°\text{C}$

(iv)   When the rotational speed is reduced to 5 rev/min we have

$$(\dot{m}c)_M/(\dot{m}c)_{min} = 0.5 \times 1.99 = 0.995$$

and
$$E = 0.636 \left\{ 1 - \frac{1}{9(0.995)^{1.93}} \right\} = 0.565$$

Therefore

$$\text{exit temperature of air} = 10 + \{0.565(300 - 10)\} = 173.8\,^\circ\text{C}$$

It can be seen that in this case for rotational speeds above 10 rev/min there is very little change in the heat recovery, and hence exit air temperature. When the rotational speed is decreased to 5 rev/min there is a larger change in the exit temperature (from 189 °C to 173.8 °C), and the heat recovery is reduced from 1799 kW to 1646.2 kW, a reduction of 8.5 %.

Figure 5.24 shows that when $(\dot{m}c)_M / (\dot{m}c)_{min}$ is less than unity then the effectiveness falls away rapidly, i.e. for a rotary regenerator the rotational speed, N, should be at least $(\dot{m}c)_{min}/Mc_M$. It can also be seen that changing the rotational speed is not a very effective way of controlling the exit temperature of the fluid being heated; control is best effected by a by-pass on the cold or hot fluid ducting.

Regenerative Burners

The use of the energy of the furnace gases to pre-heat the air in an oil or gas burner has been referred to earlier (see Fig. 4.11). A more recent innovation is the regenerative burner for gas firing developed by British Gas.[5.5]

There are two burners operating with two ceramic matrixes in a similar way to that shown earlier in Fig. 5.21. In Fig. 5.25 it can be seen that for each burner, for one part of the cycle the hot gases are fed back through the burner and through a matrix to exhaust (Fig. 5.25(a)); for the second part of the cycle, air is drawn through the matrix and supplied with gas to the burner where combustion takes place (Fig. 5.25(b)); two burners are used in tandem so that continuous combustion can take place.

Recent developments have seen the introduction of microprocessor controlled firing periods; the Energy Efficiency Office has verified in tests on a heat treatment furnace (2.4 m × 3.6 m) operating at 1000 °C that energy savings of up to 50 % are achievable using this type of regenerative burner.

Figure 5.25 Gas-fired regenerative burners

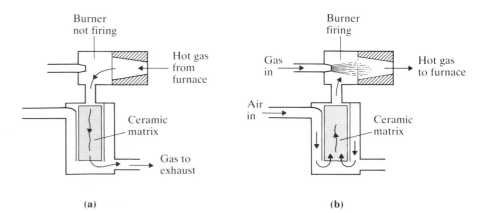

(a)   (b)

5.5 HEAT PUMPS

The theory of heat pumps is covered in Section 4.5 and in the same section there is a discussion on the possible uses including the hierarchy of use as defined by the Building Services Research and Information Association, BSRIA. BSRIA also summarizes the use of heat pumps for heat recovery in buildings.[5.6]

The heat pump is particularly suited to heat recovery from air to air in the heating and air conditioning of buildings since the coefficient of performance (see Section 4.5), is greater when the temperature difference between the condenser and evaporator is small. This applies to all cases where waste fluids at a relatively low temperature are used as the source of heat supply for a heat pump evaporator to provide a heat supply from the condenser for, say, water heating. Such heat pump systems are efficient and relatively cheap, making an electrically driven compressor a feasible proposition.

One other major use for the heat pump in energy recovery is in dehumidification and heating combined. For warehouses, laundries, and houses where humidity is high and condensation on the fabric can cause degradation the use of a heat pump as shown in Fig. 5.26 is a good solution.

Figure 5.26 Heat pump for dehumidification and heating

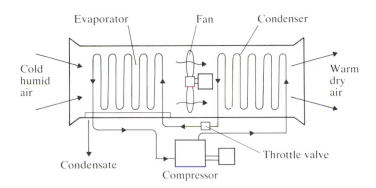

In swimming pools the air leaving the pool hall is very humid and the heat pump is a very suitable form of energy recovery system since it can recover the enthalpy of vaporization of the moisture in the air. A typical example of the heat pump used in combination with a run-around coil for a swimming pool is shown in Fig. 5.27, and a similar system using a plate-fin type heat exchanger is shown in Fig. 5.28.

Figure 5.27 Heat recovery for a swimming pool using a heat pump and run-around coil

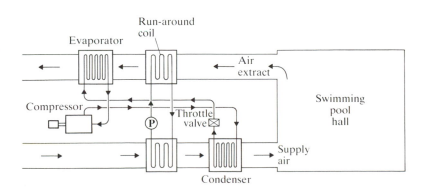

Figure 5.28 Heat
recovery from a
swimming pool using a
heat pump and plate-fin
heat exchanger

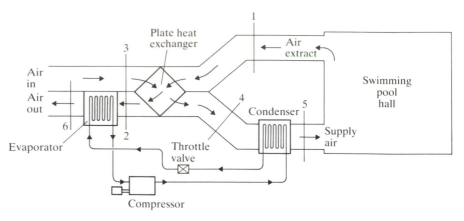

In the various heat exchange processes considered previously the effect of
condensation of vapour from an air stream has not been included. Atmospheric
air contains a certain quantity of water vapour, which is in fact vital for our
comfort, and if air is cooled below the dew point then condensation takes place.
When air is cooled but its temperature remains above the dew point this is
frequently called *sensible cooling*.

An illustration of this is Example 5.6 in which it was assumed that the two
heat transfer processes, from air to secondary fluid and from secondary fluid to
air, involved sensible cooling only; a mean specific heat was taken for the air.
In cases where cooling and de-humidification occur simultaneously a more
stringent analysis is necessary using the enthalpies of dry air and water vapour.
The psychrometric chart can be used as shown in Example 5.9.

Example 5.9
A swimming pool hall is heated by the system shown in Fig. 5.28. A volume
flow rate of air of 15 000 m³/h leaves the pool hall at 28 °C, with a percentage
saturation of 70 %, passes across a plate heat exchanger and then the coils of
the heat pump evaporator, leaving at 17 °C. The fresh air enters at 5 °C,
percentage saturation 80 %, passes across the plate heat exchanger and then
the coils of the heat pump condenser, entering the pool hall at 31 °C. Using the
data given, calculate:
(i)   the mass flow rate of refrigerant required in the heat pump circuit;
(ii) the power input required to the electric motor;
(iii) the percentage cost saving in using the energy recovery system instead of
      electric heating of the fresh air to the pool hall temperature;
(iv) sketch the psychrometric chart with all the salient points marked.

Data
Refrigerant R12: evaporator pressure, 4.233 bar; condenser pressure, 8.477 bar;
vapour leaves evaporatory dry saturated; vapour leaves compressor superheated
at 40 °C; liquid leaves condenser in the saturated state; combined mechanical
and electrical efficiency of motor, 90 %.
Effectiveness of plate heat exchanger, 0.8; air enters evaporator coils with a
percentage saturation of 90 %.

Solution
(i) Referring to Fig. 5.28 and using the psychrometric chart, at state 1

$$v = 0.85 \text{ m}^3/\text{kg dry air}$$

i.e. for dry air

$$\dot{m} = 15\,000/(0.85 \times 3600) = 4.902 \text{ kg/s}$$

Also at 28 °C, 70 % percentage saturation, $h_1 = 71.4$ kJ/kg, and at 5 °C, 80 % percentage saturation, $h_3 = 15.9$ kJ/kg, and $\omega_3 = 0.0043$.

For the plate heat exchanger:

$$E = \frac{\text{heat transferred}}{\text{maximum possible heat transfer}}$$

In this case the maximum possible heat transfer is when the air at 5 °C is heated to 28 °C at constant specific humidity. From the psychrometric chart at 28 °C and $\omega = 0.0043$ the enthalpy is 39.4 kJ/kg. Therefore heat transferred is given by

$$Q = E(39.4 - h_3) = 0.8(39.4 - 15.9) = 18.8 \text{ kJ/kg}$$
$$\therefore \qquad h_4 - h_3 = 18.8 \text{ kJ/kg}$$

i.e. $\qquad h_4 = 18.8 + 15.9 = 34.7$ kJ/kg

From chart at $h_4 = 34.7$ kJ/kg and $\omega_3 = \omega_4 = 0.0043$, we have $t_4 = 23.6$ °C. At entry to the pool hall at 31 °C and $\omega = 0.0043$ we have

$$h_5 = 42.4 \text{ kJ/kg}$$

Then, heat required from condenser coils is given by

$$Q_C = 4.902(h_5 - h_4) = 4.902(42.4 - 34.7) = 37.75 \text{ kW}$$

From tables of properties of R12 (Rogers and Mayhew[5.7]):

$$Q_C = \dot{m}_{ref}(205.2 - 69.55) = 37.75 \text{ kW}$$

i.e. $\qquad$ mass flow of refrigerant $= 37.75/135.65 = 0.278$ kg/s

(ii) Power input to compressor $= 0.278(205.2 - 191.74)/0.9 = 4.16$ kW
(iii) Without heat recovery it is required to heat fresh air from state 3 to state 5, i.e. the heat input is equal to

$$4.902(42.4 - 15.9) = 129.9 \text{ kW}$$

Assuming that the electrical heating is 100 % efficient then

$$\text{percentage saving in fuel cost} = (129.9 - 4.16) \times 100/129.9$$
$$= 96.8 \%$$

The complete process is shown on a diagram of the psychrometric chart in Fig. 5.29. State 2 is fixed by first calculating the enthalpy.

i.e. $\qquad h_2 = h_1 - 18.8 = 71.4 - 18.8 = 52.6$ kJ/kg

Then state 2 can be fixed on the chart at $h_2 = 52.6$ kJ/kg and a percentage saturation of 90 % (given). i.e. $t_2 = 19.6$ °C. Also state 6 is fixed at $t_6 = 17$ °C (given) and

$$h_6 = h_2 - \frac{(Q_E)}{4.902} = 52.6 - \frac{0.278(191.74 - 69.55)}{4.902}$$

i.e. $\qquad h_6 = 45.7$ kJ/kg

The examples in Figs 5.27 and 5.28 show 100 % fresh air intake with no

Figure 5.29
Psychrometric chart for
Example 5.9

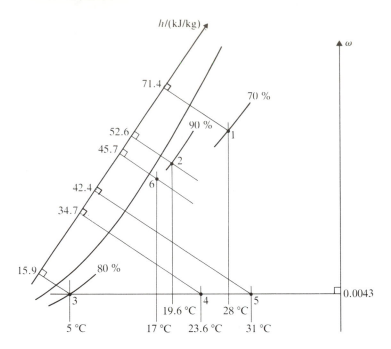

re-circulation of the exhaust air from the pool hall. This is recommended with
the type of swimming pool in which the water is chlorinated; when ozone is
used as a water disinfectant with de-humidification and filtration it is then
possible to re-circulate a proportion of the air thus making a considerable energy
saving. For swimming pools using chlorination it is sometimes claimed that
de-humidification and filtration can make re-circulation feasible but on health
grounds the case is not strong.

The systems referred to above are all air-to-air heat pump systems for heat
recovery. A more comprehensive system which includes heating the pool water
uses a water-to-air heat pump driven by a gas engine with heat recovery from
the engine. This is shown diagrammatically in Fig. 5.30.

Mechanical Vapour Re-compression

The use of a heat pump for heat recovery in cases where vapour is continuously
evaporated is becoming well-established. The vapour evaporated from the
process is compressed to a higher pressure and then condensed providing a
heating effect; the system is therefore known as Mechanical Vapour Re-
compression, MVR.

Examples of MVR applied to evaporation, distillation, and drying are shown
diagrammatically in Figs 5.31(a), (b) and (c). In the evaporator shown in
Fig. 5.31(a) a feed liquid is heated to evaporate water and produce a liquid
concentrate; the water vapour which is evaporated is compressed and passed
back to the evaporator where it is condensed by the evaporating liquid; the
heat pump is used as an alternative to the direct supply of steam to the evaporator
coil. In the distillation process shown in Fig. 5.31(b) the vapour evaporated in
the still is condensed to produce the liquid product; an indirect MVR heat
pump is used with the steam condensing in the still and evaporating in the

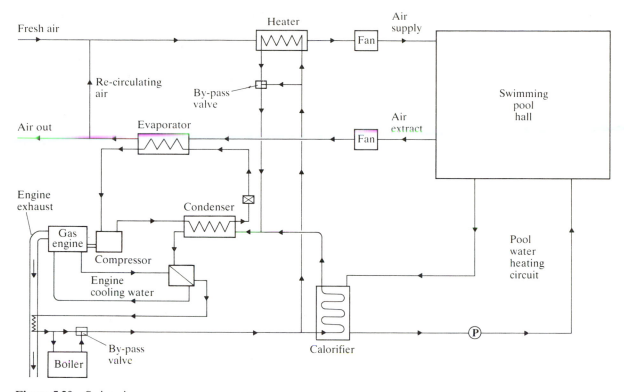

Figure 5.30 Swimming pool with gas-fired boiler and gas engine heat pump with heat recovery

Figure 5.31 Mechanical vapour recompression applied to evaporation, distillation and drying

distillation system condenser. A similar indirect MVR heat pump is shown for a drying process in Fig. 5.31(b); the vapour removed from the product being dried is passed through a heat exchanger where it condenses and thus evaporates the condensate from the heating steam in the drier; this steam is then compressed and passed to the drier.

Example 5.10
A rotary steam drier similar to that shown in Fig. 5.31(c) operates with 1 kg/s of steam entering the drier at 6 bar and condensing with no undercooling. The condensate is throttled to 0.7 bar before passing through the heat exchanger where it is evaporated leaving as a dry saturated vapour. Assuming that the

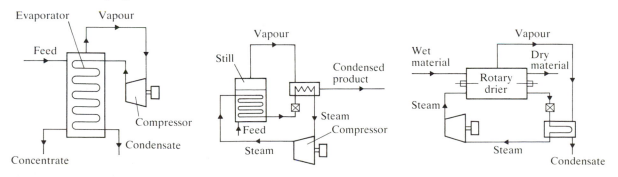

(a)   (b)   (c)

isentropic efficiency of the steam compressor is 0.65 and that the combined mechanical and electrical efficiency is 0.9, calculate:
(i)   the overall coefficient of performance of the heat pump;
(ii) the rate of cost saving compared with providing the heat for the drier with a gas boiler of 80 % efficiency; take the cost of electricity as 4 p/kW h and of gas as 1.3 p/kW h.

Solution

Figure 5.32   *T–s* diagram for Example 5.10

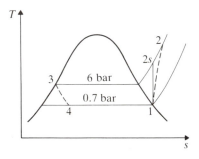

(i)   A *T–s* diagram for the steam cycle is shown in Fig. 5.32; steam enters the compressor at state 1 at 0.7 bar dry saturated, and leaves the drier at state 3 as saturated water at 6 bar. From steam tables

$$h_1 = 2660 \text{ kJ/kg} \qquad \text{and} \qquad s_1 = s_{2s} = 7.478 \text{ kJ/kg K}$$

Then interpolating from superheat tables

$$h_{2s} = 3062 + (3166 - 3062) \times \frac{(7.478 - 7.373)}{(7.546 - 7.373)} = 3125.1 \text{ kJ/kg}$$

and $$h_2 - h_1 = (3125.1 - 2660)/0.65 = 715.5 \text{ kJ/kg}$$

$$h_2 = 2660 + 715.5 = 3375.5 \text{ kJ/kg}$$

Also, the heating effect in the drier is

$$\dot{m}_s(h_2 - h_3) = 1 \times (3375.5 - 670) = 2705.5 \text{ kW}$$

and the electrical power input is

$$715.5 \times 1/0.9 = 795 \text{ kW}$$

Hence

$$\text{coefficient of performance} = 2705.5/795 = 3.4$$

(ii)   The cost of electricity is $795 \times 4/100 = £31.80$ per hour. Using a gas boiler at 80 % efficiency the cost of gas is $2705.5 \times 1.3/(0.8 \times 100) = £43.96$ per hour.

i.e.   rate of cost saving $= £43.96 - £31.80 = £12.16$ per hour

This is a percentage cost saving of $12.16 \times 100/43.96 = 27.7 \%$.

As with all forms of energy recovery a careful analysis of fuel saving against additional capital cost is necessary to determine the suitability of MVR for a particular case. Also, in a complex plant involving several processes it is important not to apply piecemeal solutions to energy recovery. In Chapter 6 the concept of Pinch Technology is introduced from which it will be seen that a heat pump may be counterproductive if it is not correctly placed within the overall system.

5.6 HEAT PIPES

A heat pipe is a device based on the concept of the thermosyphon. It consists of a sealed tube containing a fluid with a high enthalpy of evaporation. When the fluid is heated at one end of the sealed tube it evaporates, flows through the centre of the tube, condenses at the other end of the tube, and then passes back along an annular wick by capillary action to resume the cycle. A typical heat pipe is shown diagrammatically in Fig. 5.33. The heat pipe can be used in a horizontal position as shown but if the evaporator end is put below the condensing end then gravity will assist the liquid return.

Figure 5.33   Heat pipe

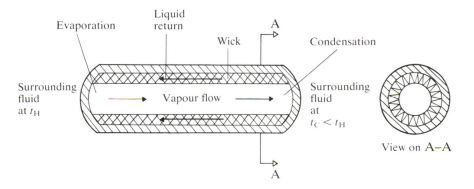

The heat pipe was invented in the 1950s but rapid development did not start until the 1970s when it was used in space technology. It is now used in a wide range of applications such as the cooling of electronic components, die casting and in the nuclear industry. Its use for heat recovery is not yet well-established in the UK although some heat pipe recovery systems have been in use since the late 1970s; in the USA heat pipes have been used to recover energy from hot waste gases, transferring heat via the heat pipe to water in a pressure vessel to raise steam.

The advantages of the heat pipe are its high heat transfer rate, robustness, and absence of pump or other moving parts; it can be used with fluids of different pressures and there is no risk of cross-contamination. The major disadvantage is its high capital cost.

A typical application in air-to-air heat recovery is shown in Fig. 5.34 where a battery of heat pipes are placed in cross-flow in two adjacent ducts. Since the heat pump performance is increased when the evaporator is below the condenser a tilting mechanism can be used to control the rate of heat transfer between the two fluids; a tilting angle of about 6° is used in air-to-air heat recovery systems for buildings.

For low temperature heat recovery systems suitable working liquids in a heat pipe are refrigerants such as ammonia, R11 and R113, as well as acetone. Where the source is at higher temperatures then water can be used, or for higher temperatures up to about 400 °C, Dowtherm A is used.

Figure 5.34 Bank of finned heat pipes for air-to-air heat transfer

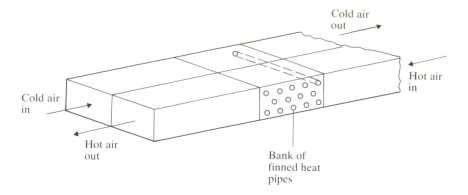

Dunn and Reay[5.8] define a Figure of Merit, M, for a heat pipe liquid as

$$M = \rho \sigma h_{fg}/\mu$$

(where ρ is the liquid density; σ is the surface tension; h_{fg} is the enthalpy of vaporization; μ is the viscosity of the liquid.)

M should be as high as possible for the most efficient operation at a given temperature range. Table 5.1 gives values of M for some typical heat pipe liquids at a normal working temperature. Various wick materials and pore designs are used depending on the fluid and the temperature range; compatibility between the fluid and the wick and the container are clearly very important. For ammonia and the freons, aluminium is suitable; for acetone, water and Dowtherm, copper can be used. Dunn and Reay[5.8] give a comprehensive coverage of practical design points.

Table 5.1 Figure of merit for some heat pipe fluids

Liquid	Mean temperature (°C)	M (10^{-6} kW/m^2)
Ammonia	20	70.24
R11	20	11.90
R113	20	6.78
Acetone	20	44.55
Water	20	177.89
Water	80	390.27
Dowtherm A	400	6.31

5.7 SELECTION OF ENERGY RECOVERY METHODS

The selection of an energy recovery system depends on the particular conditions of the plant. For a complex plant an analysis using pinch technology may be desirable before energy recovery is considered in detail (see Chapter 6).

Energy recovery may also be available from excess chemical energy as well as from excess thermal energy. In processes involving the production of combustible fumes as an unavoidable by-product it is necessary to render the products harmless in order not to pollute the atmosphere. This can be done using a thermal oxidizer (sometimes called fume incinerator), or with a catalytic converter. When a thermal oxidizer is used in conjunction with an energy

Figure 5.35 Heat recovery using a thermal oxidizer

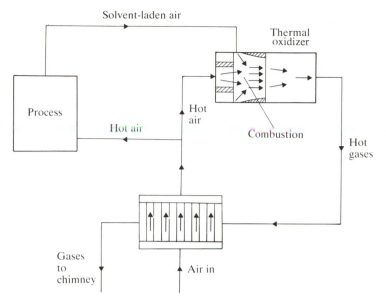

recovery system then the running costs are less and the pay-back period is shorter.

A typical example of a process in which combustible fumes are produced is a paint oven or a print baking oven. Figure 5.35 shows diagrammatically a general case of the use of a thermal oxidizer with heat recovery.

Figure 5.36 Decision chart for energy recovery

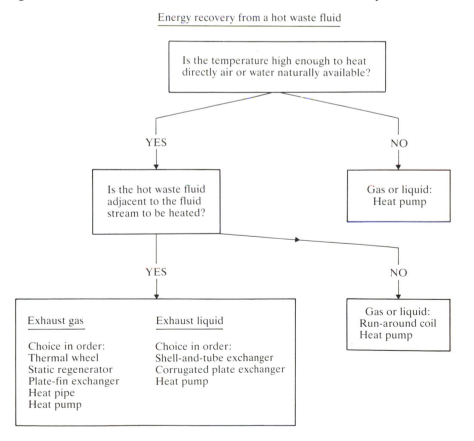

Energy recovery from high temperature exhaust gases from furnaces, ovens and other process plant has been in use for many years. The higher the temperature of the fluid going to waste the more obvious is the desirability of recovering some of the energy. Extending this type of energy recovery to a logical conclusion leads to Total Energy schemes (see Chapter 8). On the other hand considerable cost saving can be made by recovering energy from comparatively low temperature sources as some of the examples of this chapter have shown.

Choice of a heat recovery system for comparatively low temperature applications depends on a number of factors related to the plant itself – for example whether it is a new plant or an existing plant with fixed duct and pipe runs. A proper feasibility study and financial analysis must be undertaken before the final choice is made. A first step in the choice of a suitable system is given by the simple guide shown in Fig. 5.36.

Problems

5.1 A steel pipe of 60 mm external diameter and 53 mm internal diameter carries a gas flow at 150 °C through an air space at 10 °C. Using the data below, calculate:
(i)   a suitable thickness of glass fibre insulation;
(ii) the simple pay-back period for the insulation.

Data
Thermal conductivity of steel, 48 W/m K; thermal conductivity of glass fibre, 0.07 W/m K; heat transfer coefficient for inside surface of pipe, 100 W/m² K; heat transfer coefficient for outside surface of insulation, 10 W/m² K; the gas is heated by a boiler of efficiency 80 % burning gas at a cost of 1.3 p/kW h; plant in use for 3000 h per annum; insulation to be written off after five years; glass fibre thicknesses and costs: 19 mm at 476 p/m; 25 mm at 531 p/m; 32 mm at 632 p/m; 38 mm at 763 p/m; 50 mm at 1007 p/m.

(38 mm; 0.91 years)

5.2 A tank containing a hot liquid has an exposed surface area of 4 m² at a surface temperature of 80 °C. It is suggested that the surface should be insulated with a double layer of Allplas balls; tests at the National Engineering Laboratory have shown that the evaporation rate is reduced by 89 % when a double layer of balls is used at 80 °C. Using the data given calculate:
(i)   the annual cash saving by using the Allplas balls;
(ii) the simple pay-back period.

Data
Ambient air temperature, 15 °C; relative humidity of ambient air, 50 %; air velocity across tank surface, 2 m/s; thermal resistance of Allplas balls per layer, 0.1 m² K/W; heat transfer coefficient from surface of Allplas balls to ambient air, 5 W/m² K; cost of fuel used to heat tank, 0.4 p/MJ; cost per layer of Allplas balls, £30 per m²; average annual usage, 4000 h.

The heat loss per unit area, q, from the liquid surface is made up of the heat loss by evaporation plus the heat loss by convection, and is given as follows:

$$\frac{q}{[\text{W/m}^2]} = \left(9150 + 7760\frac{C_a}{[\text{m/s}]}\right)\frac{(p_g - p_s)}{[\text{bar}]} + 3.18\left\{\frac{C_a}{[\text{m/s}]}\right\}^{0.8}\frac{(t_w - t_a)}{[\text{K}]}$$

(where C_a is the air velocity; p_g is the saturation pressure of the liquid; p_s is the partial pressure of the water vapour in the ambient air; t_w is the water surface temperature; t_a is the ambient air temperature). Assume that the hot liquid has approximately the same relationship between saturation temperature and pressure as water.

(£2360.75; 1.22 months)

5.3 Air is heated by passing it across a bank of tubes containing condensing steam at 1.2 bar. The air mass flow rate is 12 kg/s and it enters the heater at 10 °C. The heat transfer coefficient on the steam side is very large, the thermal resistance of the tube walls is negligible, and fouling can be ignored. Using the data below, calculate:
(i) the exit temperature of the air when the tubes are unfinned;
(ii) the exit temperature of the air when the tubes are finned.

Data
Use Eq. [3.21] with the following properties for air:
   density, 1.01 kg/m³; viscosity, 2.08×10^{-5} kg/m s; thermal conductivity, 0.003 W/m K; specific heat at constant pressure, 1.01 kJ/kg.
Air duct size, 0.7 m by 0.7 m; tube outside diameter, 75 mm; number of tubes spanning duct, 5; number of tube rows, 10. Fins occupy half the base area in the finned tubes; fin area per tube, 0.7 m²; fin efficiency, 0.8; assume that the heat transfer coefficient for finned tubes is 80% of that for plain tubes.

(20.5 °C; 39 °C)

5.4 A shell-and-tube heat exchanger similar to the one shown in Fig. 5.6 is used to recover energy from waste water at 30 °C to heat fresh water entering at 15 °C. The mass flow rate of the waste water is 2 kg/s which is the same as that of the fresh water. Using the data given, calculate:
(i) the optimum rate of energy recovery;
(ii) the required heat transfer area;
(iii) the temperature of the fresh water at exit.

Data
Specific heat of waste water and fresh water, 4.2 kJ/kg K.
Overall heat transfer coefficient, 2500 W/m K.
For a single shell-pass, two tube-pass heat exchanger take the following characteristic when $R = 1$.

Effectiveness	0	0.46	0.53	0.56	0.57	0.58	0.58	0.58
NTU	0	1.0	1.5	2.0	2.5	3.0	3.5	4.0

(73.08 kW; 10.08 m²; 23.7 °C)

5.5 For the heat exchanger of Problem 5.4 choose a suitable nominal tube diameter and number of tubes assuming a water velocity in the tubes of about 0.3 m/s and a tube length per pass of 2 m. Take the density of water as 1000 kg/m³.

(10 mm; 80)

5.6 Air leaves a building at 20 °C at a rate of 1600 m³/h. It is proposed to recover some of the energy by using a plate-fin heat exchanger to exchange heat between the air leaving the building and the supply air. The air leaving the building passes through the heat exchanger before 50 % of it is recirculated to mix with the fresh air induced, the remainder being exhausted to atmosphere. The supply air leaving the mixing box passes across the heat exchanger before entering a heater where the air is heated to the required supply temperature to the building of 25 °C. For the heating season the mean outside temperature is 5 °C. For the data given, assuming only sensible cooling in the heat exchanger, calculate for the mean condition:

(i)   the effectiveness of the heat exchanger;
(ii) the temperature of the supply air entering the heat exchanger;
(iii) the temperature of the supply air leaving the heat exchanger;
(iv) the percentage saving in fuel cost over the heating season by using the plate heat exchanger;
(v)   the required rating of the external heater when the plate heat exchanger is used, assuming a design outside temperature of -1 °C.

Data
Heat transfer coefficient for each side of heat exchanger, 100 W/m^2 K; effective heat transfer area for each side of heat exchanger, 50 m^2; mean specific heat at constant pressure of air, 1.0043 kJ/kg K; density of air at 20 °C, 1.204 kg/m^3. For a cross-flow heat exchanger with both fluids mixed take:

NTU	1.0	2.0	3.0	4.0	5.0	6.0	7.0
Effectiveness	0.48	0.61	0.68	0.72	0.75	0.77	0.79

Assume the same percentage of air recirculation with and without the plate heat exchanger.

(0.745; 8.05 °C; 16.95 °C; 35.6 %; 5 kW)

5.7   In a particular process air is drawn in at 10 °C and leaves at 200 °C. The inlet and outlet ducts are some distance apart and heat recovery using a run-around coil is therefore suggested. The heat exchangers used in the inlet and outlet ducts are of the finned tube type with the secondary fluid flowing in counter-flow to the air. Using the data given, neglecting thermal losses, calculate:

(i)   the optimum mass flow rate for the secondary fluid;
(ii) the rate of heat recovery;
(iii) the temperatures of the heated air entering the process and the outlet air leaving the unit;
(iv) the effectiveness of the overall heat recovery;
(v)   the temperature of the secondary fluid at inlet and outlet of each heat exchanger.

Data
Mass flow rate of air, 0.5 kg/s; mean specific heat of air at constant pressure, 1.01 kJ/kg K; mean specific heat of secondary fluid, 1.37 kJ/kg K; overall heat transfer coefficient for each heat exchanger, 100 W/m^2 K; effective heat transfer area of each heat exchanger, 20 m^2.

(0.369 kg/s; 63.75 kW; 136.3 °C; 73.8 °C; 0.665; 168.1 °C, 41.9 °C)

5.8   Air enters a gas-fired furnace at 20 °C at a mass flow rate of 0.2 kg/s and is burned with an air–fuel ratio by volume of 12. The gases leave the furnace

at 350 °C. A run-around coil is installed to recover some of the energy of the exhaust gases in order to pre-heat the air entering the furnace. Using the data given, neglecting thermal losses, calculate:
(i) the required mass flow rate of secondary fluid;
(ii) the effectiveness of the overall heat recovery process;
(iii) the rate of energy recovery;
(iv) the temperature of the gases at exit and the temperature of the air at entry to the burner.

Data

Mean specific heat at constant pressure of air, 1.01 kJ/kg K; mean specific heat at constant pressure of gases, 1.15 kJ/kg K; density of air at intake, 1.204 kg/m³; density of gas at intake, 0.715 kg/m³; mean specific heat of secondary fluid, 1.6 kJ/kg K; overall heat transfer coefficient for each heat exchanger, 60 W/m² K; effective heat transfer area for each heat exchanger, 11 m².

(0.138 kg/s; 0.652; 43.45 kW; 170 °C, 235.1 °C)

5.9 A double accumulator similar to that shown in Fig. 5.22 is to be installed to recover energy from the air leaving a building. The air leaves the building at 20 °C at a rate of 2 kg/s and the mean outside air temperature for the heating season is 5 °C. Using the data below, assuming only sensible cooling, calculate:
(i) the effectiveness of the heat transfer process;
(ii) the rate of heat recovery;
(iii) the temperature of the fresh air leaving the accumulator;
(iv) the annual cash saving on fuel;
(v) the rate of heat recovery if the cycle time is increased to 50 s.

Data

Specific heat at constant pressure for air, 1.004 kJ/kg K; specific heat of matrix material, 0.9 kJ/kg K; mass of each matrix, 60 kg; effective heat transfer area of matrix, 251 m²; heat transfer coefficient between air and each matrix, 80 W/m² K; total cycle time, 30 s; length of heating season, 3000 h; cost of fuel to produce heated air, 1.4 p/kW h.
For a periodic flow heat exchanger use the following information:

NTU	1.0	2.0	3.0	4.0	5.0	6.0	7.0
Effectiveness when $(\dot{m}c)_M/(\dot{m}c)_{min} = 1.0$	0.47	0.60	0.67	0.71	0.74	0.76	0.78
Effectiveness when $(\dot{m}c)_M/(\dot{m}c)_{min} = 1.5$	0.49	0.64	0.71	0.76	0.79	0.81	0.83
Effectiveness when $(\dot{m}c)_M/(\dot{m}c)_{min} = 2.0$	0.49	0.65	0.73	0.78	0.81	0.83	0.85

(0.8; 24.1 kW; 17 °C; £1012.2; 22.5 kW)

5.10 A thermal wheel is to be used to recover energy from exhaust gases flowing at 6 kg/s at 250 °C. The air intake is at 15 °C at the same flow rate. Using the data given, calculate:
(i) the effectiveness using the tabular data given in Problem 5.9;
(ii) the effectiveness using the approximate formula given as Eq. 5.10;
(iii) the rate of heat recovery;

(iv) the temperature of the heated air leaving the wheel;

(v) the answers to (iii) and (iv) when the wheel speed is increased to 20 rev/min.

Data

Mean specific heat of both gases and air, 1.05 kJ/kg K; specific heat of matrix material, 0.4 kJ/kg K; mass of matrix, 94.5 kg; rotational speed of wheel, 15 rev/min; effective (UA) value for the wheel, 15 kW/K.

(0.667; 0.669; 987.5 kW; 171.8 °C; 1007.5 kW; 174.9 °C)

5.11 In a drying chamber a wet granular solid requires 5 m³/s of air at 40 °C. There are no heat losses from the chamber; the drying process takes place at constant wet bulb temperature with the air leaving with a percentage saturation of 70 %. The fresh air is at 10 °C with a percentage saturation of 40 %.

Currently the fresh air is heated from 10 °C to 40 °C by a gas-fired heater of efficiency 85 %. It is proposed to replace this with an electric heat pump with the condenser of the heat pump sited in the air stream entering the drier and the evaporator sited in the saturated air steam leaving the drier.

The refrigerant used is R12 with an evaporating temperature of 5 °C and a condensing temperature of 45 °C; vapour enters the compressor dry saturated and leaves the condenser as a saturated liquid; the vapour leaving the compressor is at 60 °C. The combined mechanical and electrical efficiency of the compressor is 90 %.

Assuming that the air pressure throughout is one atmosphere, neglecting heat losses, and using the psychrometric chart and the Tables of Rogers and Mayhew for R12,[5.7] calculate:

(i) the required mass flow rate of refrigerant;

(ii) the electrical power input to the motor;

(iii) the temperature of the air leaving the drying chamber;

(iv) the rate of water condensed in the heat pump evaporator if the air leaving the evaporator coils is at 8.5 °C;

(v) the fuel cost per unit time for each of the two methods of heating, assuming gas costs 1.4 p/kW h and electricity costs 4.5 p/kW h.

(1.244 kg/s; 37.42 kW; 21.2 °C; 90.8 kg/h; £2.39/h, £1.68/h)

5.12 The cellar of a large public house requires cooling due to the excessive heat generated by the beer chillers. Since there is a constant hot water demand simultaneous with the constant cellar cooling demand for 10 hours per day, it is proposed to install an air-to-water heat pump which will cool the cellar and supplement the existing hot water supply system as shown in Fig. 5.37. Using the data given and neglecting thermal losses, calculate:

(i) the electrical power input to the compressor;

(ii) the percentage reduction in fuel cost.

Data

Heat pump: refrigerant R11; enthalpy of vapour leaving compressor, 1010 kJ/kg; enthalpy of vapour entering the compressor, 962 kJ/kg; enthalpy of liquid refrigerant leaving the condenser, 822 kJ/kg; mechanical/electrical transmission efficiency, 94 %; cost of electricity, 4 p/kW h.

Cellar: volume flow of air entering evaporator, 3000 m³/h; temperature of air entering evaporator, 16 °C; temperature of air leaving evaporator, 10 °C; density of air at entry to evaporator, 1.24 kg/m³; mean specific heat of air at constant pressure, 1.011 kJ/kg K.

Figure 5.37 Plant layout diagram for Problem 5.12

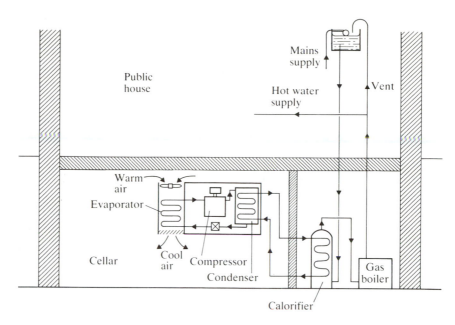

Gas boiler: efficiency, 80 %; daily gas consumption before installation of the heat pump, 29.5 m³; calorific value of gas, 32.6 MJ/m³; cost of gas, 1.3 p/kW h.
(2.26 kW; 13.32 %)

5.13 The air condition at extract from the pool hall of a chlorinated swimming pool is 27 °C, percentage saturation, 60 %; background heating is provided in the hall so that the ventilation air is supplied at 27 °C. The design condition for the outside fresh air is −1 °C, saturated.

Currently the fresh air is heated from the outside temperature to 27 °C by a gas-fired air heater. It is proposed to replace the existing heater with a heat pump energy recovery system; since the inlet and outlet ducts are spaced a distance apart the system chosen is similar to that shown in Fig. 5.27 using a run-around coil.

Using the additional data below and the psychrometric chart, neglecting thermal losses, calculate:
(i)   the air temperature entering the condenser coils;
(ii) the mass flow rate of refrigerant;
(iii) the total mass flow rate of condensation from the extract air when the air temperature leaving the evaporator coils is 13 °C;
(iv) the electrical input to the heat pump motor;
(v)   the percentage saving in fuel at the design outside conditions.

Data
Enthalpy of refrigerant leaving evaporator, 190 kJ/kg; enthalpy of refrigerant vapour leaving compressor, 209 kJ/kg; enthalpy of refrigerant liquid leaving condenser, 70 kJ/kg; mechanical/electrical transmission efficiency of compressor motor, 92 %; effectiveness of run-around coil, 0.6; ventilation rate at pool hall extract, 4.75 m³/s; efficiency of gas heater, 85 %.
(15.8 °C; 0.446 kg/s; 0.0268 kg/s; 9.21 kW; 95 %)

5.14 The wash still in a whisky distillery uses mechanical vapour re-compression as shown in the simplified diagram of Fig. 5.31(b). The vapour from the still is at 80 °C. Take a minimum temperature difference between the vapour leaving the still and the steam in the heat exchanger of about 20 K, and a minimum temperature difference between the steam in the coil and the liquid in the still also of about 20 K. Assuming a compressor isentropic efficiency of 0.65, and a combined mechanical and electrical transmission efficiency for the compressor motor of 0.9, choose suitable pressures for the steam circuit and calculate the COP of the heat pump.

(0.2 bar; 1 bar; 5.13)

References

5.1   Kays W M, London A L 1984 *Compact Heat Exchangers* 3rd edn McGraw-Hill

5.2   Emerson W H 1984 Making the most of run-around coil systems. *Heat Recovery Systems* **4** no 4: 265–70

5.3   Emerson W H 1983 Designing run-around coil systems. *Heat Recovery Systems* **3** no 4: 305–9

5.4   Bayley F J, Owen J M, Turner A B 1972 *Heat Transfer* Nelson

5.5   *Research and Development Digest* Issue no 8, British Gas 1985/86

5.6   Hamilton G, Kew J 1983 Heat Recovery and Heat Pumps in Buildings BSRIA TN 4

5.7   Rogers G F C, Mayhew Y R 1989 *Thermodynamic and Transport Properties of Fluids* 4th edn Basil Blackwell

5.8   Dunn P D, Reay D A 1982 *Heat Pipes* 3rd edn Pergamon Press

Bibliography

CIBSE *The Selection and Application of Heat Pumps* TM11 1985

CIBSE *Design Guidance for Heat Pump Systems* TM15 1988

Energy Efficiency Technologies for Swimming Pools Energy Efficiency Office, Energy Technology Series 3 1985

Energy Recovery in Process Plants IMechE Conference Proceedings, London 1975

Hamilton G 1986 *Selection of Air-to-Air Heat Recovery Systems* BSRIA TN 11

Heaton A V 1986 Compressed pay-back times with mechanical vapour re-compression. *Chartered Mechanical Engineer* April 1986

Heaton A V, Benstead R 1984 *Steam Recompression Drying* 2nd BHRA International Symposium on Large Scale Applications of Heat Pumps, York Sept. 1984

Reay D A 1977 *Industrial Energy Conservation* Pergamon Press

Sucec J 1985 *Heat Transfer* Wm C Brown

Wong H 1977 *Heat Transfer for Engineers* Longman

6 PROCESS INTEGRATION: PINCH TECHNOLOGY

The previous chapter discussed energy recovery, in which the monetary value of the energy savings was compared with the capital cost of achieving the saving. A simple example of this was the application of a recuperative burner using exhaust gas to preheat incoming air. The thermodynamic design of such a system is a straightforward matter because there is only one stream of hot gas donating heat and one stream of cold air absorbing the heat.

Many manufacturing industries use processes in which there are a large number of hot and cold streams; this makes the design of a total heat recovery scheme a more complicated affair with a large network of heat exchangers to consider.

For many years the approach to such networks was either by 'rule of thumb' or a systematic mathematical examination of all possible configurations to try to achieve the best layout. Both approaches led to good answers but neither claimed to generate the optimum solution because neither could identify the 'ideal' amount of heat recovery as a target to be achieved.

Another approach to network design has been developed and is gaining considerable popularity in process industries because the basic method allows the user to identify both the optimum heat recovery and the arrangement of heat exchangers which will achieve this recovery. The approach is given the name Process Integration, or Pinch Technology; it has been developed by Professor Bodo Linnhoff and his co-workers of ETH Zurich, Leeds University, ICI, and now UMIST.[6.1] The technique of Pinch Technology has been successfully applied to a number of process industries where energy costs represent a significant proportion of the total production cost; industries involving material transformation processes such as minerals, metals, chemicals, paper, food, drink and other organic products. In addition, the technique not only optimizes the heat recovery network but, by so doing, allows the user to determine the minimum external heating and cooling duties which are needed to complete the process demands.

Of course, cost is a major item with energy saving and the usual 'trade-off' between energy saving and capital costs applies to any heat recovery network, however well designed. Nevertheless, if an existing network is not well-designed then the application of Pinch Technology could result in the reduction of both capital and energy costs! As an example of this, examine the two circuits shown in Fig. 6.1 which are reproduced from Linnhoff *et al*[6.1] by permission of The Institution of Chemical Engineers.

Figure 6.1 (a) Initial
network design;
(b) Pinch Technology
design (reproduced by
courtesy of the Institute
of Chemical Engineers)

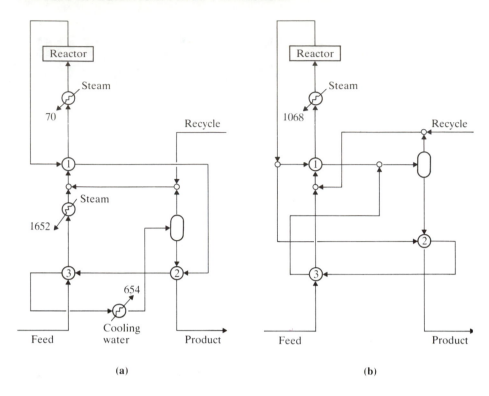

The two circuits represent diagrams for the flow of materials for part of a speciality chemicals process. In the traditional design (Fig. 6.1(a)), the heat exchanger and external utilities (heating and cooling) requirements are:

number of heat exchangers = 6 (including 2 for steam heating and 1 for
cooling water)
total external heating load = 1652 + 70 = 1722 units
total external cooling load = 654 units

For the alternative design (Fig. 6.1(b)) using Pinch Technology, the heat exchanger and external utility requirements are:

number of heat exchangers = 4 (including 1 for steam heating)
total external heating load = 1068 units
total external cooling load = zero

The alternative design requires fewer heat exchangers and reduced external utility loads. This has been achieved by increasing the amount of heat exchange between the various hot and cold streams. This means that, although there are fewer heat exchangers, the remaining ones are likely to be larger and therefore of greater capital cost. Nevertheless, the basic idea prevails: the minimum number of units means minimum cost.

6.1 BASIC CONCEPTS OF PINCH TECHNOLOGY

The circuit shown in Fig. 6.1 may at first sight seem complicated and therefore to introduce the basic ideas of Pinch Technology it is easier to examine a simpler

Figure 6.2 Heat exchanger temperature profiles

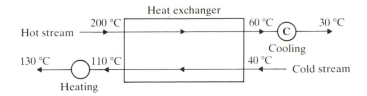

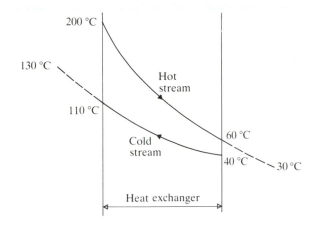

flow problem initially. Such an arrangement is shown in Fig. 6.2; a hot stream (stream 1) is required to be cooled from 200 °C to 30 °C whilst the cold stream (stream 2) requires to be heated from 40 °C to 130 °C. The two streams have different thermal properties and mass flow rates given by:

for the hot stream

$$C_{\mathrm{H}} = \text{mass flow rate} \times \text{specific heat} = 1 \text{ kW/K}$$

for the cold stream

$$C_{\mathrm{C}} = \text{mass flow rate} \times \text{specific heat} = 2 \text{ kW/K}$$

By using a counterflow heat exchanger, the energy exchange will help the streams to approach their desired final temperatures. We shall specify that the design of the heat exchanger is based on the minimum allowable temperature difference between the two streams being 20 K. Figure 6.2 shows the variation in temperature of the two streams across the exchanger.

Since the minimum temperature difference between the two streams is specified as 20 K then $t_{\mathrm{H2}} = 60\,°\mathrm{C}$. The heat exchanger duty can therefore be calculated from the heat lost by the hot stream, i.e.

$$\text{duty} = C_{\mathrm{H}} \times \text{temperature change} = 1 \times (200 - 60) = 140 \text{ kW}$$

Therefore the outlet temperature of stream 2 is

$$(\text{Duty}/C_{\mathrm{C}}) + 40 = 110\,°\mathrm{C}$$

The temperatures achieved in the heat exchanger are somewhat short of the desired values and therefore additional heating is required to raise stream 2 to its desired temperature and extra cooling is required by stream 1. The magnitude of the additional energy loads can be calculated as follows. The additional heating load for the cold stream is

$$C_C \times (\text{final desired temperature} - \text{temperature leaving exchanger})$$
$$= 2 \times (130 - 110) = 40 \text{ kW}$$

Similarly the cooling load required by the hot stream is

$$1 \times (60 - 30) = 30 \text{ kW}$$

The whole system of heat exchanger, external heating and cooling is represented in Fig. 6.3.

This system can also be represented on a graph as shown in Fig. 6.4. The horizontal axis of the graph corresponds to the rate of enthalpy change of the streams; the two streams are shown as straight lines drawn between the limits

Figure 6.3 Simple heat recovery scheme

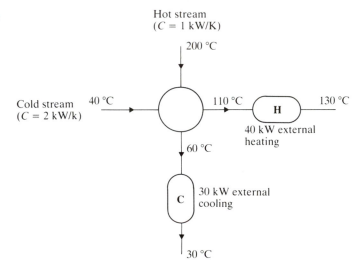

Figure 6.4
Temperature-heat load representation of heat recovery scheme

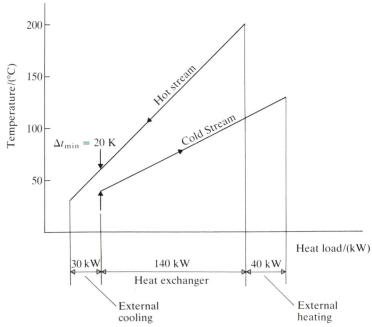

of their respective starting and finishing temperatures and having a gradient appropriate to the values of C_C and C_H. The two lines representing the streams are positioned so as to show a region of overlap which represents the action of the heat exchanger in transferring 140 kW. The hot stream overshoots the heat exchanger on the left-hand side of the graph; this means that the hot stream has left the heat exchanger above its desired final temperature and therefore requires cooling by external means to the extent of 30 kW. Similarly, the cold stream leaves the heat exchanger on the right-hand side of the graph below its desired final temperature and therefore requires external heating to the extent of 40 kW. The minimum temperature difference occurs where the two lines are nearest together – this point is called the *Pinch point*. A most important principle is illustrated by Fig. 6.4 – a principle it is vital to understand.

If we take the cold stream line as an example, then the slope of the line is determined by the value of C_C. The line must lie between the temperature limits of 40 °C and 130 °C because these temperatures are the stated inlet and outlet temperatures, but within these restrictions the line could be placed on the graph in any position such as shown in Fig. 6.5. Both of the lines A–A and B–B represent the behaviour of the cold stream. The position of each line is simply determined by how near it needs to be to the hot stream line to achieve the desired temperature difference at the Pinch. In Fig. 6.5 the cold stream line A–A is positioned to achieve a minimum temperature difference of 50 K, and line B–B to achieve a difference of 100 K. Figure 6.5 shows that the effect of increasing the Pinch temperature difference is twofold; the amount of heat exchange between the two fluids is reduced and the external duties are increased.

Exactly the same argument could be used for positioning the hot stream line. Thus we can say that the lines can be moved *horizontally* within the limits of temperature and gradient until the nearest points are separated by the minimum allowable temperature difference, that is the Pinch temperature difference.

Figure 6.5 Effect of 'moving' the cold stream

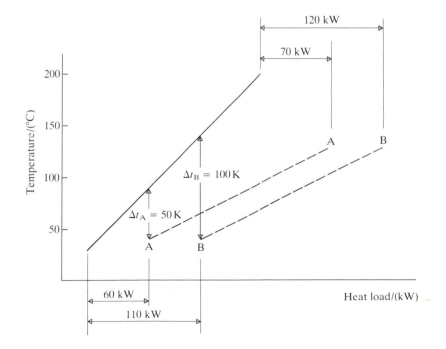

Figure 6.4 also illustrates another important point: to achieve the targets for the external heating and cooling duties, *there must be no external cooling of the hot stream above the Pinch*. If there is, then the amount of heat absorbed by the cold stream above the Pinch will be reduced, thereby increasing the external heating required.

A similar argument applies to the external heating of the cold stream below the Pinch: *there must be no external heating of the cold stream below the Pinch*. In this particular example, the cold stream starts its life at the Pinch itself and therefore could not be heated below the Pinch. In later examples, it will be seen to be a common occurrence for the cold stream to exist below the Pinch. As before, if this rule is disobeyed then the external cooling duty will be increased because of the reduced amount of heat transfer from the hot stream. This straightforward problem has been used to illustrate some of the basic ideas of Pinch Technology but to explain the points about circuit design we need a slightly more complicated network of fluid streams.

6.2 STREAM NETWORKS

We shall now consider the design of a system of heat recovery between two hot streams and two cold streams to illustrate some finer points of Pinch Technology.

Example 6.1

The heat flow capacities and temperatures of four streams are shown in the table below. For the purpose of definition, a hot stream is defined as one which requires cooling to reach its final temperature and a cold stream is one which requires heating to reach its final temperature. The minimum allowable temperature difference between the streams is 20 K.

Stream number	Type	Thermal capacity rate, C (kW/K)	Initial temperature (°C)	Final temperature (°C)	Rate of enthalpy increase ($C \times \Delta T$) (kW)
1	Hot	2	200	60	-280
2	Hot	4	170	70	-400
3	Cold	3	40	175	$+405$
4	Cold	4.5	100	150	$+225$
					-50

Solution

The negative sign opposite the heat loads of streams 1 and 2 is used to distinguish the different requirements of the cold streams to the hot, in that cold streams need heating whereas the hot streams need cooling to achieve their final temperatures. In total, the cooling requirements of the hot streams exceed the heating requirements of the cold streams by 50 kW.

The first step is to construct the curves for the temperature–enthalpy load ($t-H$) graph from which the Pinch conditions and the minimum external heating and cooling duties can be found. Although there are now four streams involved, the $t-H$ graph will again comprise two lines; one representing the combined behaviour of both hot streams and one representing both cold streams. In effect the data is simplified by combining streams into two composites; one for both the hot streams, the other for both cold streams. A composite stream is derived as shown in Fig. 6.6. For the two hot streams, for example, stream 1 exists between 200 °C and 60 °C and stream 2 between 170 °C and 70 °C. The two streams both exist in the temperature range 170 °C to 70 °C; the thermal capacity of the composite stream is therefore the sum of the two individual stream values. For the remainder of the temperature range (i.e. between 200 °C and 170 °C and between 70 °C and 60 °C) the composite stream will have the same properties as stream 1, as only this stream exists at these temperatures. The same logic applies to the evaluation of the cold stream composite and these calculations are also shown in Fig. 6.6.

The hot and cold stream composite curves are shown in Fig. 6.7. Unlike the previous example considered, the curves now display several changes of gradient,

Figure 6.6 Composite stream heat flow capacities

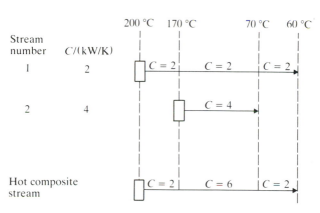

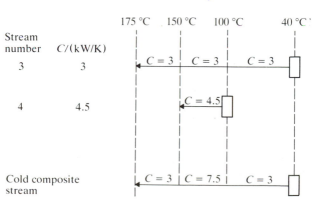

Figure 6.7 Composite curves

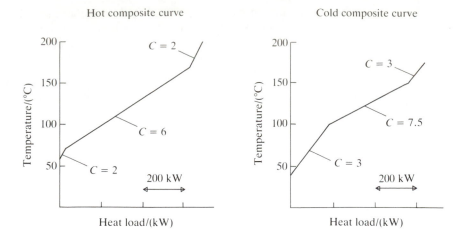

caused by the variations in thermal capacity. The two composite curves can now be placed on a single $t–H$ graph in such positions that the minimum temperature difference between the two curves is 20 K (as specified in the example). Figure 6.8 shows this condition.

Figure 6.8 Composite curves

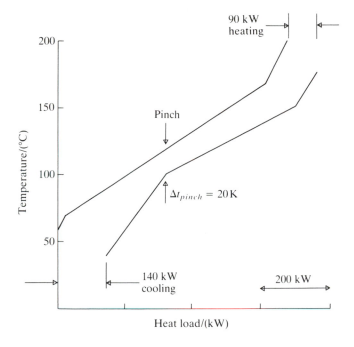

From the graph the following information can be derived:

hot stream temperature at the Pinch $= 120\,°\text{C}$
cold stream temperature at the Pinch $= 100\,°\text{C}$
target external heating load $= 90\;\text{kW}$
target external cooling load $= 140\;\text{kW}$

These figures confirm earlier calculations that the cooling load will exceed the heating load by 50 kW. The next step is to use the Pinch stream temperatures to design a system of heat recovery which will satisfy the targets for minimum external duties. Before doing that we need to pause and consider the thermodynamic significance of the Pinch as this will help to formulate another vital design criterion.

6.3 THE SIGNIFICANCE OF THE PINCH

In the previous example, by representing the streams as shown in Fig. 6.8, the lines have identified the region in which the hot streams exchange heat with the cold streams. Beyond the area of overlap the curves identify the need for external cooling (below the Pinch) and external heating (above the Pinch). Figure 6.8 can also be used to illustrate another major principle of Pinch Technology: *do not transfer heat across the Pinch*! Well, what happens if heat is transferred across the Pinch?

Imagine that an amount of heat Q is transferred across the Pinch from the hot stream above the Pinch to the cold stream below the Pinch. This will mean that the cold stream above the Pinch is not heated by an amount Q and so the external heating load will be increased by Q. In addition, the cold stream below the Pinch will provide less cooling (by an amount Q) to the hot stream below the Pinch and so the external cooling will also increase by Q. The minimum targets for external heating and cooling have now both been exceeded by an amount of heat Q. The only way that these minimum targets could be restored would be for heat to be transferred from the hot stream below the Pinch to the cold stream above the Pinch. This is not a permitted arrangement because the heat transfer would have to occur at a temperature difference between hot and cold streams of less than the Pinch temperature difference, which was stated to be the minimum. It is thus not possible to achieve the minimum external heating and cooling targets unless there is no heat transfer across the Pinch.

The question arises: what is the best Pinch temperature difference to choose? The short answer is that it is the one which leads to the lowest total cost. This aspect is covered in detail later in Section 6.5.

Summary

There are three rules which must be observed in the design of the optimum heat recovery scheme:

(a) no heat transfer across the Pinch;
(b) no external cooling above the Pinch;
(c) no external heating below the Pinch.

6.4 DESIGN OF ENERGY RECOVERY SYSTEMS

The design of an energy recovery system is based on the three basic rules previously stated *plus* a knowledge of the stream temperatures at the Pinch.

Example 6.2
The thermal capacity rates and temperatures of the four streams analysed in Example 6.1 are repeated together with the Pinch conditions. They will be used

Stream number	Type	Thermal capacity rate, C (kW/K)	Initial temperature (°C)	Final temperature (°C)
1	Hot	2	200	60
2	Hot	4	170	70
3	Cold	3	40	175
4	Cold	4.5	100	150

to illustrate the design procedure. Minimum allowable temperature difference between streams = 20 K.

Hot stream temperature at the Pinch   = 120 °C
Cold stream temperature at the Pinch = 100 °C
Target external heating load   = 90 kW
Target external cooling load   = 140 kW

Solution
The first step is to construct a design chart as shown in Fig. 6.9, where the streams are represented as horizontal lines drawn between their respective temperature limits and broken by vertical lines representing the Pinch. The temperatures of the streams at the Pinch will be 120 °C for the hot streams and 100 °C for the cold streams. The design process will therefore consist of linking the hot streams to the cold streams by heat exchangers. Rule (b) (Section 6.3) requires that both hot streams above the Pinch are cooled only by heat exchange with the cold streams (i.e. there is no external cooling here). Similarly, Rule (c) requires that, below the Pinch, the cold streams are heated only by heat exchange with the hot streams (i.e. there is no external heating here). Note that in Fig. 6.9 stream 4 is not represented below the Pinch because its initial temperature is 100 °C, which is the Pinch condition for the cold streams.

In Fig. 6.9, each stream is labelled with the rate of change of enthalpy between the temperature limits at the ends of each line. The amount is obtained simply by evaluating the product ($C \times \Delta t$). As an example, for stream 1, between 200 °C

Figure 6.9 Energy design chart

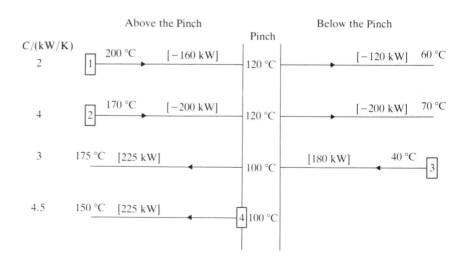

and the hot stream Pinch temperature of 120 °C the rate of increase of enthalpy is equal to

$$C_1 \times (120 - 200) = 2 \times -80 = -160 \text{ kW}$$

At first sight it seems that all that is required is to link the hot and cold streams with heat exchangers; whatever requirement that remains unsatisfied can be supplied by external heating or cooling. Before doing this we should consider one other limitation, which is imposed by the initial statement that the minimum allowable temperature difference between streams is 20 K.

Examination of Fig. 6.10 shows that if the thermal capacity rate of the hot stream (C_H) is less than that of the cold stream (C_C) then the minimum temperature difference will occur at the point where the hot stream leaves the heat exchanger and the cold stream enters. If on the other hand, C_H is greater than C_C the opposite will apply with the minimum temperature difference occurring at the point where the hot stream enters the heat exchanger and the cold stream leaves. When C_H is equal to C_C then the temperature profiles will be two parallel lines. It follows that if we wish to link a hot and cold stream *immediately adjacent to the Pinch*, the following rules should be followed:

$C_H \leqslant C_C$ above the Pinch
$C_H \geqslant C_C$ below the Pinch

These criteria will immediately limit the number of acceptable heat exchanger links between hot and cold streams. This limitation need not necessarily be applied to stream links some of whose temperatures are not at the Pinch; each heat exchanger will have to be checked with Fig. 6.10 in mind.

The correct design for the whole system, which obeys all the rules and limitations imposed by values of C_H and C_C is shown in Fig. 6.11. The duty of each heat exchanger link is shown below the link and the external duties (labelled C for cooling and H for heating) total 140 kW for the cooling and 90 kW for the heating.

If the reader has the feeling that there must be other arrangements which will satisfy all the rules then consider one alternative scheme for conditions below the Pinch as shown in Fig. 6.12. Examination of the temperature profiles in the two heat exchangers reveals that the one which links streams 1 and 3 has a temperature difference of zero at one end. This is clearly unacceptable. It is left to the reader to satisfy himself that if the heat exchanger which links streams 2

Figure 6.10 Heat exchangers at the Pinch

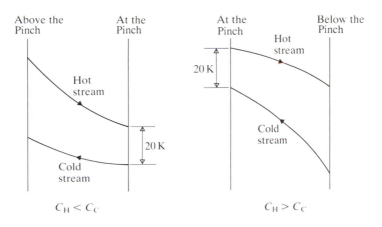

Figure 6.11 Energy design chart

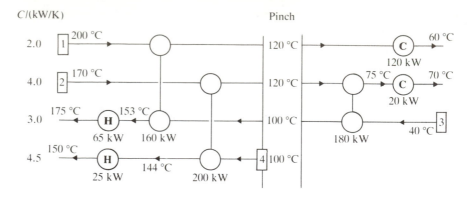

Figure 6.12 Incorrect design below the Pinch

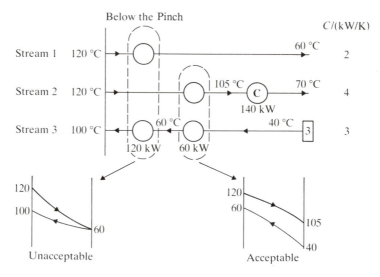

and 3 was placed immediately adjacent to the Pinch as shown in Fig. 6.11 then the arrangement would satisfy all the rules and limitations.

All that remains is to translate the correct design in Fig. 6.11 to a line diagram of the heat recovery network linking hot and cold streams. As an example, if we consider stream 3 in Fig. 6.11, then it links first with stream 2, then with stream 1 after which it is externally heated to its final temperature of 175 °C. Similarly, stream 2 first links with stream 4, then with stream 3 and is then externally cooled to its final temperature of 70 °C. Incorporating all the stream connections and external duties leads to the network design shown in Fig. 6.13. (Note: it is assumed throughout that all heat exchangers are the counterflow type.)

6.5 SELECTION OF PINCH TEMPERATURE DIFFERENCE

The heat recovery circuit of Fig. 6.13 certainly achieves the targets for the minimum external heating and cooling duties for a Pinch temperature difference of 20 K. If this difference is chosen as 40 K then the position of the composite curves can be adjusted to show that the external duties are 210 kW of cooling

Figure 6.13 Heat
recovery circuit

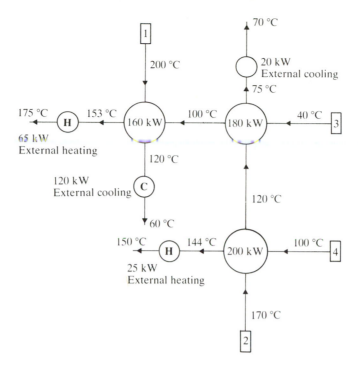

and 260 kW of heating; each duty has increased by 120 kW. The amount of
heat exchange between the streams is thus decreased by 120 kW, which in turn
implies a reduction in the total surface area of the heat exchangers. Herein lies
the problem of choosing the 'best' temperature difference for the Pinch: a small
value will reduce the external duties but incur the penalty of large heat
exchangers; a large value will require smaller heat exchangers but generate the
need for increased external duties. It thus becomes a matter of cost, with the
running costs of the system being weighed against the capital costs.

The running costs can be allocated to the following items:

(a) external heating;
(b) external cooling;
(c) pumping power;
(d) maintenance/repairs;
(e) labour.

The capital costs will derive from the following areas:

(a) heat exchangers for linking streams;
(b) heat exchangers for external heating and cooling;
(c) pumps and piping;
(d) instrumentation and control devices.

The total cost over a specified period will be given by (running costs + capital
costs).

The variation in total cost with Pinch temperature difference is likely to be
as shown in Fig. 6.14. The graph shows the predominance of capital costs at
low values of Pinch temperature difference and the greater influence of running
costs (particularly external heating and cooling duties) at higher values.

Figure 6.14 Location of optimum Pinch Δt

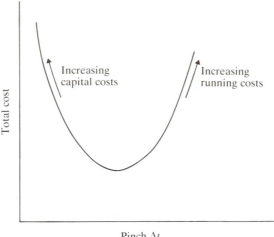

Pinch Δt

The type of curve shown in Fig. 6.14 is unlikely to appear so continuous for a real industrial problem due to the fact that the required number of heat exchangers varies with the Pinch conditions. Each heat exchanger will require certain fittings which make the total cost only partly a function of the surface area. This aspect is of interest to designers because the minimum number of heat exchangers (including the ones required for the external heating and cooling duties) corresponds to a large degree with minimum cost. Hohmann[6.3] developed an expression for the minimum number of heat exchangers for a simple system:

$$N_{min} = S - 1 \qquad\qquad\qquad [6.1]$$

(where N_{min} = minimum number of heat exchangers (including heaters and coolers) and S = total number of streams (including external cooling and heating)). This expression is applied separately to both sides of the Pinch.

As an example of the application of Eq. [6.1], we can analyse the design chart shown in Fig. 6.11, which has five streams (four process, one type of external heating) above the Pinch and four streams (three process, one type of external cooling) below the Pinch. Applying Eq. [6.1]:

above the Pinch

$$N_{min} = 5 - 1 = 4 \text{ exchangers}$$

and below the Pinch,

$$N_{min} = 4 - 1 = 3 \text{ exchangers}$$

A total of seven heat exchangers is therefore required. As can be seen from Fig. 6.11, the design contains four exchangers (including two for external heating) above the Pinch, and three heat exchangers (including two for external cooling) below the Pinch. This is in agreement with Eq. [6.1].

It is of interest to note that if Eq. [6.1] is applied to the *overall* system, then

$$N_{min} = 6 - 1 = 5 \text{ exchangers}$$

The implication of this is that fewer heat exchangers can be used if heat is transferred across the Pinch and increased external heating and cooling duties are acceptable. The decision as to which arrangement to use is a matter of cost; that is whether or not the value of the energy recovered exceeds the capital costs of extra heat exchangers.

6.6 TABULAR METHOD

The previous sections have discussed a graphical method for determining the conditions at the Pinch and the minimum external heating and cooling duties. It is equally possible to represent the composite curves and the way in which they are used by a tabular method; this method will also reveal the Pinch temperatures of the hot and cold streams and the minimum external duties.

When the composite curves were drawn in Fig. 6.8, both hot and cold composites contained several changes of gradient; each change occurred at a temperature which corresponded to the start or finish of a particular stream. The Pinch itself always occurs at a point where there is a change in gradient. By subdividing each composite into intervals of constant gradient it should be possible, by evaluating the heat flow from hot to cold stream in that interval, to determine in which interval this heat flow is zero. This interval must then contain the Pinch because of the basic rule about no heat flow across the Pinch for minimum external duties. To illustrate the tabular method we shall consider the same stream data as the previous problem so that the two methods can be seen to give exactly the same answers.

Example 6.3
The stream data are repeated here for convenience.

Stream number	Type	Thermal capacity rate, C (kW/K)	Initial temperature (°C)	Final temperature (°C)
1	Hot	2	200	60
2	Hot	4	170	70
3	Cold	3	40	175
4	Cold	4.5	100	150

Solution
The tabular method starts by representing the streams as vertical lines drawn between their initial and final temperatures as shown in Fig. 6.15. The temperatures indicated on the right-hand edge are the actual stream temperatures as shown in the above table. Since the minimum allowable temperature difference between hot and cold streams was stated to be 20 K then for Fig. 6.15 to be useful for heat flow calculations, a horizontal temperature line involving a hot and cold stream must somehow allow for the fact that the hot and cold streams must be at least 20 K apart. The easiest way to achieve this is to 'adjust' the stream temperatures; say by adding 10 K (10 K is half the minimum allowable difference of 20 K) to the cold stream temperatures and subtracting 10 K from the hot stream temperatures. This addition and subtraction process generates the column of 'adjusted' temperatures on the left-hand edge of Fig. 6.15. This

Figure 6.15 Stream
temperature adjustment

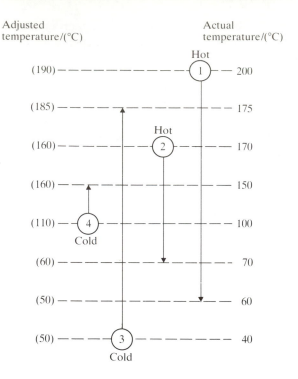

figure is then re-drawn with the streams now represented as vertical lines between 'adjusted' temperatures, as shown in Fig. 6.16.

It can be seen for example that stream 3 starts from the same horizontal temperature line (of 50 °C) as the one in which stream 1 finishes. Since the *actual* temperatures of streams 1 and 3 are 40 °C and 60 °C respectively, then the temperature difference of 20 K does permit heat transfer between the two streams and so it is appropriate to bring them together on the same horizontal

Figure 6.16 Adjusted
temperature intervals
and stream data

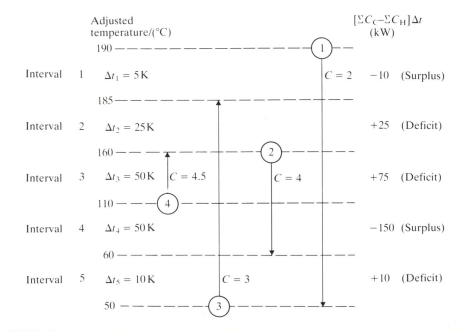

line of 'adjusted' temperature. One of the horizontal lines will represent the Pinch; the objective of the tabular method is to determine which line it is.

Figure 6.16 shows five intervals of temperature. Some intervals contain only one fluid stream (e.g. interval 1), others contain several streams (e.g. interval 3 contains all four streams). Starting with interval 2, we shall evaluate the rate of increase in enthalpy of all streams in the interval, indicating an increase in enthalpy of a cold stream as positive and that of a hot stream as negative. In fact, there is only one stream in interval 1, i.e. stream 1 (a hot stream). Therefore the rate of increase of enthalpy of stream 1 in interval 1 is

$$-C_1 \times \Delta t_1 = -2 \times (190 - 185) \text{ kW} = -10 \text{ kW}$$

There is a decrease in enthalpy which therefore represents a *surplus* of energy available for heat transfer.

Repeating this type of calculation for interval 2, which contains stream 1 and stream 3 gives the rate of increase in enthalpy of both streams in interval 2 as

$$(C_3 - C_1) \times \Delta t_2 = (3 - 2) \times (185 - 160) \text{ kW} = +25 \text{ kW}$$

The positive sign indicates that there is a net increase in enthalpy and there is thus a *deficit* of energy requiring a transfer of heat.

For interval 3 the rate of increase of enthalpy of all streams in interval three is

$$\{(C_3 + C_4) - (C_1 + C_2)\} \times \Delta t_3$$
$$= \{(3 + 4.5) - (2 + 4)\} \times (160 - 110) \text{ kW} = +75 \text{ kW}$$

The positive sign again indicates a *deficit* of energy. The calculation is repeated for intervals 4 and 5, giving a surplus of -150 kW for interval 4 and a deficit of $+10$ kW for interval 5. All the surpluses and deficits are shown in the right-hand column in Fig. 6.16.

Starting again with interval 1, we have a surplus of energy; this is energy not absorbed by cold streams in the interval and hence available to be passed to the next interval. This idea of transferring surplus energy from one interval to the next is always possible because the surplus energy is always at a higher temperature than the next interval. By the same reasoning, it would *not* be possible to transfer a deficit to the next interval because of the adverse temperature gradient; the only way to offset the deficit is to supply enough external heating to reduce the deficit to zero.

Returning to interval 1, there is a surplus of 10 kW of energy which is passed to interval 2. This interval has a deficit of 25 kW, which, together with the surplus of 10 kW from interval 1, generates a net deficit of 15 kW. This deficit is cancelled by the application of 15 kW of external heating; zero deficit is now passed to interval 3. In turn, interval 3 has a deficit of 75 kW, which together with zero deficit from the previous interval generates a net deficit of 75 kW. Again, this deficit is cancelled by the application of 75 kW of external heating and thus zero deficit is passed to interval 4. This interval has a surplus of 150 kW which, together with the zero deficit from the previous interval, will generate a net surplus of 150 kW which can be passed to interval 5. Since interval 5 has a deficit of 10 kW, the addition of a 150 kW surplus from interval 5 will generate a net surplus of 140 kW. This surplus has no other interval below it and the 140 kW is therefore removed by external cooling. The external heating was used in two intervals requiring inputs of 15 kW and 75 kW; the total external heating duty is therefore 90 kW. We have thus determined the minimum

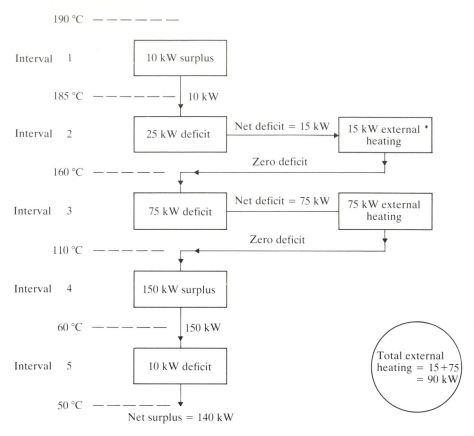

Figure 6.17 Deficit and surplus heat loads

external heating and cooling duties to be 90 kW and 140 kW respectively. This whole process of surpluses and deficits is shown in Fig. 6.17.

The Pinch is determined by passing the total external heating duty of 90 kW to interval 1 and then repeating the calculation of surpluses and deficits for each of the intervals. Referring to Fig. 6.18, interval 1 has a surplus of 10 kW which, together with the 90 kW from the external heating, will generate a net surplus of 100 kW, which is passed to interval 2. The deficit of 25 kW in interval 2 will reduce the net surplus to 75 kW which is passed to interval 3. The deficit in interval 3 is 75 kW which is exactly cancelled by the surplus passed from the previous interval. There is now zero surplus (or zero deficit) passing from interval 3 to interval 4. This condition of zero heat flow is that of the Pinch – the 'adjusted' temperature which is common to intervals 3 and 4 will therefore represent the Pinch. From Fig. 6.18 this temperature is seen to be 110 °C.

By reversing the process by which the 'adjusted' temperatures were obtained, we can now determine the actual hot and cold stream temperatures at the Pinch as

$$110 - 10 = 100\,°C \quad \text{and} \quad 110 + 10 = 120\,°C$$

respectively. Further consideration of the net surpluses of intervals 4 and 5 will again reveal the external cooling duty of 140 kW. The tabular method thus generates exactly the same information as the graphical method but has the obvious advantage that it lends itself much more easily to a computer algorithm.

Figure 6.18 Final table for Pinch determination

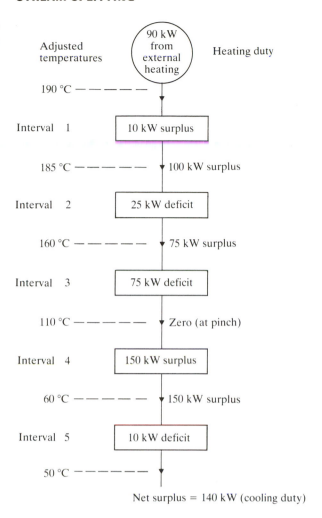

Adjusted temperatures

90 kW from external heating

Heating duty

190 °C — — — — —

Interval 1   10 kW surplus

185 °C — — — — ↓ 100 kW surplus

Interval 2   25 kW deficit

160 °C — — — — ↓ 75 kW surplus

Interval 3   75 kW deficit

110 °C — — — — ↓ Zero (at pinch)

Interval 4   150 kW surplus

60 °C — — — — ↓ 150 kW surplus

Interval 5   10 kW deficit

50 °C — — — — — ↓

Net surplus = 140 kW (cooling duty)

6.7 STREAM SPLITTING

In the previous section a number of rules and guidelines were stated which facilitate the design of the target energy recovery system. Occasionally it appears impossible to link streams because one or other of the design criteria cannot be satisfied.

Example 6.4

The energy design chart (Fig. 6.19) shows conditions above the Pinch with two hot streams supplying energy to one cold stream. The thermal capacities and temperatures of the streams are detailed in Fig. 6.19.

Solution

The stream data reveals that the two hot streams (1 and 2) have a total available heat load of $(300 + 400) = 700$ kW, exactly the amount which is required by the cold stream (3). At first sight it seems a straightforward matter to link streams 1 and 3, and then streams 2 and 3. In this way the heat loads are matched and the criterion that for above the Pinch, C_H is less than or equal to

Figure 6.19 Incorrect exchange above the Pinch

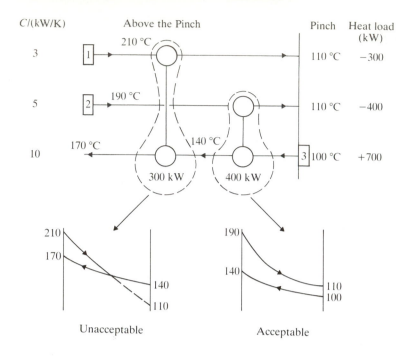

C_C is satisfied for all streams. Unfortunately, the arrangement is unacceptable because the temperature profiles in the heat exchanger which links streams 1 and 3 contravene the Laws of Thermodynamics. The other heat exchanger is operating with the correct minimum temperature difference of 10 K and is therefore acceptable.

The only solution to the problem is to split stream 3 into two separate streams. The thermal capacity rate of each sub-stream must be so chosen as to satisfy the criterion that C_H is less than or equal to C_C for each heat exchanger link. A suitable arrangement is shown in Fig. 6.20. The choice of thermal capacity rate for each sub-stream is influenced by the requirement to have as few heat exchangers as possible to reduce cost. Obviously the combined thermal capacity rate of the sub-streams must equal the thermal capacity rate of the original single stream; sub-streams of capacity 4.0 and 6.0 kW/K would equally satisfy this requirement. If these had been chosen in preference to the ones in Fig. 6.20 then the design would have required three heat exchangers to achieve the energy

Figure 6.20 Stream-splitting to obtain exchanger links

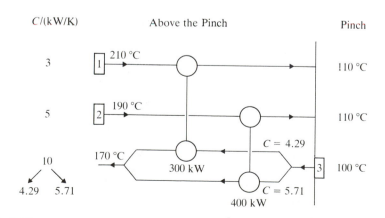

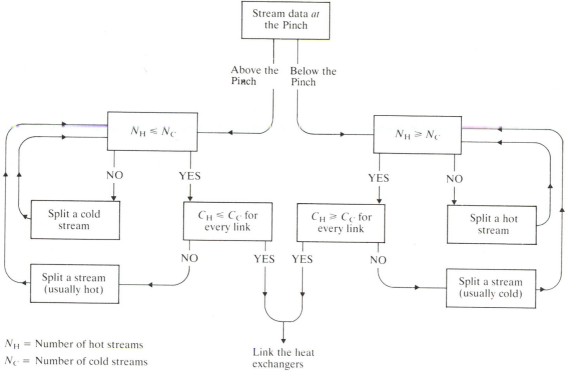

N_H = Number of hot streams

N_C = Number of cold streams

Figure 6.21 Decision diagram for stream splitting

transfer, whereas the suggested design requires only two. This aspect of system design is a matter of thoughtful trial and error, but the decision diagram shown in Fig. 6.21 will be helpful in identifying the need to split streams.

6.8 PROCESS RETROFIT

The previous material has usually exemplified the Pinch Technology approach by specifying the stream temperatures and thermal capacity flow rates and designing new circuits which adhere strictly to the Pinch rules (Section 6.3), and hence produce targets for the external duties.

In this section, we shall consider an existing recovery system. The basic approach will be to consider the stream data, evaluate the Pinch conditions, and then examine the existing circuit to see how well it compares to the Pinch rules and targets. The capital cost of completely replacing the existing circuit with the 'ideal' Pinch circuit, and the running costs which this circuit incurs will then be evaluated. A further assessment is made, based on the idea of using as many of the existing exchangers as possible as well as some extra ones to reduce the external duties. This assessment will quantify the extra capital costs and the savings in running costs over the existing scheme. In effect, this is a situation where a company may wish to retrofit additional heat exchangers to achieve moderate energy savings rather than spend a great deal more capital to achieve much bigger savings by replacing a whole system. Such a retrofit may not *exactly* produce a circuit which obeys all the design rules of Pinch Technology but may for a relatively small capital outlay achieve worthwhile savings.

For further discussion on process retrofit see Tjoe and Linnhoff.[6.2]

Case Study 6.1

An existing system is shown in Fig. 6.22 with the stream data as summarized in the table, where the stream heat transfer coefficients are specified so as to permit the calculation of heat exchanger areas. The cost data which will be used for the calculation is:

external heating cost $= £0.04/kW\,h$
external cooling cost $= £0.02/kW\,h$
heat exchanger cost $= £\{9000 + 700(\text{area in m}^2)^{0.8}\}$

The running costs will be based on continuous running, that is a yearly total of $(24 \times 365) = 8760$ hours.

Stream data

Stream number	Type	Thermal capacity rate, C (kW/K)	Initial temperature (°C)	Final temperature (°C)	Surface heat transfer coefficient (W/m² K)
1	Hot	3	350	110	250
2	Hot	1.2	260	80	350
2	Hot	9	160	90	200
3	Cold	12	105	210	130
4	Cold	5	30	140	450

Figure 6.22 Existing recovery scheme

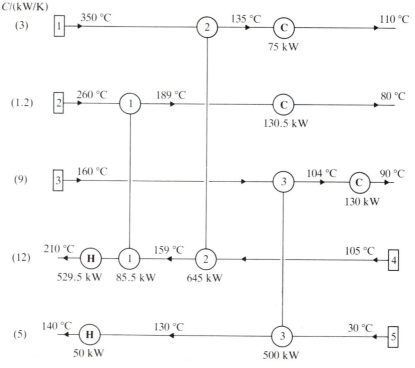

Solution

EXISTING SCHEME The first step is to calculate the areas of the heat exchangers so as to provide later calculations with data which will allow a decision to be taken on which heat exchangers to retain. As an example of the area calculation, we shall select heat exchanger 2, which links streams 1 and 4. The exchanger is shown in Fig. 6.22 as having a duty of 645 kW. The overall heat transfer coefficient U_2 (neglecting tube resistance) is calculated from

$$\frac{1}{U_2} = \frac{1}{h_1} + \frac{1}{h_4} = \frac{1}{250} + \frac{1}{130}$$

i.e. $\qquad U_2 = 85.5 \text{ W/m}^2 \text{ K}$

The logarithmic mean temperature difference is equal to

$$\frac{(350 - 159) - (135 - 105)}{\ln(191/30)} = 87 \text{ K}$$

The heat exchanger duty $= U_2 \times A_2 \times (LMTD)$. Therefore

$$645 \times 1000 = 85.5 \times A_2 \times 87$$

i.e. $\qquad A_2 = 86.71 \text{ m}^2$

Similar calculations for A_1 and A_3 give $A_1 = 16.08 \text{ m}^2$ and $A_3 = 73.79 \text{ m}^2$.

The running costs are based on continuous running, that is a yearly total of 8760 hours. Therefore

$$\begin{aligned}
\text{external heating cost} &= \text{external heating duty} \times 8760 \times \text{tariff} \\
&= (529.5 + 50) \times 8760 \times 0.04 \\
&= £203\,057 \text{ per annum}
\end{aligned}$$

Similarly

$$\begin{aligned}
\text{external cooling cost} &= (75 + 130.5 + 130) \times 8760 \times 0.02 \\
&= £261\,837 \text{ per annum}
\end{aligned}$$

PINCH DESIGN Using the stream data of the table given, the Pinch conditions and targets for a minimum allowable temperature difference between streams of 30 K are:

hot stream temperature at the Pinch $\quad = 135\,°\text{C}$
cold stream temperature at the Pinch $= 105\,°\text{C}$
external heating duty $\qquad\qquad\qquad = 415 \text{ kW}$
external cooling duty $\qquad\qquad\qquad = 171 \text{ kW}$

The design chart for this arrangement is shown in Fig. 6.23. The areas of the heat exchangers and their installed capital costs are detailed in the table below.

Heat exchanger and capital costs

Heat exchangers	P_1	P_2	P_3	P_4	P_5
Area (m^2)	26.88	95.19	122.76	16.83	60.71
Cost (£)	18 742	35 790	41 835	15 698	27 694

Therefore

$$\text{total capital cost} = £139\,759$$

and

$$\begin{aligned}
\text{total running cost} &= (415 \times 8760 \times 0.04) + (171 \times 8760 \times 0.02) \\
&= £175\,375 \text{ per annum}
\end{aligned}$$

Figure 6.23 Pinch design chart

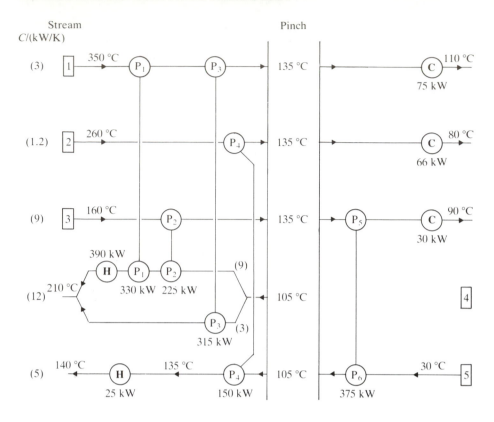

The above calculation is based on a minimum allowable temperature difference between streams of 30 K. For the purpose of this example, this is presumed to be near the optimum value. Space permitting, we should examine a number of temperature differences to determine the true thermo-economic value.

RETROFIT DESIGN   First, knowing the Pinch temperatures of the hot and cold streams (135 °C and 105 °C respectively), we can represent the existing circuit shown in Fig. 6.22 on a design chart to see how closely the circuit performs relative to the ideal Pinch design. This is shown in Fig. 6.24. The figure shows that the circuit disobeys two rules of Pinch design:

(a) there is an external cooling operation *above* the Pinch on stream 2;
(b) heat exchanger 3 transfers energy *across* the Pinch.

It may not be possible to retain all of the exchangers in their current positions and also avoid cooling above the Pinch, but it may be possible to substitute or add exchangers to achieve more energy recovery and hence obtain enough savings to offset the extra capital cost in a reasonable payback time. Figure 6.25 shows an attempt to do this. The retrofit design still has some cooling above the Pinch and requires more heat exchangers, whilst retaining exchangers 1 and 2 from the existing circuit. For the moment we shall assume that exchanger 3 has been discarded. It may be possible, by evaluating the areas required for the new exchangers 4, 5 and 6, to use exchanger 3 in a different position.

Evaluating the areas, assuming that the stream heat transfer coefficients apply as before, even if the stream is split, gives the following table with relevant exchanger areas and capital costs.

Figure 6.24 Existing scheme related to Pinch conditions

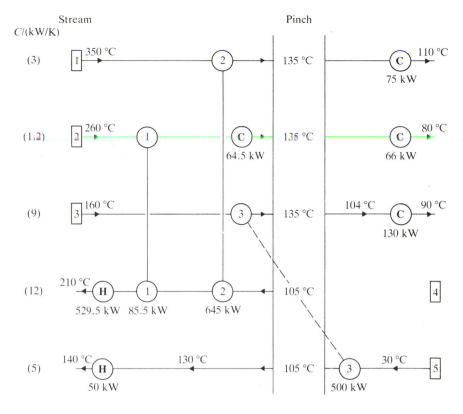

Figure 6.25 Modified circuit

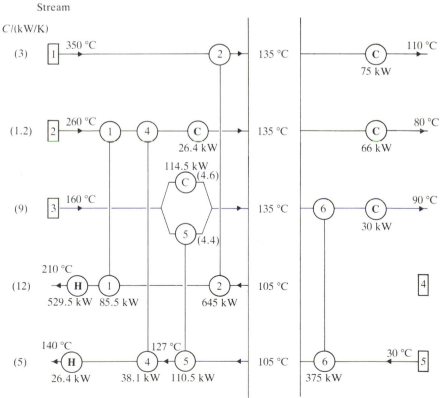

Heat exchanger areas
and capital costs

Heat exchangers	4	5	6
Area/(m^2)	5.72	25.35	60.71
Cost/(£)	11 825	18 296	41 835

Examining the areas reveals that exchanger 6 requires an area ($60.71 \, m^2$) – somewhat less than the one it replaced, i.e. exchanger 3 (area $73.79 \, m^2$). We will specify that the company decides to modify exchanger 3 (either by blocking-off tubes in a tube and shell type or by removing plates in a plate type exchanger) to achieve the area required by exchanger 6. The total capital cost is now only that due to the introduction of exchangers 4 and 5. Therefore,

$$\text{total capital cost} = 11\,825 + 18\,296 = £30\,121$$

$$\text{Total external heating duty} = 529.5 + 26.4 = 555.9 \, \text{kW}$$

and

$$\text{total external cooling duty} = 26.4 + 75 + 66 + 30 = 311.9 \, \text{kW}$$

Therefore

$$\text{total running cost} = (555.9 \times 8760 \times 0.04) + (311.9 \times 8760 \times 0.02)$$
$$= £249\,432 \text{ per annum}$$

Summary: Costs and Payback Times

Costs and payback
times

Scheme	Capital costs (£)	Annual running costs (£)	Savings in running costs over existing scheme (£)	Payback time (years)
Existing	—	261 837	—	—
Pinch	139 759	175 375	86 462	1.62
Retrofit	30 121	249 432	12 405	2.43

On these figures, the decision is not a straightforward choice of the lesser payback time as the level of investment and therefore the associated interest charges are far greater for the 'total replacement' Pinch scheme. A smaller level of investment achieves significant savings albeit over a longer payback time. Much depends on the relative importance of energy costs and the energy-intensiveness of the process and therefore the importance attached to energy conservation in the company. This point is discussed in further detail in Chapter 9.

6.9 INSTALLATION OF HEAT PUMPS

Heat pumps are a well-proven method for transferring heat from cold to hot environments; Pinch Technology will guide the designer on the most effective placement of heat pumps in heat recovery scheme. The following discussion details the underlying logic.

Figure 6.26
(a) Condensing below
the Pinch;
(b) condensing above
the Pinch

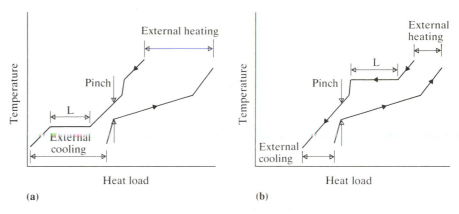

(a)   (b)

In the composite curves shown in Fig. 6.26(a) there appears a horizontal section 'L' below the Pinch which represents the condensation of a vapour at constant temperature. If the pressure, and hence the saturation temperature, could be increased the horizontal section would then appear above the Pinch as shown in Fig. 6.26(b). The effect of moving the condensation process is twofold:

(a) the hot stream duty below the Pinch is decreased thereby reducing the external cooling requirement;
(b) the hot stream duty above the Pinch is increased thereby reducing the external heating requirement.

If this argument is applied to the cold stream, then a horizontal section would represent evaporation at constant temperature. Moving such a process from above to below the Pinch would simultaneously reduce the external heating and cooling duties. These phenomena illustrate a general principle which is of particular relevance to the installation of heat pumps. External energy targets *decrease* if:

(a) hot stream duties are increased above the Pinch;
(b) hot stream duties are decreased below the Pinch;
(c) cold stream duties are decreased above the Pinch;
(d) cold stream duties are increased below the Pinch.

If we include a heat pump in the energy exchange process described by the composite curves shown in Figs 6.26(a) and (b), then the following considerations will apply. The heat pump will receive an amount of heat Q_1 at a low temperature t_1 in the evaporator. The heat pump will then have an input of power W, and will reject Q_2 units of heat at a higher temperature t_2.

$$Q_2 = Q_1 + W \qquad \text{by energy balance}$$

The effect of the heat pump evaporator will be to increase the cold stream duty below the Pinch, because the working fluid can be considered as an additional cold stream. If temperature t_1 is below the Pinch then this is compatible with the criterion: external energy targets *decrease* if cold stream duties are *increased* below the Pinch. Similarly, the heat rejected at temperature t_2 represents a new hot stream duty by an amount Q_2 because the working fluid can now be considered as an additional hot stream. If t_2 is above the Pinch then this is compatible with the criterion that: external energy targets are *decreased* if hot stream duties are *increased* above the Pinch.

Therefore to be appropriately placed, the heat pump must straddle the Pinch, that is, it must pump energy across the Pinch from the cold stream to the hot stream.

6.10 INSTALLATION OF HEAT ENGINES

A heat engine accepts heat from an external source, converts some of the heat to work and finally rejects heat to an external reservoir, usually the atmosphere. The appropriate placement of the heat engine is based on the concept that the external energy targets decrease if either hot stream duties are increased above the Pinch or cold stream duties are increased below the Pinch.

If it could be arranged for a hot stream from a process to be a source of heat input to the engine then this could have the double benefit of reducing the fuel input to the engine as well as reducing the need for external cooling of the process fluid. This situation would apply if the hot stream temperature was below the Pinch where the engine working fluid can be envisaged as an extra cold stream requiring heat input. The net effect would be to reduce the external cooling target.

Similarly, if we consider the exhaust heat from the engine, then by using this energy to heat the cold stream above the Pinch, the exhaust gas, which is being cooled, can be considered as an extra hot stream above the Pinch. The net effect here would be to reduce the external cooling requirement.

Therefore to be appropriately placed, the heat engine must either be above the Pinch donating exhaust energy to the cold stream, or below the Pinch accepting energy from the hot stream.

What happens if the heat engine operates across the Pinch? Remember that the temperature achieved by the energy input must exceed the exhaust temperature for the heat engine to produce work. Since the correct placement of the device requires the exhaust energy to be donated above the Pinch or the energy input below the Pinch, then it is clearly not feasible for the heat engine both to accept and donate energy to the same composite streams. That is to say, it cannot operate across the Pinch.

Installation of CHP (Cogeneration) Plant

The previous reasoning on the placement of heat engines can be used to make a similar decision about the placement of the heat available from CHP plant. We can say immediately that the exhaust fluid from the prime mover can be considered as part of the hot composite stream and therefore its energy must be supplied above the Pinch. This follows from the previous criterion: that external energy targets *decrease* if hot stream duties are *increased* above the Pinch. Any exhaust heat that is supplied below the Pinch will increase the external cooling duty. Only by supplying heat above the Pinch will the external heating duty be reduced.

Further comments on the suitable placement of plant is made in Section 6.11.

6.11 THE GRAND COMPOSITE CURVE

Most of the preceding material has been centred on the design of the heat recovery system; the external heating and cooling systems have been simply expressed as heat loads of a certain magnitude. Pinch Technology can help to evaluate the suitability of different types of heating and cooling systems by using what Linnhoff[6.1] calls 'the grand composite curve'. This curve may be plotted directly from the information in Section 6.6 about the tabular method for evaluating the Pinch conditions, in particular the table of surpluses shown in Fig. 6.18. For ease of reference, the information contained in Fig. 6.18 is repeated in Fig. 6.27. This figure shows that the grand composite curve is obtained by plotting energy surplus against adjusted temperature; the Pinch appears at the position of zero surplus at the adjusted temperature of 110 °C.

In one sense the grand composite curve above the Pinch represents the ever-increasing demand of the cold composite stream for external heating. It would follow from this statement that the heat load should continually increase from the Pinch up to the upper limit of 90 kW at the adjusted temperature of 190 °C. Examination of Fig. 6.27 shows that the curve reverses gradient at the shaded area marked A, which is inconsistent with the concept of a gradually increasing load. At the right-hand extreme point area A, the heat load is 100 kW. Yet it is known that only 90 kW of external heating is required; there

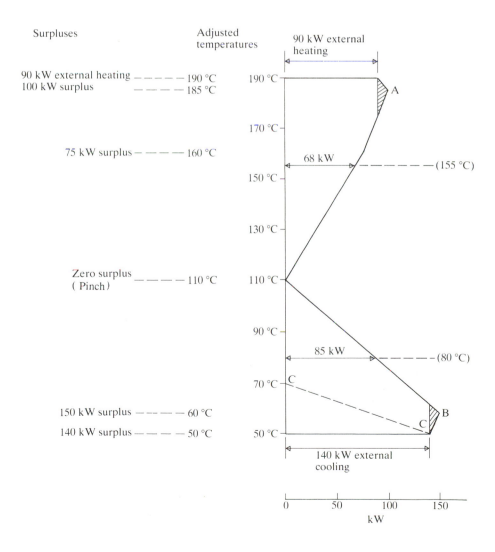

Figure 6.27 Grand compositive curve

appears to be an excess of 10 kW of heating. This apparent excess is nullified within the fluid itself and does not feature in the assessment of the external heating duties. A similar effect appears below the Pinch at the shaded area B.

In that the target external heating and cooling duties are always 90 kW and 140 kW respectively and are never exceeded, the shaded areas of Fig. 6.27 can be ignored so far as the placement of extra utilities is concerned.

The curve emphasizes the point that the design of a heat recovery system is really the sum of two separate designs: one for conditions above the Pinch, the other for below the Pinch. Above the Pinch, the grand composite curve shows how the demand for external heating increases and at what temperature the heating would need to be supplied. Similarly, below the Pinch, the curve shows how the demand for external cooling increases and at what temperature the cooling would need to be supplied. At first sight it may appear strange to discuss the external duties in this way because, as Fig. 6.11 (repeated as Fig. 6.28(a)) shows, the external utilities to heat or cool the streams to specified temperatures must themselves have specified temperatures governed by the minimum allowable temperature difference (in this case 20 K).

The circuit shown in Fig. 6.28(a) is based on the premise that high-grade energy is available and therefore all external heating can occur as the final operation on cold streams, and that low temperature cooling is available as the single method of cooling all the hot streams to their final temperatures. It will now be demonstrated that the grand composite curves can be used to show how low-grade heat can be used for some of the external heating and how a medium temperature coolant can be used for some of the external cooling duty.

Considering the external heating system as an example, imagine that we attempt to use some saturated steam at 165 °C as a source of external heat. Since the minimum allowable temperature difference between streams was stated to be 20 K then such a source would not be able to raise the temperature of a cold stream above 145 °C. This would prohibit any heating of stream 3 and provide only 1 K of temperature rise for stream 4, *if the external heating is restricted to being the final operation on each stream.* The point is that the external heating does not necessarily have to be the final operation. Before considering where to place the steam heating, we first need to examine the grand composite curve to determine the *maximum* amount of steam heating that could be used. The word 'maximum' is emphasized for two reasons: firstly, because of costs and simplicity of design, it may not be appropriate to use all of the low-grade

Figure 6.28 (a) Energy design chart; (b) low-grade steam use above the Pinch; (c) medium temperature coolant use below the Pinch

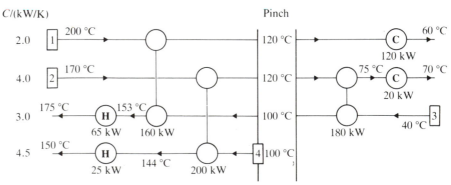

(a)

Figure 6.28 *Continued*

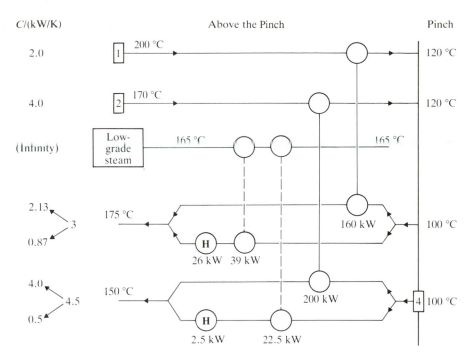

Total low grade steam load = 22.5 + 39 = 61.5 kW
Remaining external heating load = 26 + 2.5 = 28.5 kW

(b)

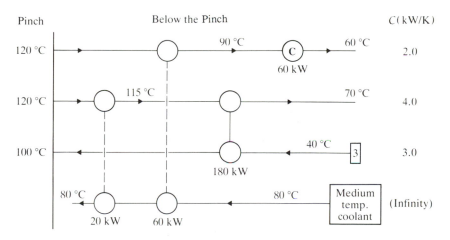

Total medium temperature coolant load = 20 + 60 = 80 kW
Remaining external cooling load = 60 kW

(c)

heat; secondly, the cold stream temperatures will eventually rise above 145 °C, thus limiting the amount of low-grade heat actually needed. Since the steam is at a temperature of 165 °C then on an adjusted temperature scale the temperature will appear as 155 °C (all hot steams are adjusted down by half the Pinch temperature difference). From Fig. 6.27 the *maximum* amount of energy available would be 68 kW. The energy design chart shown in Fig. 6.28(a) now has to be modified to include an extra hot stream (low-grade steam). For this example, the steam will be assumed to be condensing to a saturated liquid at 165 °C and so the thermal capacity of the stream will be infinity, as shown in Fig. 6.28(b).

As there are now three hot streams and only two cold streams, the decision diagram for stream-splitting (Fig. 6.21) reveals that we will have to split at least one of the cold streams. In fact, the circuit design in Fig. 6.28(b) shows that both cold streams are split. The rationale for this can be explained by reference to streams 2 and 4. Previously, stream 2 donated 200 kW to stream 4; this amount raised the temperature of stream 4 from 100 °C to 144 °C, leaving a requirement for 25 kW of external heating to achieve the desired final temperature of 150 °C.

If stream 4 is split then one of the substreams can be raised directly to 150 °C by exchange with stream 2 and the other substream can accept energy from the low-grade stream. The split in thermal capacity flow rate is determined as follows.

For the substream linking directly with stream 2 the rate of energy transfer is the product of the thermal capacity flow rate and the temperature change.

i.e. $200 \text{ kW} = \text{thermal capacity flow rate}/(\text{kW}/\text{K}) \times (150 - 100)(\text{K})$

$\therefore$ thermal capacity flow rate $= 4 \text{ kW}/\text{K}$

Since the total thermal capacity flow rate of stream 4 is 4.5 kW/K then the other substream will take a value of $(4.5 - 4) = 0.5$ kW/K, as shown in Fig. 6.28(b). This second substream can accept energy from the low-grade steam until the temperature difference between the streams is the minimum specified, i.e. 20 K. Therefore, the temperature of the second substream is given by

$$t_2 = (165 - 20) = 145 \,^{\circ}\text{C}$$

The rate of energy transfer from the low-grade steam is the product of the thermal capacity flow rate and the temperature change, which is

$$0.5 \times (145 - 100) = 22.5 \text{ kW}$$

A similar argument applies to the splitting of stream 3, giving the figures for thermal capacity flow rates and rates of energy transfer shown in Fig. 6.28(b). Therefore the total rate of energy transfer from the low-grade steam to the two substreams is

$$39 + 22.5 = 61.5 \text{ kW}$$

When compared to the one in the Fig. 6.28(a) above the circuit in Fig. 6.28(b) appears more complicated and certainly uses more heat exchangers; six as against four previously. On the other hand, the requirement for high-grade heating is reduced from 90 kW to 28.5 kW because of the contribution of 61.5 kW of low-grade steam heating. It thus becomes a matter of costs as to whether the savings accrued by using low-grade heating offset the use of more heat exchangers and ancillary equipment.

A similar situation could be specified for the use of medium temperature cooling below the Pinch. Imagine that a saturated liquid at 80 °C is required to be evaporated to a saturated vapour; could this liquid be used as part of the external cooling system of the previous problem? By adjusting the temperature to 90 °C (all cold streams are adjusted up by half the Pinch temperature difference), the *maximum* cooling contribution can be read from the grand composite curve, Fig. 6.27, as 85 kW. A circuit which will use 80 kW of this capacity is shown in Fig. 6.28(c). Again the new circuit uses more heat exchangers (four as against three previously), but the requirement for low temperature cooling has been reduced from 140 kW to 60 kW; it is again a matter of cost as to whether or not the new circuit is to be preferred.

In this discussion, both the external heating and cooling fluids were represented by evaporating or condensing processes, with no superheating or sub-cooling involved. This allowed the processes to be represented by horizontal lines on the grand composite curve. In situations where the cooling and heating fluids are not undergoing phase changes but simply changes in temperature, then they would be represented as inclined lines drawn within the grand composite curve between appropriate temperature limits (as shown by line C–C on Fig. 6.27). The maximum heating and cooling duties can then be determined from the distance between the adjusted temperature axis and the line C–C at whatever temperature interval is appropriate.

This discussion also illustrates the importance of applying Pinch Technology to the whole plant, including any low-grade steam or evaporating liquids.

Further Comments about the Placement of Heat Pumps, Heat Engines and CHP Plant

The Grand Composite Curve provides a convenient way to illustrate diagrammatically the most suitable placement of various devices relative to the Pinch. Table 6.1 summarizes the discussion of Sections 6.9 and 6.10. The table is illustrated in Figs 6.29(a), (b) and (c), (page 236), where the devices are shown in relation to the grand composite curves.

Table 6.1 Placement of devices in relation to the Pinch

Device	Stream	Placement relative to Pinch
Heat pump	Evaporator	Part of cold stream below Pinch
Heat pump	Condenser	Part of hot stream above Pinch
Heat pump	Whole device	Can operate across Pinch
Heat engine	Heat input	Part of cold stream below Pinch
Heat engine	Exhaust	Part of hot stream above Pinch
Heat engine	Whole device	Can operate only on one side of Pinch, either extracting heat below Pinch or donating heat above Pinch
CHP plant	Exhaust	Can operate only on one side of Pinch, i.e. donating heat above Pinch

6.12 GENERAL COMMENTS ABOUT PROCESS INTEGRATION

The technology of process integration has proved of great value to process industries in reducing energy costs; case studies from the Department of Energy[6.4] are testimony to this. The basic ideas of generating achievable

Figure 6.29 (a) Correct placement of heat pump across the Pinch; (b) correct placement of heat engine accepting heat input below the Pinch; (c) correct placement for heat engine donating heat above the Pinch

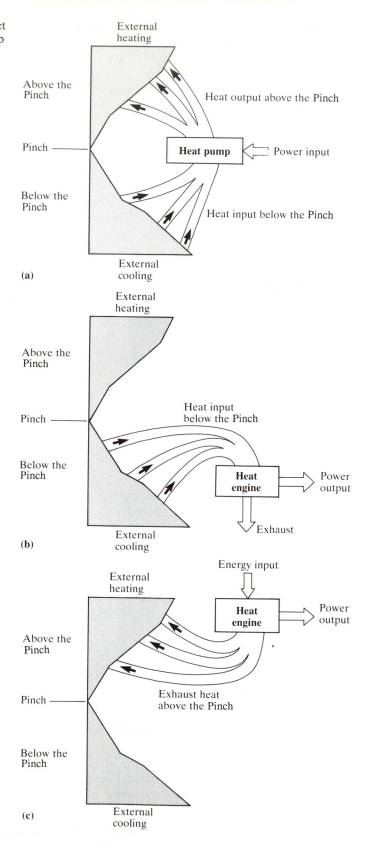

minimum energy targets for a process gives considerable confidence in the design of new networks and the re-design of existing ones.

The basic method is ideally suited to continuous processes which operate in the same way throughout the year, though it has been successfully applied to batch processes by the inclusion of suitable storage capacity. The point to emphasize is that a system designed with Pinch Technology inherently suffers the same general drawbacks as any other energy recovery scheme in the way it responds to changing flow conditions. The difference is that the Pinch Technology design will operate more economically.

The methodology of process integration not only generates information on energy targets but also has the facility for capital cost targeting and the reduction of total operating costs.

This chapter has tended to concentrate on small energy recovery systems involving a few streams. In an industrial situation the technology should be applied to the whole manufacturing plant on the site. As a result, management could have greater confidence that the optimum solutions are being proposed and energy conservation would then be seen as part of a company's investment strategy.

Problems

(Note: diagrams for the solutions to the following problems are given at the end of the Chapter.)

6.1 In a process, there are four fluid streams having thermal capacity flow rates and temperatures as shown in the table below. Calculate:
(i) the rate of external heating and cooling required if there is no heat recovery between streams;
(ii) the rate of external heating and cooling required if a heat recovery scheme is used with the minimum allowable temperature between streams being 25 K;
(iii) the Pinch temperatures of the hot and cold streams.

Stream	Type	Thermal capacity rate, $\dot{m}c_p$ (kW/K)	Temperatures (°C) Initial	Final
1	Hot	0.2	350	50
2	Hot	0.4	250	50
3	Cold	0.1	20	200
4	Cold	0.3	100	300

Construct a design chart which shows the heat exchanger connections and the external cooling and heating duties.
(140 kW cooling, 78 kW heating; 64.5 kW cooling, 2.5 kW heating; 250 °C, 225 °C)

6.2 Repeat Problem 6.1 with a minimum allowable temperature difference between streams of 50 K. Also, calculate the value of (UA) for each heat recovery exchanger.

(140 kW cooling, 78 kW heating; 72 kW cooling, 10 kW heating; 250 °C, 200 °C; 0.487 kW/K, 0.206 kW/K, 0.307 kW/K)

6.3 For the process data below, establish the Pinch conditions and hence construct a design chart to show the external heating and cooling duties based on the minimum allowable temperature between streams being 30 K.

Stream	Type	Thermal capacity rate, $\dot{m}c_p$ (kW/K)	Temperatures (°C) Initial	Final
1	Hot	1.0	400	50
2	Hot	0.75	300	100
3	Cold	0.5	20	370
4	Cold	3.0	100	200

Figure 6.30 Circuit diagram for Problem 6.4

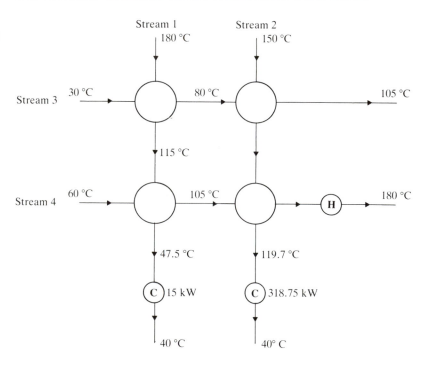

6.4 The circuit diagram in Fig. 6.30 shows a heat recovery scheme using two hot streams and two cold streams. The external heating and cooling duties and the stream temperatures have the values indicated on the diagram. Calculate:

(i) the thermal capacity flow rates of each stream;

(ii) the minimum rates of external heating and cooling required if a heat recovery scheme is used with the minimum allowable temperature between streams being 10 K;

(iii) the minimum number of heat exchangers required above and below the Pinch.

Draw a circuit diagram to represent the recovery scheme of (ii) above.

(Stream 1: 2 kW/K. Stream 2: 4 kW/K. Stream 3: 2.6 kW/K. Stream 4:

3 kW/K. 225 kW cooling, 60 kW heating; two heat exchangers above Pinch, four below Pinch: includes one for external heating and two for external cooling.)

Figure 6.31 Circuit diagram for Problem 6.5

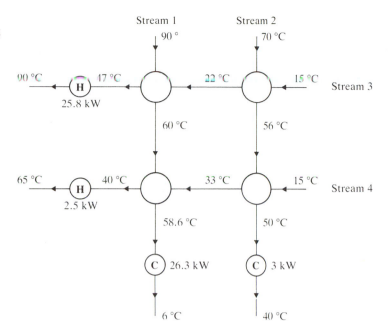

6.5 Figure 6.31 shows a circuit layout for a heat recovery scheme. The external heating and cooling duties are 28.3 kW and 29.3 kW respectively.
(a) Design and draw a circuit with a Pinch temperature difference of 20 K, which will reduce the external heating and cooling duties to 15.5 kW and 16.5 kW.
(b) Use the following data to determine the capital and annual running costs of the two schemes. Assume that the plant is running continuously.

Data
External heating cost, £0.04/kW h; external cooling cost, £0.05/kW h.
Heat exchanger costs: heat recovery exchangers, £$\{10 \times (UA)/(W/K)\}$; external heating exchangers, £100/kW; external cooling exchangers, £120/kW.
(1st scheme: running cost, £22 750 per annum; capital cost, £11 939;
 2nd scheme: running cost, £12 658 per annum; capital cost, £19 860)

6.6 In a chemical plant, a number of fluid streams require to be heated or cooled during the manufacturing process. Currently, each stream is heated or cooled individually by external means. The company is considering a heat recovery scheme in which hot and cold streams are linked with heat exchangers; the effect will be to reduce the amount of external heating and cooling and so reduce the running costs of the plant. The stream data is shown below.
(a) Estimate the external heating and cooling requirements of the plant with no heat recovery.
(b) Design a heat recovery circuit which will reduce the external duties to a minimum, given the constraint that no heat exchanger can operate with a temperature difference between streams of less than 20 K.
(c) Using the data from parts (a) and (b) previously and the cost data below,

Stream	Type	Thermal capacity rate, $\dot{m}c_p$ (kW/K)	Temperatures (°C) Initial	Final
1	Hot	6	120	40
2	Hot	300	80	79
3	Hot	3	80	40
4	Cold	8	20	135
5	Cold	2	20	70

calculate the payback times of the heat recovery scheme. Assume that the plant is operating continuously throughout the year.

External cooling cost $\qquad = £0.02$ per kW h
External heating cost $\qquad = £0.03$ per kW h
Total cost of heat recovery exchangers and ancillaries

$$= £\{400 \times (\text{total duty kW})\}$$

(900 kW cooling, 1020 kW heating; 11.4 months)

6.7 A heat recovery scheme is to be designed for a Pinch temperature difference of 15 K. Construct a design chart for the stream data shown in the table below. Draw the grand composite curve, and hence show that 90 kW of external heating can be obtained from low-grade steam condensing at 120°C, and that 55 kW of external cooling can be absorbed by an evaporating fluid whose saturation temperature is 60°C.

Stream	Type	Thermal capacity rate, $\dot{m}c_p$ (kW/K)	Temperatures (°C) Initial	Final
1	Hot	6	110	50
2	Hot	250	90	89
3	Hot	3	90	40
4	Cold	7	15	130
5	Cold	3	15	70

Figure 6.32 Solution Problem 6.1

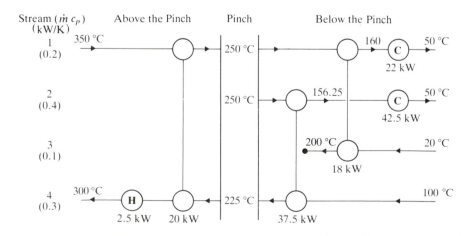

Figure 6.33 Solution
Problem 6.2

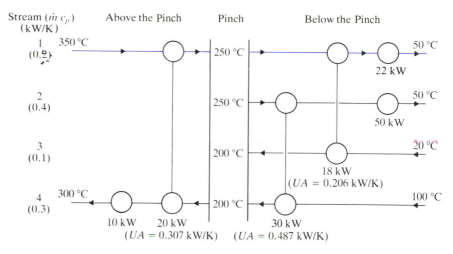

Figure 6.34 Solution
Problem 6.3

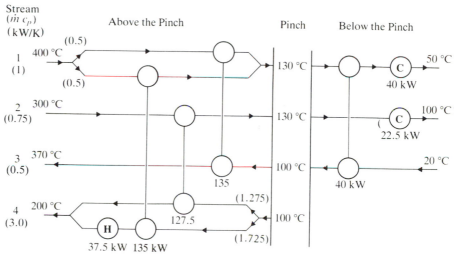

Figure 6.35 Solution
Problem 6.4

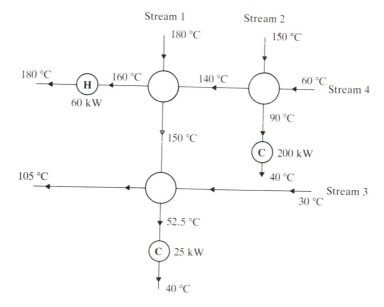

Figure 6.36 (a) Solution
Problem 6.5
(b) Alternative solution
Problem 6.5

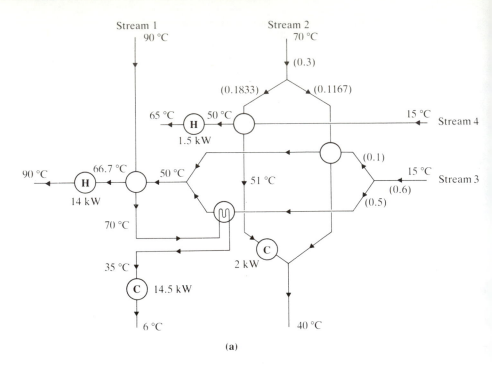

(a)

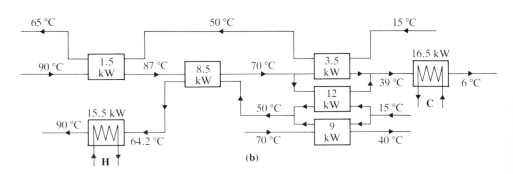

(b)

Figure 6.37 (a) Solution Problem 6.6 (b) Alternative solution Problem 6.6

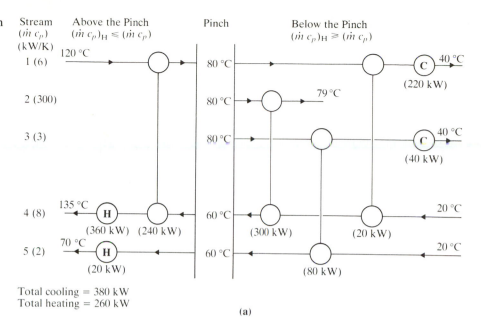

Total cooling = 380 kW
Total heating = 260 kW

(a)

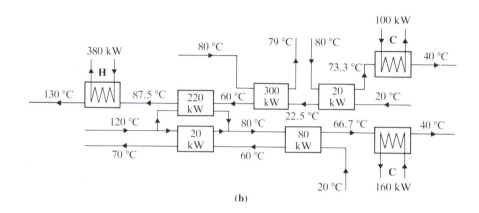

(b)

Figure 6.38 (a) Solution Problem 6.7 (b) Problem 6.7 grand composite curve; (c) Problem 6.7 solution with low-grade steam and evaporating fluid; (d) Problem 6.7 alternative solution with low-grade steam and evaporating fluid

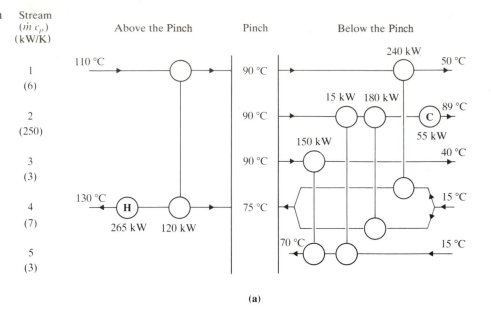

(a)

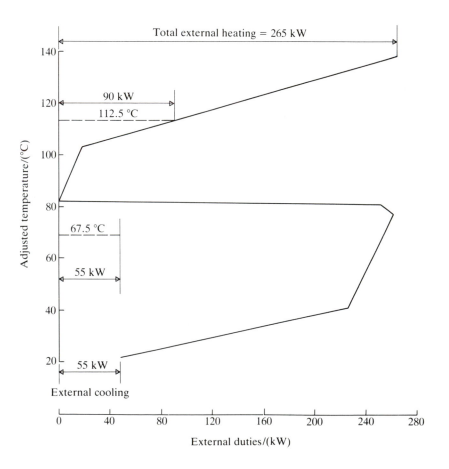

(b)

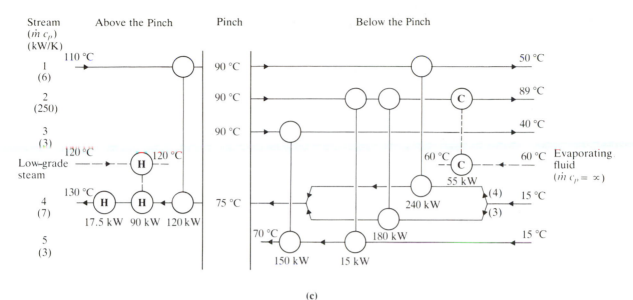

(c)

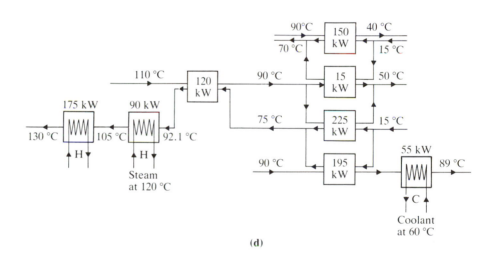

(d)

References

6.1 Linnhoff B, Townsend D W, Boland D, Hewitt G F, Thomas B E A, Guy A R, Marsland R H 1982 *User Guide on Process Integration for the Efficient Use of Energy* Inst. of Chem. Engrs, Rugby, UK

6.2 Tjoe T N, Linnhoff B 1986 Using pinch technology for process retrofit. *Chemical Engineering* April 1986: 47–60

6.3 Hohman E C *Optimum Networks for Heat Exchangers* PhD Thesis 1971, Dept. of Chem. Eng. Univ. of S. California, Los Angeles, California

6.4 Department of Energy Publications:
   (i)   *Energy Management – Focus,* Issue numbers 9 (1986), 13 (1987)
   (ii) *Research and Development Reports on Process Integration Studies in UK Industry*

Sources of Further Information

Linnhoff-March Ltd, Tabley Court, Moss Lane, Over Tabley, Knutsford, Cheshire, WA16 0PL, UK

National Engineering Laboratory, East Kilbridge, Scotland, UK

The Energy Efficiency Office, Thames House, South Millbank, London SW1P 4QJ, UK

The Enquiries Bureau, Energy Technology Support Unit, Harwell Laboratory, Didcot, Oxon OX11 0RA, UK

7 ENERGY IN BUILDINGS

A major use of energy is in the heating, air conditioning and lighting of buildings and there is considerable scope for energy saving in these areas.

A building is essentially a modifier of the micro-climate, a space isolated from climatic temperature and humidity fluctuations, sheltered from prevailing winds and precipitation, and with enhancement of natural light by artificial means. Conditions within the building are determined by the type of use; for example, the number of people, the type of activity (e.g. sedentary activity, heavy manual work, etc.), water vapour generated (e.g. from kitchens, wash-rooms, laundries etc.), process use (e.g. furnaces, ovens, computer terminals etc.).

The creation of a successful building with adequately designed and controlled building services is a complex process involving a wide range of engineering disciplines. In this chapter the major design features will be considered briefly in order to examine possible ways in which energy, and hence money, can be saved. For a proper study of building services engineering the reader is recommended to read the books listed in the Bibliography at the end of the chapter.

7.1 STEADY STATE LOADS AND COMFORT

Unlike most other engineering design situations, the external design conditions for the heating and air conditioning of buildings vary widely both in position and with time. The building must be designed to provide sufficient heating capacity for the most likely external winter design temperature for its location, but must also be able to provide heating at a high efficiency when the outside temperature is much higher than the design minimum. The CIBSE *Guide*[7.1] gives a summary of winter, summer and annual weather data for the UK, and a table of world weather data; there is also a complete set of data for world solar radiation; the information is mainly taken from the records of the UK Meterological Office.

Heat losses from a building fall into two categories: the fabric loss (i.e. through the walls, windows, roof and floor); and the ventilation loss due to air leakage into and out of the building.

Fabric Loss

The heat loss through the fabric depends on the thermal resistances of the various elements making up the walls, roof, floors etc. and the thermal resistances of the inside and outside surfaces of the building due to convection through the fluid film on the surface and thermal radiation from the surface to the surroundings. Section 3.2 covers the basic heat transfer theory and Problem 3.11 is a simple example of heat transfer through the wall of a building.

The overall heat transfer coefficient, or *thermal transmittance*, U, in this case is given by

$$\frac{1}{U} = \frac{1}{h_i} + \Sigma R_w + \frac{1}{h_o} \qquad [7.1]$$

(where the heat transfer coefficients, h_i and h_o, include both convection and radiation effects; the summation term is the sum of the thermal resistance of the elements making up the wall or roof etc.). The thermal transmittance is sometimes known as the *U-value* in building services design.

It can be seen from Eq. [7.1] that the thermal transmittance depends on the weather conditions outside since the value of h_o will vary, mainly with air speed; it is necessary therefore to modify the thermal transmittance for a particular wall construction depending on whether it is under exposed or sheltered conditions.

The UK Building Regulations set down minimum thermal transmittances to be achieved for new houses and offices; these are revised downwards as technology and materials improve. At April 1990 the minimum values laid down are 0.45 W/m² K for walls in dwellings, and walls in industrial and commercial buildings; 0.25 W/m² K for roofs in dwellings; 0.45 W/m² K for roofs in industrial and commercial buildings; 0.45 W/m² K for floors. Table 7.1 shows some typical values for different types of construction given by the CIBSE *Guide*.[7.1]

Table 7.1 Thermal transmittance of some typical constructions

Construction	Thermal transmittance (W/m² K)
Cavity wall with plaster on inside: 105 mm thick bricks, 50 mm air gap, 13 mm plaster	1.5
As above with glass wool in cavity	0.5
Pitched roof with no loft insulation	2.6
Pitched roof with 100 mm thick loft glass-fibre insulation	0.35
Solid ground floor with four edges exposed: 10 m by 10 m, edges uninsulated	0.62
As above with only two perpendicular edges exposed	0.36
Solid ground floor with four edges exposed and insulated to a depth of 25 mm with insulation of thermal resistance 0.25 m² K/W	0.58
Single glazed window: aluminium frame	6.40
Double glazed window: aluminium frame	4.30

(Note: The thermal transmittance for a floor is to be used with the overall temperature difference from room to outside air as for the other fabric elements.)

Emphasis on more efficient housing design has been given by such initiatives as the 1986 Energy World run by the Milton Keynes Development Corporation; the Canadian R2000 system of housing has also been used as a model by some builders. By using a brick and block construction with a 100 mm cavity filled with fibreglass batts the thermal transmittance for a wall can be reduced to about 0.25 W/m² K.

The roof thermal transmittance can be reduced to about 0.20 W/m² K by using lined roof tiles and a 200 mm fibreglass quilt over the plasterboard ceiling, and by insulating the floor with 50 mm of rock fibre below the top screed of concrete the thermal transmittance can be reduced to about 0.40 W/m² K. Double glazed windows with a low emissivity coating and with the gap between the glass filled with argon gas can have a thermal transmittance as low as 1.60 W/m² K.

Ventilation Loss

A building may be naturally ventilated by air movement through doors, windows and ventilators, or mechanically ventilated by fans, or air conditioned by providing air through ducts at controlled conditions and removing the air through outlet ducts. In all cases the building should be constructed to be as air tight as possible; air which leaks into and out of a building through fortuitous cracks, inadequate sealing etc. is known as *infiltration*.

Fresh air requirement for spaces occupied by people is governed by the need to remove body odours and tobacco smoke; carbon dioxide from the normal breathing process can be removed by extremely low ventilation rates. The CIBSE *Guide*[7.1] gives recommended ventilation rates for different types of buildings; factories and certain public buildings are governed by statutory requirements and local bye-laws.

When infiltration has been estimated for a particular building then, if this is less than the air change rate required, mechanical ventilation must be provided to satisfy the requirements. For a building of volume, V, with n air changes per unit time, ventilation heat loss rate is given by

$$Q_V = \rho c_p n V (t_{ai} - t_{ao}) \qquad [7.2]$$

(where ρ = mean density of air; c_p = mean specific heat of air at constant pressure; t_{ai} = inside air temperature; t_{ao} = outside air temperature). The mean specific heat and the mean density of air over the normal temperature range are approximately 1000 J/kg K and 1.2 kg/m³, and hence the product $\rho c_p = 1200$ W s/m³ K = (1/3) W h/m³ K, and we can write:

$$\frac{Q_v}{[\text{W}]} = \frac{1}{3}\frac{n}{[\text{h}^{-1}]}\frac{V}{[\text{m}^3]}\frac{(t_{ai} - t_{ao})}{[\text{K}]} \qquad [7.3]$$

Environmental Temperature

Environmental temperatures are equivalent temperatures introduced to allow for radiation effects, either externally from solar radiation on the outside surfaces or internally from radiant interchange between internal surfaces.

The effect of solar radiation in winter gives an external environmental temperature higher than the outside air temperature; this is known as the sol-air temperature, t_{eo}. The amount of heating required is therefore reduced but this is normally ignored and the outside environmental temperature taken to be equal to the mean outside air temperature, i.e. $t_{eo} = t_{ao}$.

The inside environmental temperature depends on the temperatures and emissivities of the various room surfaces, the shape of the room, and the convective heat transfer coefficients for the various inside surfaces. The CIBSE[7.1] take a cubical enclosure with a mean value of the convection heat transfer coefficient for all surfaces of 3 W/m^2 K, and a mean value of the radiation heat transfer coefficient of 5.13 W/m^2 K (based on Eq. [3.38]). The analysis then gives an approximate expression for inside environmental temperature, t_{ei}, as follows:

$$t_{ei} = \frac{2}{3} t_{mr} + \frac{1}{3} t_{ai} \qquad [7.4]$$

The mean radiant temperature, t_{mr}, is the temperature of a small sphere completely surrounded by a number of surfaces at different temperatures in the absence of convection. For a cubical enclosure at the temperatures normally encountered in buildings the mean radiant temperature is approximately equal to the mean surface temperature, t_{ms}, given by:

$$t_{ms} = \Sigma(A_s t_s)/\Sigma(A_s) \qquad [7.5]$$

(where A_s = area of any surface at temperature t_s). The fabric loss for the building is then given by:

$$Q_F = \Sigma(A_o U)(t_{ei} - t_{ao}) \qquad [7.6]$$

(where A_o is the area of any element of the fabric which transmits heat to the outside air).

To find the heat loss from a particular room it is necessary to take into account the heat transfer to adjacent rooms if these are at a different environmental temperature. In such cases the (UA) value for the internal wall is multiplied by the difference in environmental temperatures across the internal wall.

Example 7.1

An office has one external wall, 5 m long, containing a window of area 4 m^2; the width of the office from the window to the wall adjacent to the corridor is 4 m, and the ceiling height is 3m. There are similar offices above, below, and on either side. The environmental temperature in the corridor is 16 °C, the environmental temperature in the room is 20 °C, the internal air temperature in the room is 19 °C, and the external design temperature is −1 °C. Taking thermal transmittances for the external wall, window, and internal wall as 1.0 W/m^2 K, 5.6 W/m^2 K, and 2.7 W/m^2 K respectively, and an air change rate of one per hour, calculate the required rate of heat input to the room.

Solution

A tabular method of laying out the data is useful in more complex cases and is used below as an illustration of the method. The heat loss is obtained using Eq. [7.6].

Item	Area (m^2)	U (W/m^2 K)	Δt (K)	Heat loss (W)
External wall	$15 - 4 = 11$	1.0	$20 - (-1) = 21$	231.0
Window	4	5.6	21	470.4
Corridor wall	15	2.7	$20 - 16 = 4$	162.0
				$Q_F = 863.4$

From Eq. [7.3]

$$Q_V = 1 \times (4 \times 5 \times 3)(19 + 1)/3 = 400 \text{ W}$$

Therefore

$$\text{total heat input required} = 863.4 + 400 = 1263.4 \text{ W}$$

Steady State Network

A model of the heat transfer process using an electrical resistance analogy can be used as shown in Fig. 7.1. Part of the heat input, Q_a say, is taken to be at the air temperature, t_{ai}, and part, Q_e say, at the environmental temperature, t_{ei}. It is assumed that convective heating is input at the air point, and that at the environmental point the heat input is partly radiant and partly convective.

Figure 7.1 General steady state network

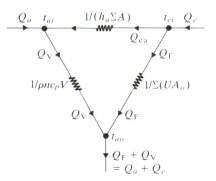

The thermal resistances for each leg of the network are as shown in the figure: the resistance $1/\Sigma(UA_o)$ follows from Eq. [7.6]; the resistance $1/\rho c_p nV$ follows from Eq. [7.2]; the resistance $1/h_a\Sigma A$ can be derived from the method used to obtain Eq. [7.4] (see the CIBSE *Guide*[7.1]). (Taking the values used previously for convective and radiative heat transfer coefficients then h_a can be shown to be approximately 4.5 W/m^2 K.) Note: it is important to note that ΣA is the summation of all the room surface areas whereas in the term $\Sigma(UA_o)$ only the areas of the surfaces conducting to the outside are taken.

Ducted warm air heating is purely convective in which case $Q_e = 0$ and the network is modified as shown in Fig. 7.2. No form of heating is purely radiative since there is always some convection from a hot surface, but since a heat input at the environmental point is partly radiative and partly convective the values of Q_a and Q_e must be modified accordingly.

Figure 7.2 Steady state network for warm air heating

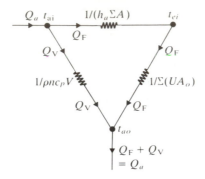

In the CIBSE method the heat input at the environmental point (in the absence of indirect radiation), is taken to be

$$Q_e = 1.5\gamma Q_i \qquad [7.7]$$

where γ is the proportion of the heat input which is due to direct radiation. In the steady state, $(Q_a + Q_e) = Q_i$, and $\gamma + \delta = 1$, where δ is the proportion of the heat input by convective heating. Then

$$Q_a = Q_i - (1.5\gamma Q_i) = \{(\gamma + \delta) - 1.5\gamma\} Q_i$$

i.e. $$Q_a = (\delta - 0.5\gamma)Q_i \qquad [7.8]$$

It can be seen that in this model the value of Q_a is negative when δ is less than 0.5γ. For example, for vertical radiant panels or ceiling panels the proportions of convective and radiant heat are taken to be 33.33 % and 66.67 % respectively; this gives Q_e equal to Q_i, and a value of Q_a of zero. For any greater proportion of radiant to convective heating then Q_a becomes negative (e.g. for high temperature radiant strip panels or quartz linear heaters).

The network method is detailed in the CIBSE *Guide*[7.1] and is based on work by Danter, and by Loudon.[7.2] The basis of the model has recently been questioned by Davies.[7.3] There are so many approximations and assumptions built into any simple model that the one proposed by CIBSE using environmental temperature is probably in its end result as good as any other. It is easy to use, gives results which are more accurate than neglecting the effects of internal radiation, and can be also used for transient response (see later).

Comfort

Many research workers have attempted to define comfort criteria and numerous comfort indices have been proposed. Comfort is a subjective state but it is generally agreed that air temperature, mean radiant temperature, air velocity, and relative humidity are the main factors governing thermal comfort. Of these the first two are by far the most important with air velocity becoming important if it rises above about 0.1 m/s. There is a wide range of relative humidities that people find acceptable (say 40 % to 70 %).

Fanger's work on thermal comfort[7.4] is highly regarded and is recommended for a more extensive treatment. A concise coverage of all the factors involved is given by the CIBSE *Guide*;[7.1] they recommend the use of *dry resultant temperature*, t_c. This is defined as the temperature recorded at the centre of a blackened globe of 100 mm diameter in still air. It can be taken to be given by the following approximate expression:

$$t_c = (t_{mr} + t_{ai})/2 \qquad [7.9]$$

As before, it can be assumed that the mean radiant temperature, t_{mr}, is equal to the mean surface temperature, t_{ms}. Combining Eqs [7.9] and [7.4] we then have

$$t_c = \frac{3}{4} t_{ei} + \frac{1}{4} t_{ai} \qquad [7.10]$$

It can be seen from Eq. [7.10] that the temperature, t_c, can be placed in the network diagram at a point $1/3$ of the way from t_{ei} to t_{ai} as shown in Fig. 7.3. The equivalent heat transfer coefficients can be easily shown to be $h_{ac} = 4h_a/3 = 6$ W/m² K, and $h_{ec} = 4h_a = 18$ W/m² K.

Figure 7.3 General steady state network with dry resultant temperature

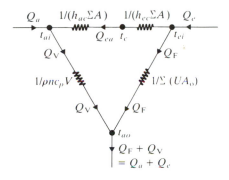

Table 7.2 Some recommended dry resultant temperatures

Type of building	Dry resultant temperature, t_c (°C)
Factories: heavy work	13
Stores and appliance rooms	15
Factories: light work	16
Class-rooms, hospital wards	18
Offices, laboratories, museums	20
Living rooms, public rooms	21
Hotel bedrooms, sports changing rooms	22
Swimming pool halls	26

Table 7.2 gives some recommended values of dry resultant temperature taken from the CIBSE *Guide*.[7.1] For velocities within the space greater than 0.1 m/s there is a recommended addition to the temperatures in Table 7.2 (e.g. for a velocity of 0.2 m/s, 0.9 K; for a velocity of 0.4 m/s, 1.6 K; for velocity of 0.6 m/s, 2.1 K).

For any given space, heated in a particular way, the recommended value of t_c can be placed in the network and knowing the outside air temperature and all the resistances in the network the required heat input can be calculated. The CIBSE *Guide*[7.1] derives formulae which allow a computer programme to be drawn up but the procedure can be illustrated by a step-by-step method as shown in the following worked example.

Example 7.2

(a) Using the data given calculate the fabric heat loss, ventilation heat loss, and total heat load for a small workshop when it is heated:

(i) by ducted warm air;

(ii) by vertical radiant panels;

(iii) by high temperature radiant panels.

(b) Re-calculate the heat loads for (i), (ii), and (iii) when the air change rate is reduced by a factor of 4 by installing more energy efficient loading doors.

Data

Size of workshop: 15 m by 10 m by 4.5 m high.

Thermal transmittances: walls (all external), 0.9 W/m² K; windows (area 50 m²), 5.6 W/m² K; door (area 5 m²), 3.0 W/m² K; floor, 0.7 W/m² K; flat roof, 0.9 W/m² K.

Air change rate, 2 per hour; dry resultant temperature required, 16 °C; outside design temperature, -1 °C; for vertical panels take 33.33 % convective and 66.67 % radiant heating; for high temperature panels take 10 % convective and 90 % radiant heating.

Solution

(a) Using a tabular method:

Item	Area (m²)	U (W/m² K)	(UA) (W/K)
Walls	$225 - 50 - 5 = 170$	0.9	153
Windows	50	5.6	280
Door	5	3.0	15
Floor	150	0.7	105
Roof	150	0.9	135
	$\Sigma A = \overline{525}$ m²		$\Sigma(UA) = \overline{688}$ W/K

The equivalent heat transfer coefficient between the inside air temperature point and the dry resultant temperature point can be taken as $h_{ac} = 6$ W/m² K, and the equivalent heat transfer coefficient between the dry resultant temperature point and the equivalent temperature point can be taken as $h_{ec} = 18$ W/m² K. The thermal resistances of these two links are then given by

$$1/(h_{ac}\Sigma A) = 1/6 \times 525 = 1/3150 \text{ K/W}$$

and $$1/(h_{ac}\Sigma A) = 1/18 \times 525 = 1/9450 \text{ K/W}$$

The thermal resistance of the fabric heat loss link from internal environmental temperature to the outside temperature is $1/\Sigma(UA) = 1/688$ K/W. For the ventilation heat loss the thermal resistance between the inside and outside air points is given by Eq. 7.3 as

$$3/(nV) = 3/(2 \times 15 \times 10 \times 4.5) = 1/450 \text{ K/W}$$

(i) For the case of ducted warm air heating all the input is at the air point and Q_e is zero. The heat input, Q_i, is equal to Q_a, and is also equal to $(Q_F + Q_V)$. In this case the fabric loss, Q_F also flows along the top arm of the network and

Figure 7.4 Steady state network for Example 7.2 part (i)

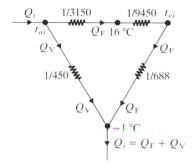

hence, referring to Fig. 7.4

$$(16 - t_{ei}) \times 9450 = (t_{ei} + 1) \times 688$$
$$\therefore \quad t_{ei} = 150\,512/10\,138 = 14.85\,°C$$

Then

$$\text{fabric heat loss} = (14.85 + 1) \times 688 = 10\,902 \text{ W}$$

Also

$$(t_{ai} - 16) \times 3150 = Q_F = 10\,902$$
$$\therefore \quad t_{ai} = 19.46\,°C$$

and

$$\text{ventilation heat loss} = (19.46 + 1) \times 450 = 9207 \text{ W}$$

i.e.

$$\text{heat input} = 10\,902 + 9207 = 20\,109 \text{ W} = 20.11 \text{ kW}$$

(ii)   In this case the proportion of the heat input at the environmental point is given by Eq. [7.7] as

$$Q_e = \frac{(1.5 \times 2)Q_i}{3} = Q_i$$

and the heat input at the air point is given by Eq. [7.8] as

$$Q_a = \left(\frac{1}{3} - 0.5 \times \frac{2}{3} \right) = 0$$

The ventilation loss, Q_V, flows along the top arm of the network and hence, referring to Fig. 7.5

$$(16 - t_{ai}) \times 3150 = (t_{ai} + 1) \times 450$$
$$\therefore \quad t_{ai} = 49\,950/3600 = 13.88\,°C$$

Figure 7.5 Steady state network for Example 7.2 part (ii)

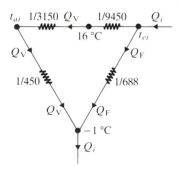

Then

$$\text{ventilation heat loss} = (13.88 + 1) \times 450 = 6694 \text{ W}$$

Also

$$(t_{ei} - 16) \times 9450 = Q_v = 6694$$
$$\therefore \quad t_{ei} = 16.71\,^\circ\text{C}$$

and hence

$$\text{fabric heat loss} = (16.71 + 1) \times 688 = 12\,183 \text{ W}$$

Hence

$$\text{heat input} = 6694 + 12\,183 = 18\,877 \text{ W} = 18.88 \text{ kW}$$

(iii)   For this case, using Eqs [7.7] and [7.8]:

$$Q_e = 1.5 \times 0.9 Q_i = 1.35 Q_i$$
and $$\quad Q_a = (0.1 - 0.5 \times 0.9) Q_i = -0.35 Q_i$$

Figure 7.6 Steady state network for Example 7.2 part (iii)

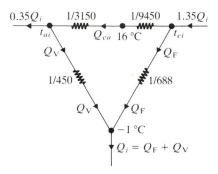

The network is now more complex, as shown in Fig. 7.6, and a solution involving simultaneous equations is required.

i.e. $$Q_{ea} = (t_{ei} - 16) \times 9450 = 1.35 Q_i - Q_F$$

Therefore $$(t_{ei} - 16) \times 9450 = 1.35 Q_i - (t_{ei} + 1) \times 688$$

or $$10\,138\, t_{ei} - 150\,512 = 1.35\, Q_i \qquad [1]$$

Also

$$Q_{ea} = (16 - t_{ai}) \times 3150 = 0.35\, Q_i + (t_{ai} + 1) \times 450$$

or $$49\,950 - 3600\, t_{ai} = 0.35\, Q_i \qquad [2]$$

And

$$(16 - t_{ai}) \times 3150 = (t_{ei} - 16) \times 9450$$

or $$t_{ai} = 64 - 3\, t_{ei} \qquad [3]$$

From [3] in [2]

$$49\,950 - (3600 \times 64) + (3600 \times 3) t_{ei} = 0.35\, Q_i$$

or $$10\,800\, t_{ei} - 180\,450 = 0.35\, Q_i \qquad [4]$$

From [1] $\quad t_{ei} = \dfrac{1.35\ Q_i + 150\,512}{10\,138}$

Substituting in [4]

$$\frac{(10\,800 \times 1.35\ Q_i)}{10\,138} + \frac{(10\,800 \times 150\,512)}{10\,138} - 180\,450 = 0.35\ Q_i$$

i.e. $\qquad$ heat input $= 18\,481\ \text{W} = 18.48\ \text{kW}$

Then substituting into Eqs [1] and (2) we have

$$t_{ei} = 17.31\,°\text{C} \qquad \text{and} \qquad t_{ai} = 12.08\,°\text{C}$$

Hence

$$\text{fabric heat loss} = (17.31 + 1) \times 688 = 12.60\ \text{kW}$$

and

$$\text{ventilation heat loss} = (12.08 + 1) \times 450 = 5.89\ \text{kW}$$

(b)   Reducing the air change rate by a fraction of four (i.e. from 2 per hour to 0.5 per hour) increases the resistance for the ventilation loss to $4/450 = 1/112.5\ \text{K/W}$.

(i)   The network is the same as in Fig. 7.4 but with the resistance of the left arm changed to $1/112.5\ \text{K/W}$. Hence, by inspection of the network it can be seen that the fabric heat loss, environmental temperature, and internal air temperature are as before; only the ventilation heat loss is different.

i.e. $\qquad$ ventilation heat loss $= (19.46 + 1) \times 112.5 = 2.30\ \text{kW}$

and

$$\text{heat input} = 2.30 + 10.90 = 13.20\ \text{kW}$$

(ii)   The network is now as in Fig. 7.5 with the ventilation resistance changed to $1/112.5\ \text{K/W}$. In this case the air temperature is altered since the ventilation loss depends also on the resistance of the top arm of the network.

i.e. $\qquad (16 - t_{ai}) \times 3150 = (t_{ai} + 1) \times 112.5$
$$\therefore \qquad t_{ai} = 15.41\,°\text{C}$$

Then

$$\text{ventilation heat loss} = (15.41 + 1) \times 112.5 = 1.85\ \text{kW}$$

Also

$$(t_{ei} - 16) \times 9450 = 1850$$
$$\therefore \qquad t_{ei} = 16.20\,°\text{C}$$

i.e. $\qquad$ fabric heat loss $= (16.20 + 1) \times 688 = 11.83\ \text{kW}$

and

$$\text{heat input} = 1.85 + 11.83 = 13.68\ \text{kW}$$

(iii)   In this case the network is as in Fig. 7.6 with the ventilation resistance changed to $1/112.5\ \text{K/W}$. Using the same method as before Eq. [1] is unchanged as

$$10\,138\ t_{ei} - 150\,512 = 1.35\ Q_i$$

Equation [2] is now modified to

$$50\,287.5 - 3262.5\ t_{ai} = 0.35\ Q_i$$

Equation [3] is unchanged and hence Eq. [4] becomes

$$9787.5\ t_{ei} - 158\,512.5 = 0.35\ Q_i$$

Substituting for t_{ei} from Eq. [1] in the above we have

$$\text{heat input} = 13.85\ \text{kW}$$

Then

$$t_{ei} = 16.69\,^{\circ}\text{C} \qquad \text{and} \qquad t_{ai} = 13.93\,^{\circ}\text{C}$$

Hence

$$\text{fabric heat loss} = (16.69 + 1) \times 688 = 12.17\ \text{kW}$$

and

$$\text{ventilation heat loss} = (13.93 + 1) \times 112.5 = 1.68\ \text{kW}$$

The complete results of the example are summarized in the table below.

Conditions	Q_i (kW)	Q_F (kW)	Q_V (kW)	t_{ai} (°C)	t_{ei} (°C)
Warm air heating: air changes, 2 h^{-1}	20.11	10.90	9.21	19.46	14.85
Radiant heating: 66.67 % radiant; air changes, 2 h^{-1}	18.88	12.18	6.70	13.88	16.71
Radiant heating: 90 % radiant; air changes, 2 h^{-1}	18.48	12.60	5.89	12.08	17.31
Warm air heating: air changes, 0.5 h^{-1}	13.20	10.90	2.30	19.46	14.85
Radiant heating: 66.67 % radiant; air changes, 0.5 h^{-1}	13.68	11.83	1.85	15.41	16.20
Radiant heating: 90 % radiant; air changes, 0.5 h^{-1}	13.85	12.17	1.68	13.93	16.69

It can be seen that for the larger air change rate of 2 per hour the total heat input is reduced as the proportion of radiant heating is increased from zero to 90 %, but when the air change rate is reduced to 0.5 per hour the total heat input is increased as the radiant proportion increases. In the latter case the ventilation loss is a much lower percentage of the total input. Since the example is based on the recommended dry resultant temperature for thermal comfort then each of the cases would produce satisfactory working conditions. The decision then becomes one of economics with the capital cost of each method of heating set against the fuel running costs in the usual way.

In any given factory there may be additional air changes necessary due to the need for fume or dust extraction; in such cases energy recovery should be considered along the lines of one of the methods discussed in Chapter 5.

Ventilation Conductance

A ventilation conductance, C_V, is sometimes used as an equivalent to the thermal conductance (UA). It is defined by the following equation:

$$\text{steady state heat loss} = \{\Sigma(UA_o) + C_V\}(t_{ei} - t_{eo}) \qquad [7.11]$$

In winter in the UK then $t_{eo} \simeq t_{ao}$ as before. From Eqs [7.2] and [7.6] we have

$$\text{steady state heat loss} = \Sigma(UA_o)(t_{ei} - t_{eo}) + \rho c_p n V(t_{ai} - t_{ao})$$

Comparing this equation with Eq. [7.11] we have

$$C_V = \rho n c_p V \frac{(t_{ai} - t_{ao})}{(t_{ei} - t_{ao})} \qquad [7.12]$$

When all the heat input is at the air point (see for example Fig. 7.2 or Fig. 7.4) then the same heat flows through both the temperature differences in Eq. [7.12] and therefore

$$C_V = \rho n c_p V \frac{\{(1/h_a \Sigma A) + 1/\Sigma(UA_o)\}}{1/\Sigma(UA_o)}$$

i.e.
$$C_V = \rho c_p n V \left\{ 1 + \frac{\Sigma(UA_o)}{h_a \Sigma A} \right\} \qquad [7.13]$$

(where h_a is the equivalent heat transfer coefficient between the air and environmental temperatures introduced earlier; h_a is usually taken as 4.5 W/m^2 K). When all the heat input is at the environmental temperature (see for example Fig. 7.5) then from Eq. [7.12] we have

$$C_V = Q_V/(t_{ei} - t_{ao})$$

Also from the network it can be seen that in this case

$$t_{ei} - t_{ao} = Q_V\{(1/h_a \Sigma A) + (1/\rho c_p n V)\}$$

Hence substituting

$$C_V = \frac{1}{\{(1/h_a \Sigma A) + (1/\rho c_p n V)\}} \qquad [7.14]$$

For the general case where the heat input at the environmental point is eQ_i and that at the air point $(1-e)Q_i$, it can be shown that

$$C_V = \rho c_p n V \frac{\{(h_a \Sigma A) + (1-e)\Sigma(UA_o)\}}{\{(h_a \Sigma A) + (e\rho c_p n V)\}} \qquad [7.15]$$

Example 7.3

For a small building with an internal environmental temperature of 19 °C at an outside air temperature of -1 °C use the data below to plot the heat loss and the dry resultant temperature against the proportion of radiant heating in the range 0 % to 90 %.

Data

$\Sigma(UA_o) = 700$ W/K; $\Sigma A = 500$ m^2; air change rate, 0.6 h^{-1}; volume, 540 m^3.

Solution

Substituting in Eq. [7.15]

$$C_V = \frac{0.6 \times 450}{3} \frac{\{(4.5 \times 500) + (1 - e) \times 700\}}{\{(4.5 \times 500) + (e \times 0.6 \times 450/3)\}}$$

i.e.

$$C_V = \frac{(265\,500 - 63\,000\,e)}{(2250 + 90\,e)}$$

From Eq. [7.7], $e = 1.5\,\gamma$, where γ is the proportion of the heat input which is from radiant heating. Therefore in this case e varies from 0 to $(1.5 \times 0.9 = 1.35)$. C_V can then be calculated for a series of values of e and substituted into Eq. [7.11] to give Q_i. A table of values of C_V and Q_i can then be drawn up as below and plotted as shown in Fig. 7.7. The value of the dry resultant temperature is also shown; this is found referring to Fig. 7.3 as:

$$(t_{ei} - t_c)h_{ec}\Sigma A = (eQ_i - Q_F)$$

and in this case, $Q_F = (t_{ei} - t_{ao})\Sigma(UA_o) = 20 \times 700 = 14\,000$ W. Therefore

$$t_c = 19 - \frac{(eQ_i - 14\,000)}{18 \times 500}$$

(where Q_i is in watts).

	0 %	33.33 %	50 %	66.67 %	100 %
e	0	0.50	0.75	1.00	1.35
$C_V/(\text{W/K})$	118.0	102.0	94.2	86.5	76.1
$Q_i/(\text{kW})$	16.36	16.04	15.88	15.73	15.52
$t_c/(^\circ\text{C})$	20.6	19.7	19.2	18.8	18.2

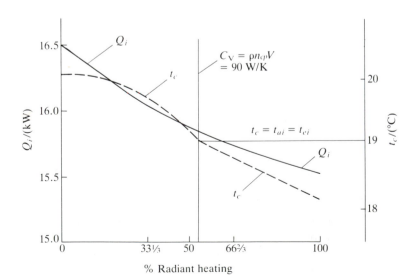

Figure 7.7 Heat input and dry resultant temperature variation for Example 7.3

When C_V from Eq. [7.15] is substituted in Eq. [7.11] an expression is obtained for Q_i with temperature difference based on environmental temperature. It is CIBSE practice to derive expressions for Q_i based on t_c for heating and on either t_c or t_{ai} for air conditioning. In the intermittent heating or air conditioning cases (considered in Section 7.2) the building control system is designed to maintain a constant internal temperature and therefore the expression used for the heat load should be the one with a temperature the same as the control parameter. Now a parameter measuring instrument will measure an equilibrium value such that the radiant heat falling on it is balanced by the heat convected from it (with conduction from it ignored); it is not immediately obvious which temperature will be sensed in any given control system. This point is not important in spaces such as offices etc. where the difference between the mean surface temperature and the internal air temperature is small, but it can be significant in situations where the radiant input is high.

A method based on dry resultant temperature will be outlined here; this implies that the comfort parameter, t_c, is able to be controlled. (Note that the definition of dry resultant temperature is that measured at the centre of a 100 mm diameter blackened sphere in air velocities below 0.1 m/s.) For the steady state the following expression can be derived, proceeding from Fig. 7.3 and using the simple electrical analogy method which was used to solve Example 7.2:

$$F_V Q_a + F_U Q_e = \{F_V \rho c_p nV + F_U \Sigma(UA_o)\}(t_c - t_{ao}) \qquad [7.16]$$

where $\qquad F_V = h_{ec}\Sigma A / \{(h_{ac}\Sigma A) + \rho c_p nV\}$

and $\qquad F_U = h_{ec}\Sigma A / \{(h_{ec}\Sigma A) + \Sigma(UA_o)\}$

Internal Gains

Modern buildings have a wide range of internal thermal gains the most common of which are from lighting and people. Offices nowadays have equipment with considerable thermal energy output, e.g. computer terminals, fax machines etc., and factories have an even wider range of process equipment and machinery. In the past in buildings such as offices, housing etc. it was common practice to ignore internal gains thus slightly overestimating the heating required in winter. Nowadays it is more usual to allow for these, particularly in summer when the need for air conditioning in modern high-technology buildings becomes more and more necessary. The human body has an average surface temperature of about 35 °C and the body metabolism maintains body temperature at this value as the activity level and air temperature changes. Threlkeld[7.5] gives a simple treatment of the energy interchanges between a human body and its surroundings.

Energy from lighting is an important topic and is considered separately in Section 7.4.

For machinery driven by an electric motor the total output of the motor electrical input appears eventually as a heat transfer to the space it occupies; when there is a separate motor room then only the effective output is transferred to the space housing the machinery.

The CIBSE *Guide*[7.1] gives values of heat emission from various types of equipment and a selection is given in Table 7.3. In summer, solar radiation is

Table 7.3 Some typical
energy outputs from
people and office
equipment

People	Output (W)
Light office work	140
Light factory work	235
Heavy factory work	440
Office equipment	
intelligent VDU	500
80 Mbyte disk drive	980
small photocopier, standby	750
small photocopier, running	1500
electric typewriter	50
electronic typewriter	100
small printer	450
coffee maker	400

an extraneous undesirable gain through glazing. In winter the energy of the sun can be used positively for space heating or hot water supply; in the UK the accepted view is that solar panels for hot water supply are uneconomic (unlike in Mediterranean Europe for example) but exploiting solar radiation for space heating is an economic proposition. The use of solar radiation as part of the heat load is known as *passive solar design*; a typical way of achieving this in housing is by using south-facing conservatories with high thermal mass floors with louvred openings carrying warmed air into the rooms of the house.[7.6]

In South East England the daily mean solar irradiances on a vertical surface facing south on a clear day in December and in June are about 115 W/m^2 and 155 W/m^2 respectively; the equivalent values for a horizontal surface are 40 W/m^2 and 315 W/m^2; on a cloudy day the equivalent values are 20 W/m^2 and 80 W/m^2 respectively (vertical or horizontal).

7.2 TRANSIENT HEATING AND AIR CONDITIONING LOADS

Under steady state conditions the heat loss from a building depends on the thermal transmittance and the ventilation rate as discussed in Section 7.1. A true steady state is never attained since the outside climate varies on a 24-hour cycle as well as from day-to-day. Also most buildings are not continuously heated or cooled; the energy input or output is controlled to maintain the required comfort conditions and most buildings are not occupied continuously for 24 hours and hence require intermittent loads.

When a material is heated up or cooled down the energy change depends on the density and specific heat of the material as well as on the thermal conductivity (see Section 3.5). In Table 7.4 different materials are compared to illustrate the importance of density and specific heat; the term, t, is the thickness of material required to give a thermal resistance of 1 m^2 K/W, and the term, E, is the energy required for a surface area of 1 m^2, to heat or cool the thickness, t, through a temperature rise or fall of 1 K. It can be seen that the heavyweight concrete needs a much greater thickness than expanded polystyrene for the same insulating effect but that it needs 3110 times as much energy to heat or cool that thickness.

Table 7.4 Comparison of materials for heat transfer and energy storage

Material	Density (kg/m^3)	Specific heat (J/kg K)	Thermal conductivity (W/m K)	t (mm)	E (MJ)
Heavyweight concrete	2100	840	1.00	1000	3.110
Brick	1700	840	0.84	840	1.200
Timber	600	1210	0.14	140	0.100
Wood wool	500	1000	0.10	100	0.050
Fibreboard	300	1000	0.05	50	0.015
Expanded polystyrene	25	1000	0.03	30	0.001

For buildings which are continuously heated the type of structure is unimportant for a given thermal transmittance; for intermittently heated buildings in winter then a lightweight structure will respond much more quickly than a heavyweight structure and hence is more economic. In summer on the other hand a lightweight structure will respond too quickly to the effects of solar radiation thus increasing the air conditioning load required whereas a heavyweight structure will tend to even out the temperature fluctuations.

A heavyweight structure can be modified to reduce the response time by adding a thin layer of a lightweight material to the *inside* surface; adding a lightweight material to the outside, or to the interior of the structure (e.g. adding insulation to a wall cavity) does not have the desired effect. The reason for this is given below.

Thermal Admittance

Transient heat flow through a structure is analogous to an AC electrical circuit with a resistance and capacitance in parallel (see Section 3.3). In the electrical circuit the capacitance causes the current to be out-of-phase and in the building structure the thermal capacity of the materials causes the heat flow to be out-of-phase with the temperature changes. In the electrical circuit the impedance, Z, is introduced as the ratio of the voltage to the current; the reciprocal of the impedance, $1/Z$, is called the admittance.

The thermal admittance is given the symbol Y and since it is a reciprocal of a resistance-type term then it has the same units as thermal transmittance (i.e. W/m^2 K). It can be defined as the rate of heat flow per unit surface area between the internal surface and a space for each degree of swing in the temperature of the space. The admittance, Y, determines the rate of energy fluctuation in a structure in the same way that the thermal transmittance, U, determines the heat flow rate. A small value of Y denotes a material of lightweight construction which will produce a large temperature fluctuation.

Table A3.16 in the CIBSE *Guide*[7.1] gives values of thermal admittance for typical building constructions; some typical values of Y and U are reproduced in Table 7.5 below. On page A3–45 of the CIBSE *Guide*[7.1] the derivation of admittance is given based on one-dimensional transient heat transfer with a sinusoidal temperature variation on a 24-hour cycle.

Table 7.5 Thermal admittance and transmittance for some typical constructions

Construction	Y (W/m² K)	U (W/m² K)
External brick, 105 mm thick	4.2	3.3
External brick, 220 mm thick	4.6	2.3
External brick, 335 mm thick	4.7	1.7
Asbestos cement sheet, 5 mm thick	6.5	6.5
External brick, 105 mm thick with plaster 13 mm thick on inside	3.3	2.6
As above but with an air gap of 25 mm	3.5	1.4

(Note: In calculating values of U in the above table, air film resistances were taken of 0.12 m² K/W and 0.06 m² K/W for the inside and outside surfaces respectively.)

The following points can be noted from Table 7.5.

(1) With reference to the brickwork values, the thermal transmittance, U, of a structure reduces with thickness but the admittance, Y, tends to a constant value as thickness increases above a certain value.

(2) With reference to the asbestos sheet, for a very thin layer of a material the values of U and Y are approximately equal.

(3) Comparing the brickwork with and without a layer of plaster it can be seen that adding a layer of a low admittance material to the inside surface decreases the admittance of the structure.

(4) Adding a second layer of bricks with an air cavity, although substantially reducing the value of U, has a very small effect on the admittance, Y, which is largely determined by the layer of plaster on the inside surface.

The CIBSE *Guide*[7.1] uses a harmonic method based on assuming a sinusoidal fluctuation of the outside temperature over a 24-hour cycle. It also assumes that the fabric comes into equilibrium with its surroundings; this is achieved if the weather conditions remain stable over, say, a five-day period. This is a reasonably good assumption for calculating heating loads where solar radiation is neglected. For air conditioning, the method is not so accurate; solar radiation, although cyclic, does not follow a 24-hour time period. Also, other internal gains occur only during occupancy and again do not conform to a sinusoidal variation.

Practice in the USA is to use a response factor method which allows for an hour by hour variation in parameters; the method is complex and lengthy but can be easily handled by a computer. For further details reference should be made to the handbook of the American Society of Heating, Ventilating, Refrigerating and Air Conditioning Engineers, ASHRAE.

Other methods used are based on a fundamental finite difference analysis of the heat transfer processes involved, leading to complete computer simulation of the structure's thermal behaviour. A simple summary of dynamic simulation methods in use is given in a series of articles.[7.7]

The CIBSE admittance method was originally developed at the Building Research Establishment by Millbank and Harrington-Lynn,[7.8] based on work by Danter and Loudon, and was designed to be used for manual calculations. Computer packages are available using the method but more accurate computer software now exists. The admittance model is used in this book since it provides a quick manual check on the required loads and is reasonably accurate particularly for heating loads.

The transient model is based on the same network as in Fig. 7.1 or Fig. 7.3 with swings in heat flow about the mean in place of steady state values, and swings in temperature about the mean in place of the mean values; the term $\Sigma(UA_o)$ is replaced by the term $\Sigma(YA)$; it should be noted that the latter term includes all surfaces and not just the surfaces conducting heat to the outside as in the term $\Sigma(UA_o)$.

For example, referring to the model of Fig. 7.1 and Eq. [7.11]

$$Q_i = \{(UA_o) + C_V\}(t_{ei} - t_{ao})$$

Then for the transient case

$$\tilde{Q}_i = \{\Sigma(YA) + C_V\}\tilde{t}_{ei} \qquad [7.17]$$

where $\tilde{Q}_i$ is the swing about the mean heat input, Q_i, and $\tilde{t}_{ei}$ is the swing in environmental temperature about the mean value, t_{ei}. Then

$$\frac{\tilde{Q}_i}{Q_i} = \frac{\{\Sigma(YA) + C_V\}\tilde{t}_{ei}}{\{\Sigma(UA_o) + C_V\}(t_{ei} - t_{ao})}$$

and the response factor, C_r, of the building is defined by

$$f_r = \frac{\{\Sigma(YA) + C_V\}}{\{\Sigma(UA_o) + C_V\}} \qquad [7.18]$$

The CIBSE *Guide*[7.1] defines response factor with the term $\rho nc_p V$ in place of C_V; the relationship between these terms is given by Eq. [7.15]. The numerical difference in response factor in a specific case is very small. A similar equation could be derived for a response factor based on dry resultant temperature, defined as follows:

$$\frac{\tilde{Q}_i}{Q_i} = f_r \frac{\tilde{t}_c}{(t_c - t_{ao})}$$

Then as an approximation the response factor can be written as

$$f_r = \frac{\{\Sigma(YA) + \rho nc_p V\}}{\{(\Sigma(UA_o) + \rho nc_p V\}} \qquad [7.19]$$

A building is classified as a heavyweight construction when the response factor is greater than or equal to 6, and as a lightweight construction when the response factor is equal to or less than 4.

Intermittent Heating

Assuming a constant internal design temperature over a 24-hour period then for the case of continuous heating the plant load can be found from the steady state equations considered previously. For the more common case of intermittent heating, the inside temperature will vary throughout the 24-hour period. A simple on-off system with n hours on and $(24-n)$ hours off would give the temperature and heat flow variations shown in Figs 7.8(a) and (b), where a cyclic variation in temperature is assumed. Q_{in} is the actual heat input during

Figure 7.8 Intermittent
heating

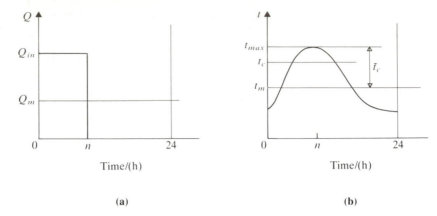

(a)   (b)

the heating period, and the average heat input over 24 hours is Q_m, where $Q_m = nQ_{in}/24$, and the fluctuation in Q_m is given by, $\tilde{Q}_i = Q_{in} - Q_m$. Similarly the mean temperature over 24 hours is t_m, and the maximum and minimum temperatures are as shown; the mean building temperature during the period when the building temperature is above the mean is, for a sine wave

$$t_c = t_m + \frac{2\tilde{t}_c}{\pi} \qquad\qquad [7.20]$$

where $\tilde{t}_c$ is the fluctuation of the temperature from the mean value, t_m. The value of the average heat input over the 24-hour period, Q_m, is given by Eq. [7.16], where $Q_m = Q_a + Q_e$.

For intermittent heating a similar equation can be written where the fluctuating heat input, $\tilde{Q}_i$, replaces Q_m, and the admittance is used instead of the thermal transmittance.

i.e. $$\tilde{Q}_i\{eF_V + (1-e)F_Y\} = F_V \rho c_p nV + F_Y \Sigma(YA)\tilde{t}_c \qquad [7.21]$$

where

$$F_Y = h_{ec}\Sigma A/\{(h_{ec}\Sigma A) + \Sigma(YA)\}$$

A plant operation ratio, r_P, is defined as the ratio of the rate of heat input during the heating period to the heat input rate required for the steady state at the same inside design temperature. In the CIBSE *Guide*[7.1] an expression is derived for the operation ratio assuming that the temperature response is as in Fig. 7.9 with the design dry resultant temperature the maximum value as shown.

Figure 7.9 Idealized
temperature response

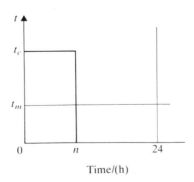

Example 7.4

A room 5 m by 4 m by 3 m high has an external wall 5 m by 3 m high with a window which occupies half the wall area. The rooms on all sides and above and below are at the same temperature as the room itself. The external wall is of cavity brick (105 mm brick) with 13 mm thick lightweight plaster on the inside surface, and the cavity filled with expanded polystyrene. The floor is 150 mm cast concrete with 50 mm screed and carpeted. The internal walls are of 100 mm heavyweight concrete block with 13 mm lightweight plaster on each side. There are 1.5 air changes per hour.

The building is heated intermittently with a 12-hour on period and a 12-hour off period. During the time period when the room temperature is above the 24-hour mean value, the mean dry resultant temperature is to be 20 °C. The outside design temperature is −3 °C.

Neglecting casual and solar gains, and assuming that the heat input is 45 % at the environmental point and 55 % at the air point, calculate:

(i)   the 24-hour mean dry resultant temperature;
(ii) the maximum and minimum dry resultant temperatures;
(iii) the rate of heat input during the heating period;
(iv) the rate of heat input required for continuous heating to obtain a dry resultant temperature of 20 °C;
(v)   the energy saved by intermittent heating when the 24-hour mean seasonal temperature is 6 °C for a 210-day heating season.

Solution

In the table above the values of U and Y have been taken from the CIBSE *Guide*;[7.1] note that a carpeted floor gives the same values as a wood block floor.

Item	Area (m^2)	U (W/m^2 K)	UA_o (W/K)	Y (W/m^2 K)	YA (W/K)
Window	7.5	5.60	42.00	5.6	42.0
External wall	7.5	0.69	5.18	3.6	27.0
Floor	20.0	—	—	2.9	58.0
Ceiling	20.0	—	—	6.0	120.0
Internal walls	39.0	—	—	4.4	171.6
	$\Sigma A = 94.0$		$\Sigma(UA_o) = 47.18$		$\Sigma(YA) = \overline{418.6}$

(i)   Using Eq. [7.16] to obtain the mean heat input, Q_m, we require F_U and F_V.
i.e.
$$F_U = h_{ec}\Sigma A /\{(h_{ec}\Sigma A) + \Sigma(UA_o)\} = 18 \times 94 /\{(18 \times 94) + 47.18\}$$
$$= 0.973$$

and
$$F_V = h_{ac}\Sigma A /\{(h_{ac}\Sigma A) + \rho c_p nV\}$$
$$= 6 \times 94 /\{(6 \times 94) + (1.5 \times 5 \times 6 \times 3/3)\} = 0.950$$

Then in Eq. [7.16]

$$Q_m = \frac{\{(0.950 \times 30) + (0.973 \times 47.18)\}(t_m + 3)}{(0.55 \times 0.950) + (0.45 \times 0.973)} = 77.48(t_m + 3)$$

[1]

For the intermittent case

$$F_Y = h_{ec}\Sigma A /\{(h_{ec}\Sigma A) + \Sigma(YA)\}$$
$$= 18 \times 94 /\{(18 \times 94) + 418.6\} = 0.802$$

Then in Eq. [7.21]

$$\tilde{Q}_i = \frac{(0.950 \times 30) + (0.802 \times 418.6)}{(0.55 \times 0.950) + (0.45 \times 0.802)}\tilde{t}_c = 412.32\tilde{t}_c \qquad [2]$$

Assuming a sine wave for the temperature variation then from Eq. [7.20] we have

$$t_c = 20 = t_m + 2\tilde{t}_c/\pi \qquad [3]$$

The heating period is 12 hours. Therefore for this case we have

$$Q_m = 12Q_{in}/24 \qquad \therefore \qquad Q_{in} = 2Q_m$$

Then, since $\tilde{Q}_i = Q_{in} - Q_m$

$$\therefore \qquad \tilde{Q}_i = 2Q_m - Q_m = Q_m$$

Hence in Eq. [1] above

$$\tilde{Q}_i = 77.48(t_m + 3)$$

Substituting from Eq. [3] in the above

$$\tilde{Q}_i = 77.48\{20 - (2\tilde{t}_c/\pi) + 3\} = 1782 - 49.33\tilde{t}_c \qquad [4]$$

Then substituting from Eq. [2] in Eq. [4]

$$412.32\tilde{t}_c = 1782 - 49.33\tilde{t}_c$$
$$\therefore \qquad \tilde{t}_c = 3.86 \text{ K}$$

Therefore, from Eq. [3]

$$\text{mean dry resultant temperature} = 20 - (2 \times 3.86/\pi) = 17.54\,°\text{C}$$

(ii)   The maximum and minimum temperatures are then given by

$$t_{max} = 17.54 + 3.86 = 21.40\,°\text{C}, \qquad t_{min} = 17.54 - 3.86 = 13.68\,°\text{C}$$

(iii)   The heat input during the heating period, Q_{in}, is given by $2Q_m$. Therefore from Eq. [1]

$$Q_{in} = 2 \times 77.48 \times (17.54 + 3) = 3182.9 \text{ W} = 3.183 \text{ kW}$$

(iv)   At the steady state for continuous heating with a dry resultant temperature of 20 °C the heat input is given by Eq. [1] as

$$Q_i = 77.48 \times (20 + 3) = 1782 \text{ W} = 1.782 \text{ kW}$$

Note that for a complete building a similar method could be used and a plant load found equivalent to Q_{in} above. The plant operation ratio would then be, for example

$$r_P = 3182.9/1782 = 1.786$$

(v)   The energy saved by intermittent heating is given by:

$$\{(1.782 \times 24) - (3.183 \times 12)\}\frac{(20-6)}{(20+3)} \times 210 = 584 \text{ kW h}$$

It can be seen from Figs 7.8(a) and (b) that the method is an approximation; the true temperature response is not a sine wave and it is necessary to start the heating before the period of occupancy to allow the temperature to attain an acceptable value before people arrive for work.

The period of heating before occupancy is known as the *pre-heat period* and this can be varied according to the severity of the weather outside using an optimum start control system.

An intermittently heated building with a plant capacity greater than the steady state capacity required for continuous 24-hour heating has a greater capital cost although the fuel running costs are less. The decision to oversize the heating plant in order to heat the building intermittently must be taken after a proper economic assessment.

It should also be remembered that a plant sized on a steady state basis for a low design outside temperature will have spare capacity in milder weather and can be used in an intermittent way with optimum start control. On the other hand the trend to highly insulated buildings is causing an increase in the response factor since $\Sigma(UA_o)$ is decreased without any significant change in $\Sigma(YA)$; larger plant margins may therefore become necessary.

Air Conditioning

For an air conditioned building the equations for the steady state and transient state are as before (i.e. Eqs [7.16] and [7.21]). In this case as the outside conditions vary cyclically in the model the inside temperature, say t_c, is held constant by providing the necessary air conditioning load; it is a good assumption to take the air conditioning input at the air point.

The inputs which cause the need for air conditioning are as follows.

Solar radiation

This is assumed to have two mean components, one at the air point and one at the environmental point, and is dealt with in the CIBSE model by using solar gain factors, i.e.

$$\text{solar gain at environmental point} = S_e I A_G$$
$$\text{solar gain at the air point} = S_a I A_G \qquad [7.22]$$

(where S represents solar gain factor defined as the ratio of the mean solar gain to the mean solar irradiance on the glazing; I is the mean solar irradiance; A_G is the area of the glazing). The fluctuating component can be included by equations similar to Eq. [7.22] with a factor $\tilde{S}$ in place of S, and the cyclic solar gain, $\tilde{I} = I' - I$, replacing the mean solar irradiance I. The fluctuating component lags the instantaneous component, I', by between 0 and 2 hours approximately depending on the response factor of the structure, and therefore I' is taken at a time one or two hours before the peak time.

There is also a cyclic solar gain from radiation due to the fluctuation in sol-air temperature which appears at the environmental point after a time lag and with a value less than the instantaneous value,

i.e. $\qquad Q_f = \Sigma(f U A_o \tilde{t}_{eo}) \qquad [7.23]$

For the fabric this term appears some hours after the peak load; $\tilde{t}_{eo}$ is the difference between the sol-air temperature at a time before the peak, and the mean sol-air temperature; f is the *decrement factor* defined as the ratio of the heat flow through the structure per degree in swing of outside temperature, to the steady state heat flow.

Casual gains internally

These are the gains due to people, office equipment, machinery etc. and the mean total value can be denoted by ΣQ_C and will be assumed to be an input at the air point. The total fluctuating component can be taken as

$$\Sigma \tilde{Q}_C = \Sigma \{ Q_C - (n Q_C / 24) \} \qquad [7.24]$$

(where n is the number of hours of application of each gain).

Total load

We therefore have mean heat input at the air point given by

$$Q_a = \Sigma Q_C + \Sigma (S_a I A_G) - Q_{AC} \qquad [7.25]$$

(where Q_{AC} is the air conditioning load)
and, mean heat input at the environmental point given by

$$Q_e = S_e I A_G \qquad [7.26]$$

Also, the fluctuating component at the air point is given by

$$\tilde{Q}_a = \Sigma \tilde{Q}_C + \Sigma (\tilde{S}_e \tilde{I} A_G) + \Sigma (f U A_o \tilde{t}_{eo}) + \Sigma (U_G A_G) \tilde{t}_{ao} \qquad [7.27]$$

(note: the last term in the above represents the fluctuating component conducted through the glass due to the swing in the outside air temperature; it is assumed that there is no time delay with this term and that the decrement factor is unity) and, the fluctuating component at the environmental point is given by

$$\tilde{Q}_e = \Sigma (\tilde{S}_e \tilde{I} A_G) \qquad [7.28]$$

In a sealed, air conditioned building the infiltration rate should be zero or negligibly small and hence no allowance has been made in the above treatment for fluctuating ventilation gains. For the case of a constant infiltration rate the fluctuating gain due to ventilation is given by $\rho c_p n V \tilde{t}_{ao}$, and this term should be included in Eq. [7.27].

From Eq. [7.16] for the steady state, using mean values and putting the air infiltration to zero, we have:

$$Q_a + F_U Q_e = F_U \Sigma \{ (U A_o)(t_c - t_{ao}) \}$$

(Note that $F_V = 1$ in this case since $\rho c_p n V = 0$.) In the above equation the appropriate sol-air temperature, t_{eo}, should be used in place of t_{ao} for outside surfaces of the fabric, but not for the glazing. From Eq. [7.21] using $\tilde{Q}_a$ and $\tilde{Q}_e$ and with $\rho n c_p V = 0$ as before we have

$$\tilde{Q}_a + F_Y \tilde{Q}_e = F_Y \Sigma (Y A) \tilde{t}_c$$

When the dry resultant temperature is controlled at a steady value then $\tilde{t}_c$ is zero in the above equation. Then, adding the two equations

$$Q_a + \tilde{Q}_a + F_U Q_e + F_Y \tilde{Q}_e = F_U \Sigma \{ (U A_o)(t_c - t_{ao}) \} \qquad [7.29]$$

The expressions from Eqs [7.25], [7.26], [7.27], and [7.28] can then be substituted in Eq. [7.29] to enable the mean air conditioning load, Q_{AC}, to be calculated.

Example 7.5

A room 5 m by 4 m by 3 m high has one external wall (5 m by 3 m) which faces south and which has a window, 3 m by 2 m. The window is single-glazed with normal exposure and white venetian blind fitted internally. There are two occupants who do light office work for 8 hours per day. For a typical sunny day in June using the data below calculate the required air conditioning load to maintain the inside dry resultant temperature at 18 °C. Assume that the rooms above, below, and at the sides of the room are at the same conditions as the room. Assume also that the room is well-sealed.

Data

Mean total solar irradiance 155 W/m²; peak total solar irradiance at solar noon, 535 W/m²; 24-hour mean external temperature, 16.5 °C; mean sol-air temperature, 20 °C; external air temperature at 1300 h, 20.5 °C; decrement factor for external wall, 0.31; time lag for glazing, 1 h; time lag for external wall, 9 h; sol-air temperature at 0400 h, 10 °C; heat gains for light office work per person, 140 W; heat gains from office equipment, 2000 W; average period of use of equipment, 4 h; $S_a = 0.16$; $S_e = 0.31$; $\tilde{S}_a = 0.17$; $\tilde{S}_e = 0.24$.

Take the values of U and Y as in the table in the solution below.

Solution

Item	A (m²)	U (W/m² K)	UA_c (W/K)	Y (W/m² K)	YA (W/K)
External wall	9	0.73	6.57	3.6	32.4
Window	6	5.60	33.60	5.6	33.6
Internal walls	39	—	—	3.5	136.5
Floor	20	—	—	2.9	58.0
Ceiling	20	—	—	6.0	120.0
$\Sigma A = 94$			$\Sigma(UA_o) = 40.17$		$\Sigma(YA) = 380.5$

Then

$$F_U = 18 \times 100/\{(18 \times 100) + 40.17\} = 0.978$$

and

$$F_Y = 18 \times 100/^E\{(18 \times 10) + 380.5\} = 0.826$$

From Eq. [7.25]

$$Q_a = \left\{\frac{(2 \times 140) \times 8}{24} + \frac{(2000 \times 4)}{24}\right\} + (0.16 \times 155 \times 3 \times 2) - Q_{AC}$$

i.e.

$$Q_a = 575.47 - Q_{AC} \qquad [1]$$

From Eq. [7.26]

$$Q_e = 0.31 \times 155 \times 6 = 288.3 \text{ W} \qquad [2]$$

From Eq. [7.27]

$$\tilde{Q}_a = \left\{(2 \times 140) - \frac{(2 \times 140 \times 8)}{24}\right\} + \left\{2000 - \frac{(2000 \times 4)}{24}\right\}$$

$$+ 0.17 \times 3 \times 2(535 - 155) + 0.31 \times 6.57(10 - 20)$$
$$+ 33.6(20.5 - 16.5)$$

i.e. $\qquad\qquad \tilde{Q}_a = 2355 \text{ W}$ [3]

From Eq. [7.28]

$$\tilde{Q}_e = 0.24 \times 3 \times 2(535 - 135) = 547.2 \text{ W} \qquad\qquad [4]$$

Then substituting Eqs [1], [2], [3], and [4] in Eq. [7.29]:

$$575.47 - Q_{AC} + 2355 + (288.3 \times 0.978) + (547.2 \times 0.826)$$
$$= 0.978\{6.57(18 - 20) + 33.6(18 - 16.5)\}$$

i.e. $\qquad\qquad$ air conditioning load $= 3627 \text{ W}$

The above can be applied to a complete building for continuous air conditioning. When intermittent air conditioning is used then an additional load can be added to the calculated plant load for continuous air conditioning to give the new plant load for the on-period. For a well-sealed building this is given by the following expression:

$$\Delta Q_P = \frac{\{F_Y \Sigma(YA) - F_U \Sigma(UA_o)\}}{\dfrac{(24 - n)}{24} F_U \Sigma(UA_o) + \dfrac{n}{24} F_Y \Sigma(YA)\}} \times \begin{array}{l} \text{(24-hour mean load} \\ \text{in the off period)} \end{array}$$

(where n is the number of hours of plant operation). For example, for a 12-hour on and off period for the room of Example 7.5

$$\Delta Q_P = \frac{\{(0.826 \times 380.5) - (0.978 \times 40.17)\}}{\{(0.5 \times 0.978 \times 40.17) + (0.5 \times 0.826 \times 380.5)\}}$$

$$\times \begin{array}{l} \text{(24-hour mean load} \\ \text{in the off period)} \end{array} = 1.56 \times \begin{array}{l} \text{(24-hour mean load in} \\ \text{the off period)} \end{array}$$

Then

$$Q_{AC} = 3672 + \Delta Q_P$$

7.3 THERMAL PERFORMANCE MONITORING

In Section 7.2 methods are described for obtaining plant loads for space-heating based on a chosen design outside temperature. During the heating season in the UK (September to May) the weather conditions vary from day to day and from month to month, and each year is different from the previous one. The heat required to maintain a building at a specified comfort temperature is directly proportional to the difference between the fixed inside temperature and the outside temperature and therefore the fuel used in heating the building varies with the 24-hour mean outside temperature over the heating season.

Degree Days

One convenient way of comparing fuel heating costs throughout the year and from year to year is by defining the weather conditions by *degree days*. This is defined as the daily difference in temperature between a base temperature and the 24-hour mean outside temperature when the base temperature is higher than the maximum daily temperature.

The *base temperature*, t_b, is the outside temperature above which no space-heating is required; the value normally chosen for the UK is 15.5 °C. It is assumed that internal gains from people etc. allow the required internal temperature to be maintained when the outside temperature is 15.5 °C and above. For certain buildings (e.g. hospitals) a higher base temperature is selected because a higher inside temperature must be maintained.

The 24-hour mean outside temperature is taken to be the average between the maximum and minimum outside temperatures. Hence the number of degree days in any one month is given by:

$$D = \Sigma\{t_b - \tfrac{1}{2}(t_{max} + t_{min})\} \qquad [7.30]$$

On any day for which $t_{max} > t_b$ then the degree days for that day are taken as *either* (i)   when $t_b > \tfrac{1}{2}(t_{max} + t_{min})$

$$D = \tfrac{1}{2}(t_b - t_{min}) - \tfrac{1}{4}(t_{max} - t_b)$$

or (ii)   when $t_b < \tfrac{1}{2}(t_{max} + t_{min})$

$$D = \tfrac{1}{4}(t_b - t_{min})$$

Example 7.6
Calculate the degree days for each of the three days shown below for the normal base temperature of 15.5 °C:
(i)   maximum temperature, 10 °C; minimum temperature, 2 °C;
(ii) maximum temperature, 16 °C; minimum temperature, 10 °C;
(iii) maximum temperature, 20 °C; minimum temperature, 14 °C.

Solution
(i)   In this case, $t_{max} < t_b$. Therefore

$$\text{degree days} = 15.5 - \tfrac{1}{2}(10 + 2) = 9.5$$

(ii) In this case, $t_{max} > t_b$, and $t_b > \tfrac{1}{2}(t_{max} + t_{min})$. Therefore

$$\text{degree days} = \tfrac{1}{2}(15.5 - 10) - \tfrac{1}{4}(16 - 15.5) = 2.875$$

(iii) In this case, $t_{max} > t_b$, and $t_b < \tfrac{1}{2}(t_{max} + t_{min})$. Therefore

$$\text{degree days} = \tfrac{1}{4}(15.5 - 14) = 0.375$$

Methods of correcting data for different base temperatures are given in Fuel Efficiency Booklet no. 7.[7.9]

The Department of Energy publishes monthly totals for degree days for 17 areas of the UK including an average value over the previous 20-year period. These are also published monthly in the CIBSE Journal *Building Services*, and are also available from the BSRIA Information Centre.

Table 7.6 gives degree day values for the Midlands and the South East of England for each month of 1986 and 1988, with the 20-year average values for 1968–1988. The years 1986 and 1988 are chosen as examples of a severe winter and an exceptionally mild winter respectively, this can be seen by comparing the values with the 20-year averages. It can be seen that degree days for the Midlands are higher than for the South East for each month, indicating that

Table 7.6 Degree day values for the Midlands and South East England

| Month | Degree days | | | | | |
| | 1986 | | 1988 | | 20-year average | |
	Midlands	SE	Midlands	SE	Midlands	SE
January	386	368	325	306	379	365
February	482	467	316	316	347	331
March	335	335	284	267	315	306
April	296	277	235	217	244	234
May	142	141	134	108	162	151
June	77	64	72	63	80	73
July	40	30	52	43	44	40
August	78	61	54	43	47	44
September	150	132	88	72	87	79
October	164	130	172	132	173	160
November	249	227	332	304	275	267
December	301	300	252	259	343	334
Total	2700	2532	2316	2130	2496	2384

the weather is milder in the South East and hence fuel heating costs are less in proportion. The largest number of degree days in the heating season is recorded in the North East of Scotland and the lowest number in the South West of England. In a typical year the ratio of degree days in the heating season for those two areas of the country is about 1.5 and hence the fuel costs are 50 % greater for a factory sited in the North East of Scotland compared with an identical factory in the South West of England, assuming the cost of fuel is the same in each region.

Thermal Energy Consumption

A typical well-run building has a space-heating load which varies with the outside temperature on a controlled basis; in addition the building has a hot water load and perhaps process thermal energy demands all of which remain substantially constant over the year.

If the weekly fuel consumption for thermal energy is plotted against the weekly mean outside temperature then two straight lines should be obtained as shown in Fig. 7.10. The base load represents the fuel used for the combined weekly hot water and process loads; any process demand which varies markedly during the year must be subtracted to avoid distorting the lines. It can be seen that the sloping line should cut the base load line at a temperature of 15.5 °C which is the temperature above which no heating is required (see earlier). The greater the number of readings taken the more accurate is the sloping line which can be drawn to give the best fit; plotting over a complete heating season will give the best results.

A least squares method can be used to obtain the best straight line through the points obtained. Representing fuel consumption by F and temperature by t, then the best straight line is

$$F = at + b \qquad [7.31]$$

Figure 7.10 Weekly fuel consumption against weekly mean outside temperature

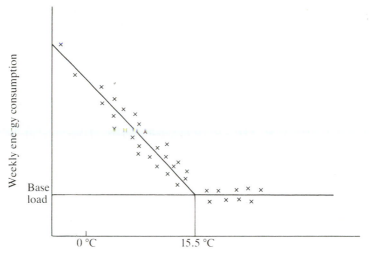

Then if individual readings are given the subscript i and mean values are given the subscript m we have:

$$a = \frac{\Sigma(F_i t_i) - nF_m t_m}{\Sigma(t_i)^2 - n(t_m)^2} \quad \text{and} \quad b = F_m - at_m$$

The closeness of correlation between F and t can be defined by a correlation coefficient, r, given by

$$r = \frac{\Sigma(F_i t_i) - nF_m t_m}{\sqrt{\{[\Sigma(t_i)^2 - n(t_m)^2][\Sigma(F_i)^2 - n(F_m)^2]\}}} \qquad [7.32]$$

If all points lie on the straight line then $r = 1$. (Many hand calculators have provision for performing these simple calculations.)

In the previous section degree days were defined and taken to be a convenient way of allowing for variations in weather conditions when comparing fuel consumption. Since degree days are more readily available than weekly mean outside temperatures then it is usually more convenient to plot fuel consumption against degree days on a monthly basis.

Figure 7.11 Fuel consumption for each month against degree days for that month

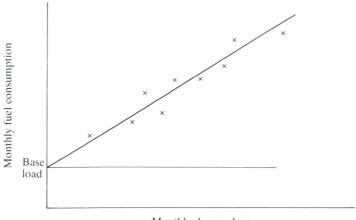

Fig. 7.11 is a typical graph of fuel consumption for thermal energy against degree days for a nine month heating season; a straight line can be fitted to the points by the least squares method described above (use Eq. [7.31] with t replaced by D, for degree days). The sloping line should cut the vertical axis at the monthly base load fuel consumption figure.

When a complete heating season has been plotted in this way then any subsequent heating season can be compared with the graph. Comparison will show whether changes to the building (e.g. new controls or improved insulation), have produced a true reduction in fuel consumption.

Example 7.7

The monthly gas consumption figures for a given year for a building are as shown in the table below with the degree days corresponding to the heating season for the region in which the building is sited.

(a) Plot gas consumption against degree days and hence estimate the base thermal load; check the likely reliability of the graph obtained by calculating the correlation coefficient.

Month	Gas consumption (therm)	Current year degree days	20-year average degree days
September	6 050	79	87
October	5 740	149	173
November	8 950	239	275
December	8 660	298	346
January	11 340	467	379
February	11 440	368	347
March	10 530	346	315
April	7 490	240	244
May	6 770	189	162

(b) For the building described, the fabric heat loss is found to be 75 % of the total heat loss with the ventilation loss 25 % of the total heat loss. It is proposed to upgrade the insulation of the building with all other factors unchanged and it is estimated that this measure will reduce the fabric heat loss by 10 %. Calculate the expected annual cash saving by introducing this measure and plot the new performance line on the graph; take a gas price of 37 p/therm.

Solution

(a) Solving for Eq. [7.31] using a simple programmable calculator gives

$$F/[\text{therm}] = 17.23 \, D/[\text{degree days}] + 4005$$

The base load is therefore approximately 4000 therm. From Eq. [7.32] the correlation coefficient, r, is found to be 0.94.

Figure 7.12 shows a graph of gas consumption against degree days with the nine points for the nine months of the heating season. The line drawn is the best straight line fit with a slope of 17.23 therm/degree day, and an intercept on the y-axis of 4005 therm. The slope of the line defines the heat loss characteristic of the building. Once a line is drawn for a given building then in subsequent years fuel consumption figures can be compared with this norm.

Figure 7.12 Gas consumption against degree days for Example 7.7

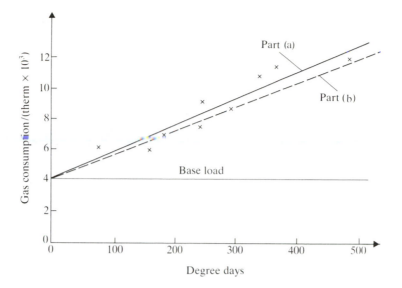

Degree days

Energy efficiency measures should reduce the slope of the line; changes to the base load without a change in the space-heating load will move the line up or down parallel to the original line.

(b) From part (a) the total gas consumption is 76970 therm for total degree days in the heating season of 2375. The gas used for hot water heating etc. is given by (9×4005) for the nine-month period, i.e. the gas used for space-heating is given by

$$76970 - (9 \times 4005) = 40925 \text{ therm}$$

The gas used for the fabric loss is 75 % of this total and hence the saving in gas by reducing the fabric loss by 10 % is given by

$$\text{annual gas saving} = 0.75 \times 40925 \times 0.1 = 3069.4 \text{ therm}$$

i.e. $$\text{annual cash saving} = (3069.4 \times 37)/100 = \pounds 1136$$

To find the new performance line it is necessary to use the 20-year average degree day total since the weather conditions for subsequent years are unknown. The degree day total for the 20-year average for the 9-month heating season is found to be 2328. Therefore the slope of the graph after the new insulation is installed is given by

$$\text{slope} = \frac{(40925 - 3069.4)}{2328} = 16.26$$

The new performance line can then be plotted as shown in Fig. 7.12.

The use of degree days as a means of compensating for weather conditions has been criticized on the grounds that the method ignores solar radiation, prevailing winds, and local outdoor temperature profiles for the building. It may be for these reasons that the correlation coefficient for the best line on the graph of fuel consumption against degree days is not always high enough.

Research at the Polytechnic of the South Bank by Levermore and Wong shows that the correlation coefficient can be improved for a building when energy efficiency methods are applied to it such as upgrading and re-setting compensator heating controls; their research therefore indicates that the degree day method of monitoring fuel consumption is reasonably accurate if the building is properly controlled. For further information on this topic consult Levermore.[7.10,7.11]

To obtain a better indication of the changes in fuel consumption as the heating season proceeds, due to efficiency measures taken, the cumulative fuel consumption can be plotted against the cumulative degree days. Since the total fuel consumption and the total degree days are known for the previous year then a straight line can be drawn for the previous year as shown in Fig. 7.13; as the next year proceeds the new line should slope below the existing line indicating the effectiveness of the measures taken.

Figure 7.13 Cumulative fuel consumption against cumulative degree days

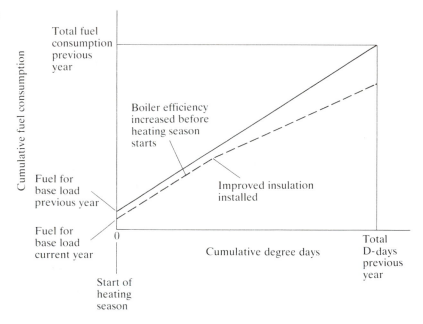

7.4 LIGHTING

In a factory the energy consumption due to lighting may be as low as 3 % of the total energy consumption but considerable cash savings can still be made by simple changes to the lighting system. In offices, shops etc. lighting may account for as much as 30 % of the total energy consumption and hence larger savings are possible.

In this section ways will be considered of reducing the annual lighting costs with comparatively small capital cost and with short pay-back periods. Savings are possible due in part to advances in lighting design and in part to the introduction of control systems and energy recovery methods.

Lighting Levels

The basic SI unit of luminous intensity is the *candela*, cd; it is defined such that the luminance of the total radiator at the centre of the temperature of the solidification of platinum (i.e. $1769\,^{\circ}$C), is 60 cd/cm^2.

Light from a point source illuminates a spherical surface surrounding it and so the light emitted is defined in terms of the solid angle; the unit of solid angle is the area of part of a spherical surface divided by the square of the radius of the sphere and is called the steradian, sr. We then define the *lumen*, lm, as the *luminous flux* emitted within unit solid angle by a point source with a luminance of 1 candela. The *lux*, lx, is then defined as an *illuminance* of 1 lumen per square metre.

CIBSE[7.12] gives some recommended values of illuminance for various interiors; some of these are quoted in Table 7.7. Daylight of a satisfactory standard precludes the need for artificial lighting and hence saves power and money. CIBSE[7.13] gives methods for estimating daylight levels in buildings. For a given shape of room with a given window area the number of hours per year for which the illuminance falls below a given value can be calculated. For example for a typical case there may be as many as 2200 hours per year when the illuminance from daylight is below 750 lx, and about 1100 hours for which it is below 320 lx.

Table 7.7 Illuminance recommended for some interior spaces

Interior space	Illuminance (lx)
Cable tunnels, walkways	50
Loading bays, switch rooms	150
General offices, engine assembly	500
Drawing offices	750
Tool rooms, electronic component assembly	1000
Assembly of minute mechanisms	2000

An electric light source producing a given amount of light from a given electrical power input has an *efficacy* (sometimes called efficiency) defined as the luminous flux in lumens divided by the power input in watts, lm/W.

For acceptable comfort light must also be of the right colour, contrast and with an absence of glare. In certain machinery applications the stroboscopic effect must be avoided; this occurs to varying degrees with certain types of lighting when the mains frequency of 50 Hz is close to the rotating speed of the machinery, i.e. 3000 rev/min.

Types of Lamps and Tubes

Incandescent

The most common type of lamp, or light bulb, is the *incandescent* type in which a filament of wire is heated electrically to a high temperature such that the radiation emitted is in the visible waveband. The wire used is made of tungsten and such lamps are therefore usually known as *tungsten lamps*; the basic tungsten lamp is frequently referred to as GLS, General Light Standard. Incandescent lamps are more efficient and have a longer life when filled with a halogen gas; various proportions of halogens are used but the type is generally known as *tungsten halogen*. In the UK the fitting for incandescent lamps is the simple bayonet type whereas in some other countries a screw-in bulb is used.

Fluorescent

In the *fluorescent tube* light is produced by the excitation of a gas (originally neon) or a metallic vapour, by an electric discharge; a coating of phosphorus, say, on the inside surface of the tube produces a white light. This type of light has a much higher efficacy than the incandescent lamp and also has a longer life. The connection is usually made by a two-pin fitting on each end of the tube.

Various improvements to the design have been made in recent years to increase the efficiency: krypton gas is replacing argon gas in the tube; tube diameters have been reduced; tri-phosphorus coatings provide more stability and better colour rendering; tubes have been made in a compact form in a size very little larger than the tungsten GLS lamp and with the same type of fitting.

The latter improvement is a major step forward since compact fluorescent tubes of high efficacy can replace tungsten incandescent lamps with no additional capital cost. Fluorescent tubes require control gear to start the discharge process and to provide the correct discharge rate. Compact lamps may have integral control gear using an electronic ballast.

High frequency fluorescent tubes have recently been introduced which operate on frequencies as high as 28 kHz instead of on the normal mains frequency of 50 Hz. Such tubes have an electronic ballast which increases the cost over the normal tube but the efficacy is increased by about 10 %. A further advantage of high frequency tubes is that the stroboscopic effect is eliminated.

Other discharge types

Lamps using the principle of discharge through a gas or metallic vapour but not necessarily using the fluorescent concept are also in common use.

The high pressure *mercury fluorescent* lamp produces light partly by discharge in the vapour and partly from ultraviolet radiation striking a fluorescent coating on the outer surface of the bulb; such lamps have been superseded by high pressure sodium lamps (see below).

The *metal halide* lamp produces light by radiation from a mixture of a metallic vapour (usually mercury) and the dissociated products of halides. Such lamps give good colour rendering and are used for floodlighting, studio lighting etc.

Low pressure sodium (SOX) lamps have a high efficacy but give an almost monochromatic yellow light; they are used for street lighting or security lighting.

High pressure sodium (SON) lamps are almost as efficient as the SOX lamps, have a much better colour quality, and are therefore very suitable for industrial use.

Such types of discharge lamp are not suitable for instant lighting since there is a warming-up period before the full illumination is obtained. For example, the mercury fluorescent lamp requires six minutes to reach full illumination with six minutes on re-start; the metal halide lamp requires 12 minutes with five minutes on re-start; the low pressure sodium requires eight minutes with about the same on re-start; the high pressure sodium requires six minutes with one minute on re-start. Table 7.8 gives some typical approximate figures of efficacy and life expectancy for a given illuminance and lamp power the examples chosen are not intended to indicate the range of use of any particular type but merely to show some simple comparisons between types. It can be seen that, for example, a compact fluorescent tube giving an illuminance of 1200 lm requires

Table 7.8 Efficiency and life expectancy for various lamps

Type	Illuminance (lm)	Power (W)	Efficacy (lm/W)	Life expectancy (h)
Tungsten GLS	1 200	100	12	1 000
Tungsten GLS	20 000	1000	20	2 000
Tungsten halogen	50 000	2000	25	4 000
Compact fluorescent	1 200	28	43	8 000
Fluorescent tube	4 500	70	64	10 000
High frequency tube	5 000	62	80	10 000
Low pressure sodium	30 000	210	143	12 000
Metal halide	160 000	2000	80	10 000
High pressure sodium	25 000	280	90	12 000

(Note: The efficacy quoted is the luminous flux divided by the total electrical input; in the case of all tubes and lamps other than the incandescent type the electrical input is greater than the rating due to the losses in the ballast; the efficacy is sometimes defined as the luminous flux divided by the rated output of the lamp.)

28 % of the power input of an equivalent Tungsten lamp yet on average it will last eight times as long.

The apparatus which fixes the lamp to the structure, protects it, and at the same time controls the distribution of light to the areas to be illuminated, is known as the *luminaire*. The normal working plane for lighting is taken to be a horizontal plane at a height of 0.85 m from the floor. A utilization factor can then be defined as the ratio of the flux on the working plane to the total flux from the lamp. Luminaires can be compared by their utilization factor.

Energy Saving

There are several ways in which the energy used for lighting may be reduced with a consequent reduction of annual running costs. The first and perhaps most obvious way is to check that each lamp is switched on only when required. Simple timers can be used and photocell switches connected in areas where daylight levels provide adequate lighting for parts of the period of use of the building. Certain types of lamp can be dimmed to lower levels of illuminance (e.g. incandescent and most fluorescent but not the high intensity discharge type). Future trends are towards microprocessor controlled lighting systems with the possibility of dealing with specific lamps or groups of lamps, and with sensors of the acoustic or infra red type which allow lamps to be switched on only when the space is occupied.

The choice of lamp and luminaire for the purpose required is obviously very important: too much light wastes energy but it has been shown by various studies that there is a strong correlation between productivity and lighting levels; it will also be clear from the previous section that choosing the right type for a specific purpose can lead to greater energy efficiency.

The cost of replacement must be considered as well as the running cost and hence the probable life of a lamp is an important factor; replacement cost includes the labour cost as well as the cost of the lamp itself and in fact the former may be the greater cost element.

Lighting costs can be reduced by proper maintenance; luminaires, tubes and bulbs must be cleaned at predetermined intervals depending on the cleanliness of the atmosphere in which they are sited; lamps and tubes should be replaced after a fixed number of hours of burning based on the probable life expectancy since if this is not done productivity may suffer when lights fail unexpectedly.

Example 7.8

A factory building of 15 000 m^2 floor area is lit by 500 twin fluorescent tube luminaires each tube having a rating of 68 W and an electrical input of 80 W with an efficacy of 60 lm/W and a life expectancy of 8000 h. It is proposed to replace the tubes with 100 high pressure sodium lamps of 560 W electrical input, efficacy 100 lm/W, and a life expectancy of 10 000 h. The building is occupied for 12 hours per day for five days per week for a 50-week year. Taking the cost of each sodium lump as £20, the cost of each fluorescent tube as £3, the labour cost for replacement as £5 per luminaire, the capital cost of installing the new system as £6000, and the price of electricity as 4.5 p/kW h, calculate:

(i) the percentage increase in illuminance;

(ii) the break-even time for the new system neglecting depreciation and inflation.

Solution

(i) Assuming that the efficacies given are based on the total electrical input then initial illuminance is

$$500 \times 2 \times 80 \times 60/15\,000 = 320 \text{ lx}$$

and final illuminance is

$$100 \times 560 \times 100/15\,000 = 373.3 \text{ lx}$$

i.e.   percentage increase in illuminance $= (373.3 - 320) \times 100/320$
$$= 16.67\%$$

(ii) The annual hours of use of lighting $= 12 \times 5 \times 50 = 3000$ h. Therefore the annual electricity cost is initially

$$2 \times 500 \times 80 \times 3000 \times 4.5/(100 \times 1000) = £10\,800$$

and finally

$$100 \times 560 \times 3000 \times 4.5/(100 \times 1000) = £7560$$

Initially the tubes have a life of 8000 h and hence should be replaced at time intervals of $8000/3000 = 2.67$ years. Therefore, the annual labour cost for replacement is

$$£(500 \times 5)/2.67 = £937.5$$

With the high pressure sodium lamps we have an annual labour cost for replacement of

$$£(100 \times 5) \times 3000/10\,000 = £150$$

Also, the annual cost of replacement tubes is initially

$$£(2 \times 500 \times 3)/2.67 = £1125$$

and finally

$$£(100 \times 20) \times 3000/10\,000 = £600$$

Hence we have an annual cash saving of

$$£\{(10\,800 - 7560) + (937.5 - 150) + (1125 - 600)\} = £4552.5$$

i.e.   break-even period $= 6000/4552.5 = 1.32$ years

Energy Recovery

The energy input to a tube or lamp appears in the space eventually as thermal energy. This can be important when calculating air conditioning loads for buildings in summer, and may be exploited to reduce the heating load required in winter. Some of the electrical input appears as short wavelength radiant energy (i.e. light) some as long wavelength radiant energy from the luminaire, some as convective heat from the luminaire to the air in the space, and some as conduction from the luminaire to the structure of the building. For a tungsten lamp the proportion is about 85 % radiation to 15 % convection plus conduction; for a fluorescent tube the proportion is about 45 % radiation to 55 % convection plus conduction; for a high pressure discharge lamp the proportion is about 62 % radiation to 38 % convection plus conduction.

In winter the thermal energy from conventional lighting is uncontrolled and may even cause discomfort to the occupants due to a high radiant flux. It is therefore desirable to design luminaires which are cooled so that the energy recovered can be used effectively while at the same time improving the comfort of the occupants.

The most frequently used heat-recovery luminaire draws air from the room across the tubes or lamps and delivers it to a plenum above the ceiling from where it can be passed to mix with the inlet air to the room. Less frequently water is used as the coolant fluid.

One point which must be taken into account when designing heat-recovery luminaires is that the efficacy of a typical fluorescent tube varies with the temperature of the tube surface. For example a typical fluorescent tube will give its maximum light output at a tube surface temperature of about 35 °C; if the tube surface temperature is cooled to 20 °C the light output will fall to about 75 % of the maximum; similarly if the tube surface temperature is allowed to rise to 70 °C the output will again be about 75 % of the maximum.

It can be seen that for the normal use of luminaires the way the luminaire is enclosed and ventilated is important in order to maintain optimum light output. For example it may be possible to use an enclosed, unventilated luminaire in spaces at low ambient temperature, but for normal room temperatures of say 20 °C it will be necessary to ensure that the luminaires are properly ventilated.

When the luminaire is designed for heat recovery then the heat transfer from the tube must not be too high otherwise the tube surface temperature could fall below the optimum value.

Example 7.9

An office is to be lit to a level of 500 lx with twin 68 W fluorescent tubes of 80 W electrical input with an efficacy of 60 lm/W. It is suggested that energy recovery luminaires could be used in which 75 % of the output from each tube can be recovered.

Estimate the total energy recovery possible from the office lighting, and the mass flow rate of air through each luminaire when the room air temperature is $20\,°C$ and the air entering the plenum is $28\,°C$. The office floor area is $115\,m^2$.

Solution

An illuminance of 500 lx requires $(500 \times 115) = 57\,500$ lm for the area of $115\,m^2$. Therefore, using twin-tubes of 80 W total electrical input, the number of luminaires required is

$$\frac{57\,500}{60 \times 80 \times 2} = 5.99, \text{ say } 6$$

Then

$$\text{total energy recovery} = 0.75 \times 6 \times 2 \times 68 = 612 \text{ W}$$

and

$$\text{air mass flow rate per luminaire} = \frac{612}{6 \times 1005(28 - 20)}$$
$$= 0.0127 \text{ kg/s}$$

(where the specific heat of air at constant pressure is taken as $1005\,J/kg\,K$).

7.5 ENERGY TARGETS

The thermal and electrical demands of a building can be calculated using the theory outlined in this chapter. Hot water requirements depend on the type of building, the nature of its use, and the number of occupants; CIBSE[7.14] gives a simple method of calculating this demand.

In the CIBSE method used to calculate energy demands[7.14] the quantities are calculated in primary energy terms by multiplying the actual energy used by the reciprocal of the overall efficiency of the generation of the energy from a primary source, (for example electrical loads are multiplied by 3.82). Any thermal demand from process heaters etc. must be subtracted from the energy consumption so that the true building thermal demand is obtained.

To calculate the lighting load it is necessary to know the hours of occupancy and the diversity factor for the lights; the latter depends on the daylight which in turn depends on the amount, size and position of glazing in relation to the whole building. Apart from lighting, the electrical demand depends on the number of electrical motors driving fans, pumps and other items of electrical equipment related to the thermal demand. Other electrical power consumption, for example for lifts, machinery etc., is not taken as part of the building energy demand.

For an electrically heated building the electrical demand for heating is part of the thermal demand and not the electrical demand as defined above.

Targets

Thermal and electrical demand targets can be calculated for particular categories of buildings; the thermal and electrical demands of a particular building can then be compared with the target value. The thermal demand target can be calculated for a particular building type using a building envelope number and various constants which define the type of structure (see for example CIBSE[7.14]). A lighting target can be calculated using recommended levels of illuminance and total floor areas; it is also recommended[7.14] that to allow for the electrical demand from fans, pumps etc. 10 % of the thermal demand should be added to the lighting target to give the total electrical demand target.

An alternative way of establishing a target is to survey a large number of buildings of different types, patterns of occupancy, and location, and to identify good and bad examples of energy efficient buildings from such a survey. Targets can then be set. The Energy Efficiency Office produced the results of such a survey in 1985[7.15] for four economic sectors: industry; offices; retail and distributive trades; hotels and catering. The figures are based on the 1985 fuel and electricity prices. In Table 7.9 some of these figures have been converted into general energy terms to give a broad guide to energy targets.

Table 7.9 Energy targets for some building types

Building		Energy target (MJ/m^2)		
		Good	Satisfactory	Very poor
Factory	electrical	<65	65–80	>430
	thermal	<550	550–830	>1380
Warehouse	electrical	<65	65–70	>325
	thermal	<430	430–460	>1170
Office	electrical	<85	85–105	>245
	thermal	<640	640–700	>1405
Shop (air conditioned)	electrical	<150	150–160	>260
	thermal	<310	310–370	>1080
Hotel (air conditioned)	electrical	<240	240–290	>430
	thermal	<1075	1075–1200	>1780
Secondary school	electrical	<135	135–170	>170
	thermal	<765	765–950	>950

Notes:

(i) The secondary school type quoted has an indoor swimming pool.
(ii) The figures for the school are adapted from the Energy Office publication on schools, (ref 7.16).
(iii) The figures shown are based on approximate annual hours of occupancy as follows: factories, 2 268 h; offices, 2 268 h; shops, 2 584 h; hotels, 8 760 h; schools, 1 500 h.
(iv) For Scotland increase the figures by 10%; for South West England decrease the figures by 10%.

Example 7.10
A multi-storey office building has a total floor area of 11 500 m². The annual gas consumption is 77 000 therm; the installed lighting has an illuminance of 500 lx with an efficacy of 74 lm/W, and the annual consumption of electricity due to fans, pumps etc. is found to be 12 000 kW h. Assuming an occupancy of 2268 hours per year, calculate the annual electrical and thermal demands and compare these with the targets in Table 7.9.

Solution
The annual thermal load is

$$77\,000 \text{ therm} = 77\,000 \times 0.1055 \text{ GJ}$$

i.e. the annual thermal load per unit area is equal to

$$\frac{77\,000 \times 0.1055 \times 10^3}{11\,500} = 706.4 \text{ MJ/m}^2$$

The illuminance is $500 \text{ lx} = 500 \text{ lm/m}^2$, and therefore the installed rating is

$$500/74 = 6.76 \text{ W/m}^2$$

i.e.   annual lighting load $= \dfrac{6.76 \times 2268 \times 3600}{1000 \times 1000} = 55.2 \text{ MJ/m}^2$

and   annual electrical load $= \dfrac{(12\,000 \times 3600)}{1000 \times 11\,500} + 55.2$

$$= 3.8 + 55.2 = 59 \text{ MJ/m}^2$$

From Table 7.9 for an office occupied for 2268 hours per year we can see that the thermal demand of 706.4 MJ/m^2 is just outside the satisfactory limit of 700 MJ/m^2 and hence attention needs to be given to improving the energy efficiency of the heating and hot water system. At least 10 % decrease in the thermal load should be possible. From Table 7.9 the electrical demand is seen to be well below the target for a good building of 85 MJ/m^2.

Estimating Thermal Demand for Space-Heating

In Sections 7.1 and 7.2 calculation methods for the required heat input (and hence plant size) are based on a low design outside temperature based on weather data for the region in which the building is sited. For example the design 24-hour mean temperature for a lightweight building is taken to be $-3\,°\text{C}$ in London and $-5\,°\text{C}$ in Birmingham.[7.1]

To estimate the average rate of supply of energy for space-heating in the heating season the outside temperature used must be the 24-hour mean averaged over the heating season (taken to be $7\,°\text{C}$ in London, and $6.5\,°\text{C}$ in Birmingham, for example). The rate of supply of energy for space-heating can be calculated using the methods outlined in Section 7.2 simply substituting the 24-hour mean outside temperature averaged over the heating season for the 24-hour mean design outside temperature.

The space-heating load in Joules is then obtained by multiplying the average rate of energy supply in Watts by (24×3600) times the number of days in the heating season; the total time is taken since the rate of energy supplied is based on the 24-hour mean inside and outside temperatures.

To obtain the annual fuel supplied for space-heating the heating load thus obtained must be divided by the seasonal efficiency of the device producing the heat supply (e.g. boiler).

The hot water demand is usually constant throughout the year and hence the rate of energy supplied for this purpose should be multiplied by the annual hours of occupancy.

In the CIBSE Energy Code[7.14] an approximate method of calculating the rate of energy supply for space-heating for an intermittently heated building is given as follows:

rate of energy supply for space-heating is

$$\{\Sigma(UA_o) + \rho n c_p V\}(t_m - t_{ao}) \qquad [7.33]$$

where t_{ao} is the 24-hour mean outside temperature averaged over the heating season; t_m is the 24-hour mean dry resultant temperature during the heating season, and is related to the design dry resultant temperature, t_c, by the following equation:

$$(t_m - t_{ao}) = \frac{nf_r}{\{(nf_r) + (24 - n)\}}(t_c - t_{ao}) \qquad [7.34]$$

(where f_r is the response factor given by Eq. 7.19).

Example 7.11

The data below has been calculated for a small factory which has a design dry resultant temperature of 13 °C and is intermittently heated for 10 hours per day by forced circulation warm air.

(a) Calculate the space-heating fuel energy load per unit floor area during the heating season.

(b) Re-calculate (a) using the approximate formulae.

Data

Total surface area, 550 m^2; floor area, 160 m^2; $\Sigma(UA_o) = 1850$ W/K; $\Sigma(YA) = 2600$ W/K; $\rho c_p nV = 180$ W/K; 24-hour mean outside temperature averaged over the heating season, 7 °C; number of days in heating season 175; overall seasonal heating efficiency, 70 %.

Solution

(a) For warm air heating the input is entirely at the air point. Using Eqs [7.16] and [7.21] we have

$$F_U = 18 \times 550/\{(18 \times 550) + 1850\} = 0.8426$$
$$F_V = 6 \times 550/\{(6 \times 550) + 180\} = 0.9483$$
$$F_Y = 18 \times 550/\{(18 \times 550) + 2600\} = 0.7920$$

and $\qquad Q_m = \dfrac{\{(0.9483 \times 180) + (0.8426 \times 1850)\}(t_m - 7)}{0.9483}$

i.e. $\qquad Q_m = 1823.79(t_m - 7)$ W $\qquad\qquad [1]$

Also $\qquad \tilde{Q} = \dfrac{\{(0.9483 \times 180) + (0.792 \times 2600)\}}{0.9483}\tilde{t}_c$

i.e. $\qquad \tilde{Q} = 2351.47\,\tilde{t}_c$ W $\qquad\qquad [2]$

The design dry resultant temperature of 13 °C (given) is:

$$13 = t_m + \tilde{t}_c \qquad\qquad [3]$$

Also

$$Q_{in} = Q_m + \tilde{Q} \qquad \text{and} \qquad Q_m = nQ_{in}/24 = 10Q_{in}/24$$

i.e. $\qquad Q_m = (Q_m + \tilde{Q})/2.4$

$\therefore \qquad Q_m = 0.714\tilde{Q} \qquad\qquad [4]$

From Eqs [2] and [4]

$$Q_m = 1678.95\,\tilde{t}_c \text{ W}$$

Substituting in [1]

$$1678.95\,\tilde{\iota}_c = 1823.79(t_m - 7)$$

i.e. $\tilde{\iota}_c = 1.086t_m - 7.604$

Substituting in (3)

$$13 = t_m + 1.086t_m - 7.604$$

i.e. $t_m = 9.88\,°C$

Then substituting in [1] the mean heat input over one day is given by

$$Q_m = 1823.79(9.88 - 7) = 5247.4 \text{ W}$$

The heating season is 175 days, the seasonal heating efficiency is 70 %, and the floor area is 160 m². Therefore

total fuel energy load for space-heating
$$= 5247.4 \times 175 \times 24 \times 3600/(160 \times 0.7 \times 10^6)$$
$$= 708.4 \text{ MJ/m}^2$$

(b) The response factor is given by Eq. [7.19]

$$f_r = \frac{2600 + 180}{1850 + 180} = 1.37$$

Then from Eq. [7.34]:

$$(t_m - t_{ao}) = \frac{10 \times 1.37}{(10 \times 1.37) + (24 - 8)}(13 - 7) = 2.77 \text{ K}$$

i.e. $t_m = 7 + 2.77 = 9.77\,°C$

(compare with previous value of 9.88 °C). Substituting in Eq. [7.33]

$$Q_m = (1850 + 180) \times 2.77 = 5623.1 \text{ W}$$

This is high compared with the previous value of 5247.4 W. Equation [7.33] will always give a high answer since the term in brackets $\{(\rho nc_p V) + \Sigma(UA_o)\}$ ignores all the 'F' terms in Eq. [7.16].

i.e.   total fuel energy load for space-heating $= \dfrac{5623.1 \times 708.4}{5247.4}$
$$= 759.1 \text{ MJ/m}^2$$

Problems

7.1 An office on the first floor of a building has one external wall, 12 m long, containing windows of total area 16 m². The width of the room is 6 m and the height is 3 m. There is a corridor adjacent to the inside wall which is parallel to the external wall, and there are similar offices above and on either side of the office; the room below is a store-room. Using the data below, calculate for the steady state:
(i) the inside air temperature;
(ii) the ventilation conductance;
(iii) the total heat input required for the office.

Data

Thermal transmittances: external wall, 0.73 W/m² K; windows, 3.6 W/m² K; internal walls, 1.8 W/m² K; floor and ceiling, 1.3 W/m² K.

Design environmental temperatures: offices, 19 °C; corridor, 16 °C; store-room, 10 °C; outside, −1 °C.

Air change rate, 1 h⁻¹.

Take the mode of heating as 100 % convective.

(21.2 °C; 79.9 W/K; 4.08 kW)

7.2   An office 6 m by 4 m by 3 m high has one external wall 6 m long, which has a window of area 10 m². There is a corridor on one side of the 6 m long internal wall, and there are similar offices on either side of the 4 m long walls as well as above and below the office. Using the data below, assuming that all the adjacent rooms other than the corridor are at the same environmental temperature as the office, calculate for the steady state:

(i)   the inside air temperature;

(ii) the rate of heat input required for the office;

(iii) the inside environmental temperature for the office.

Data

Thermal transmittances: external wall, 0.7 W/m² K; corridor wall, 1.7 W/m² K; window, 2.9 W/m² K.

Dry resultant temperature for office, 20 °C; outside environmental temperature, −1 °C; environmental temperature for corridor, 16 °C.

Air change rate, 1 h⁻¹.

Take the mode of heating as two-thirds radiant and one-third convective.

(19.25 °C; 1.35 kW; 20.25 °C)

7.3   A room which is heated entirely by convection has a total steady state heat input of 1307 W. Using the data below, calculate:

(i)   the inside air temperature;

(ii) the air infiltration rate.

Data

Inside environmental temperature, 20 °C; outside environmental temperature, −2 °C; total inside surface area, 94 m²; $\Sigma(UA_o) = 38.5$ W/K.

(22 °C; 0.016 m³/s)

7.4   A factory used for heavy work is 20 m long by 10 m wide by 5 m high and is heated by radiant panels with a 36 kW output. Using the data below and that of Table 7.2, (page 253), calculate for the steady state:

(i)   the dry resultant temperature, stating whether the factory may be deemed to be comfortable, assuming that the humidity level is satisfactory;

(ii) the inside air temperature;

(iii) the air infiltration rate.

Data

Thermal transmittances: external walls, 0.8 W/m² K; windows, 5.6 W/m² K; floor, 0.48 W/m² K; flat roof, 0.75 W/m² K.

Total window area, 100 m². Mean air velocity internally, 0.4 m/s.

Internal environmental temperature, 18 °C. External design temperature, −5 °C.

Take the mode of heating as two-thirds radiant and one-third convective.

(16.91 °C, too warm; 13.63 °C; 0.617 m³/s)

7.5 For the factory of Problem 7.4 calculate the required total heat input, the inside air temperature, and the inside environmental temperature, when the dry resultant temperature is taken at the comfort value. Assume the same infiltration rate as before.

(32.21 kW; 11.67 °C; 15.58 °C)

7.6 The mode of heating in the factor of Problem 7.4 is changed to 30 % radiant, 70 % convective (i.e. double panel radiators). Assuming the comfort value of dry resultant temperature from Table 7.2, (page 253), and with all other data unchanged, calculate:
(i)   the inside environmental temperature;
(ii) the inside air temperature;
(iii) the total heat input required.

(14.33 °C; 15.43 °C; 33.78 kW)

7.7 An office 6 m by 4 m by 3 m high has one external wall, 6 m long, which has a double-glazed window of area 10 m². All spaces adjacent to the office are at the same internal environmental temperature as the office. The building is heated with 12-hour on and off periods and it may be assumed that the dry resultant temperature follows a sinusoidal variation with time. Neglecting solar and casual heat gains and using the data below, calculate:
(i)   the 24-hour mean dry resultant temperature;
(ii) the minimum dry resultant temperature;
(iii) the design rate of heat input during the heating period;
(iv) the design rate of heat input required for continuous heating to obtain a dry resultant temperature equal to the design (i.e. maximum value) in intermittent heating;
(v)   the energy saved per year and the percentage reduction in fuel costs by using intermittent heating.

Data
Thermal transmittances: external wall, 0.66 W/m² K; window, 3.3 W/m² K.
Thermal admittances: external wall, 0.98 W/m² K; window 3.3 W/m² K; internal walls, 2.8 W/m² K; floor, 3.2 W/m² K; ceiling, 2.4 W/m² K.
Design (in this case the maximum) dry resultant temperature, 20 °C; design outside environmental temperature, −5 °C; 24-hour mean seasonal temperature, 6 °C, for a 210-day heating season. Air changes, 1 h⁻¹.
Take the mode of heating as one-third radiant and two-thirds convective.

(15.73 °C; 11.46 °C; 2.59 kW; 1.56 kW; 752 kW h, 17.08 %)

7.8 Repeat Problem 7.7 assuming that the design value of dry resultant temperature of 20 °C is the mean temperature in the part of the cycle when the temperature is above the 24-hour mean value.

(18.00 °C; 14.85 °C; 2.87 kW; 1.56 kW; 353 kW h; 8.01 %)

7.9 A small workshop, 20 m by 8 m by 3 m high at the eaves, is heated by forced circulation warm air heaters. It is adjacent, along one of the 8 m long walls, to another building with the same internal environmental temperature; the other three walls of the workshop are external and there is a total window area of 80 m² in these walls. The roof ridge is central and runs parallel to the 20 m long walls; the height of the ridge is 6 m. The building is heated intermittently with an eight-hour operating period. Using the data below, calculate:

(i) the 24-hour mean dry resultant temperature;

(ii) the required plant load.

Data

Design (maximum) dry resultant temperature, 16 °C. Design outside temperature, −2 °C.

Thermal transmittances: external walls, 0.73 W/m² K; windows, 5.3 W/m² K; pitched roof, 6.5 W/m² K; floor, 0.45 W/m² K.

Thermal admittances: external walls, 3.6 W/m² K; windows, 5.3 W/m² K; internal wall, 3.6 W/m² K; pitched roof, 6.5 W/m² K; floor, 2.9 W/m² K.

Air infiltration rate, 0.75 h⁻¹.

(5.04 °C; 27.6 kW)

7.10 A small well-sealed laboratory is continuously air conditioned at a constant dry resultant temperature. The laboratory is occupied for eight hours per day by five people and there is equipment which is in use for five of the eight-hour occupancy. There is one external wall facing South West, and the flat roof is also exposed. Using the data below, assuming that the optimum load occurs at 1400 h solar time, calculate the air conditioning plant load for the design conditions given.

Data

Thermal transmittances: external wall, 0.6 W/m² K; flat roof, 1.6 W/m² K; window, 3.3 W/m² K.

Total surface area, ΣA, 190 m²; window area, 20 m²; external wall area, 10 m²; roof area, 50 m²; $\Sigma(AY) = 760$ W/K.

Mean total solar irradiance, 180 W/m²; peak total solar irradiance, 585 W/m². Dry resultant temperature, 20 °C; mean sol-air temperature for external wall, 24.5 °C; sol-air temperature for external wall at 0500 h, 11.5 °C; mean sol-air temperature for roof, 26 °C; sol-air temperature for roof at 1200 h, 50.5 °C; external air temperature at 1400 h, 21 °C; mean external air temperature, 16.5 °C. Decrement factor for external wall, 0.28; time lag for external wall, 9 h; decrement factor for roof, 0.96; time lag for roof, 2 h.

Mean solar gain factor at environmental point, 0.26; alternating solar gain factor at environmental point, 0.21; mean solar gain factor at air point, 0.19; alternating solar gain factor at air point, 0.21.

Energy output per person, 140 W; energy output for equipment, 1500 W.

(9.84 kW)

7.11 Using the normal base temperature of 15.5 °C calculate the degree days for the following days:

(i) maximum temperature, 8 °C; minimum temperature, −2 °C;

(ii) maximum temperature, 17 °C; minimum temperature, 5 °C;

(iii) maximum temperature, 22 °C; minimum temperature, 10 °C.

(12.5; 4.875; 1.375)

7.12 A building heated by gas oil has one boiler house which provides space-heating and hot water heating; the only records of fuel consumption are of the combined consumption of oil given in the table below. The table gives the degree days for the past year's heating season and the 20-year average figures.

(a) Plot oil consumption against degree days, find the best straight line which fits the points, and hence determine the total hot water heating load during the heating season.

(b) It is proposed to reduce fuel costs by two measures: improving the insulation of the roof of the building; adjusting the boiler to improve its efficiency. Using the data below, calculate the reduction in fuel cost for the heating season and draw the expected line of oil consumption against degree days on the graph.

Data

Boiler efficiency initially, 75 %; boiler efficiency finally, 80 %. Space heating load: 20 % ventilation loss, 80 % fabric loss; heat loss through roof initially, 50 % of fabric loss; reduction in thermal transmittance of roof, 50 %. Density of gas oil, 840 kg/m^3; calorific value of gas oil, 45.5 MJ/kg; cost of gas oil, 19 p/litre.

Month	Oil consumption (l)	Current year degree days	20-year average degree days
September	10 000	69	87
October	27 000	227	173
November	35 000	246	275
December	34 000	294	346
January	41 000	340	379
February	36 000	369	347
March	33 000	243	315
April	25 000	245	244
May	20 000	147	162

(1465 GJ; £10 577)

7.13 A hotel has 60 corridor luminaires each with two tungsten light bulbs which are continuously lit throughout the year. It is proposed to replace the light bulbs with a single bent fluorescent tube which will fit into the same luminaire. Using the data below, calculate:

(i) the percentage increase in illuminance in the corridor;
(ii) the annual cash saving for the new system.

Data

Tungsten light bulbs, 60 W; fluorescent tube, 20 W lighting and 28 W circuit load; efficacy of tungsten bulb, 12 lm/W; efficacy of fluorescent tube, 60 lm/W; average life of tungsten bulb, 1000 h; average life of fluorescent tube, 8000 h; cost of tungsten bulb, 65 p; cost of fluorescent tube, £9; labour cost for replacement of lamps, £2 per luminaire; average cost of electricity, 3 p/kW h.

(16.7 %; £2462)

7.14 A warehouse of 15 000 m^2 area is illuminated by 300 twin fluorescent tube luminaires. The existing lighting is to be replaced by high intensity sodium, SON, lamps to give approximately the same illuminance. Using the data below, calculate:

(i) the required number of SON lamps;
(ii) the break-even period for the SON system.

Data

Fluorescent tube, 65 W rating and 80 W circuit load; efficacy of fluorescent tube, 60 lm/W; SON lamp, 250 W rating and 280 W circuit load; efficacy of SON lamp, 110 lm/W; average life of fluorescent tube, 5000 h; average life of SON lamp, 12 000 h; cost of fluorescent tube, £3; cost of SON lamp, £20; labour cost to replace a lamp, £5 per luminaire; capital cost of replacing fluorescent tube luminaires with SON luminaires, £7500; lighting is required for 3000 h per year; average cost of electricity, 3 p/kW h.

(94; 2.68 years)

7.15 A small specialized workshop of floor area 200 m² is to be lit to an illuminance of approximately 1000 lx by high frequency fluorescent tubes of efficacy 80 lm/W with a circuit load of 62 W. Water-cooled luminaires are installed with the water flowing in parallel through each luminaire. The heated water is pumped through a heat exchanger which acts as the heat sink for a heat pump used to heat air for building heating purposes. Using the data below, calculate:
(i)   the number of luminaires required;
(ii) the total mass flow rate of water required;
(iii) the rate of energy input to the air for building heating;

Data
Operating temperature of fluorescent tube, 30 °C; temperature of water at entry to luminaire, 10 °C; minimum temperature difference required for heat transfer to the water, 8 K; energy recovery from circuit power input, 70 %; overall coefficient of performance of heat pump, 3.

(40; 0.0364 kg/s; 2.6 kW)

7.16 For the workshop of Problem 7.9 the hot water demand through the year is at a rate of 2 kW. The 24-hour mean outside temperature during the heating season can be taken to be 6.5 °C. The annual hours of occupancy are 2268 and the length of the heating season is 175 days. The average boiler efficiency for both heating and hot water provision may be taken as 70 %.
(a) Using the approximate Eqs [7.33] and [7.34], calculate the fuel energy thermal demand per unit floor area and with reference to Table 7.9, (page 285), determine whether the factory meets the thermal energy demand target.
(b) With reference to the fabric of the workshop state any realistic improvements and estimate the reduction that might be possible in the fuel thermal energy demand. Assume that the hot water demand, hours of usage, length of heating season, boiler efficiency etc. are all unchanged.

(1202 MJ/m²; very unsatisfactory; insulate roof, double-glaze windows; 600 to 700 MJ/m²)

7.17 For the factor of Problem 7.4 the lighting level is 500 lx using lighting with an efficacy of 60 lm/W. The lighting is used continuously during the annual occupancy period of 2268 hours. The annual electrical load required for the heating and hot water system is found to be 250 kW h. Determine whether the factory meets a satisfactory electrical demand target based on Table 7.9.

(72.5 MJ/m²; satisfactory)

References

7.1 CIBSE *Guide to Current Practice* A 1986
7.2 Loudon A G 1970 Summertime temperatures in buildings without air conditioning. *J. Inst. Heat. Vent. Engrs* **37**: 280–92

7.3   Davies M G 1988 What should we do about environmental temperature? *Building Services* **10** no 5: 55–8 May

7.4   Fanger P O 1982 *Thermal Comfort* Robert E Kreiger, Florida

7.5   Threlkeld J L 1970 *Thermal Environmental Engineering* Prentice Hall

7.6   *Passive Solar Energy in Buildings* Watt Committee Series, No. 17, Elsevier, 1989

7.7   Clarke J, Holmes M J, Bowman N, Lomas K 1986 Energy simulation models. *Building Services* **8** no 3: 25–32

7.8   Millbank M O, Harrington-Lynn J 1974 Thermal response and the admittance procedure. *Building Services Engineering* **42**: 38–51 May

7.9   Fuel Efficiency Booklet No 7 *Degree Days* Energy Efficiency Office 1984

7.10 Levermore G J 1985 Energy conservation in London boroughs. *CIBSE Technical Conference Proceedings* 104–16 March

7.11 Levermore G J 1986 Motivating energy savings. *Building Services* **8** no 7: 33–4 July

7.12 CIBSE 1984 *Code for Interior Lighting*

7.13 CIBSE 1987 *Window Design Applications Manual*

7.14 CIBSE Building Energy Code:
Part 1   *Guidance Towards Energy Conserving Design of Buildings and Services* 1977
Part 2   *Calculation of Energy Demands and Targets for the Design of New Buildings and Services* 1981
Part 3   *Guidance Towards Energy Conserving Operation of Buildings and Services* 1979
Part 4   *Measurement of Energy Consumption and Comparison with Targets for Existing Buildings and Services* 1982

7.15 *Energy Efficiency in Buildings* Energy Efficiency Office 1985

7.16 *How to Bring Down Costs in Schools* Energy Efficiency Office March 1987

Bibliography

Burberry P 1988 *Environment and Services* The Mitchell Publishing Co

CIBSE 1986 *Guide to Current Practice* **B**

CIBSE 1989 *Lighting Guide*: *The Industrial Environment*

Eastop T D, McConkey A 1986 *Applied Thermodynamics for Engineering Technologists* 4th edn Longman

Energy Design Guide British Standards Institution Nov 1985

Energy Efficiency in Buildings British Standards Institution BS 8207 1985

Energy Manager's Workbook Energy Publications Cambridge 1982

Jackman P J 1987 *Specification of Indoor Environmental Performance of Buildings* BSRIA TN 3

Markus T A, Morris E N 1980 *Buildings, Climate and Energy* Pitman

McLaughlin R K, Craig McLean R, Bonthron W J 1981 *Heating Services Design* Butterworth

Reay D A 1977 *Industrial Energy Conservation* Pergamon Press

Sherratt A F C (ed.) 1987 *High Insulation: Impact on Building Services Design* CICC Publications

The Businessman's Energy Saver, section 7, *Lighting*, Formecon Publishing Nov. 1985

8 TOTAL ENERGY SCHEMES

The phrase 'total energy' has been used to describe a range of applications from, for example, an industrial gas turbine using the exhaust gas to raise steam, to much more complex plants involving a large number of processes. An example of the former is given in Section 4.4, Example 4.7.

A definition of a total energy scheme is offered as follows: *A scheme in which the total energy requirements of a plant in the form of power and heat are provided from a supply of primary fuel and in which the energy wastage is reduced to a minimum.*

In a particular case it may be more practical or more economic to import electricity from the grid for some or all of the power requirements; this does not preclude an attempt to minimize the energy wastage by, for example, the use of energy recovery methods such as those covered in Chapter 5, and by applying Pinch Technology (see Chapter 6). A total energy scheme may be more broadly defined as one in which multiple power and heat requirements are provided by optimizing the energy supplied.

A scheme which uses the minimum energy for a particular requirement of heat and power is not necessarily the most cost effective. The methods of Chapter 2 should be used to determine which scheme yields the best financial return.

The definition of total energy given above covers the use of refrigeration, heat pumps, recovery of energy from waste water and waste gases etc. as well as the use of boilers and all types of prime movers used either for electricity generation or for direct mechanical drives. More restricted schemes which combine electrical power generation with heat for space-heating and/or specific industrial processes are frequently referred to as *Combined Heat and Power* (CHP) or *cogeneration*. Schemes in which thermal energy is produced centrally and distributed via pipe networks for the heating of houses and public buildings are referred to as *district heating*. CHP schemes can incorporate district heating.

8.1 BASIC CONCEPTS OF CHP

CHP is receiving considerable attention from both industry and commerce as well as local and central government; it is a proven technology which has been used in a few energy-intensive industries for many years. The resurgence of interest in CHP in the late 1980s is due to a number of technical, economic and political factors which will be discussed later in the chapter. In the UK after the 1939–45 war the newly nationalized electricity generating industry took the decision to site the required new power stations away from centres of population, near coal fields and usually near large rivers or the sea to make use of natural cooling water. The policy was to generate electricity as efficiently as possible and to distribute it nationally via a distribution grid; the system is loaded so that the base load is taken by units with the lowest costs and the increments of load as they arise are taken by units with progressively higher generating costs. This national policy for electricity generation militates against the use of large CHP schemes in the UK.

Electricity is generally thought of as the most convenient form of energy because it can be converted into heat or mechanical energy and easily transmitted to the place where it is required. Unfortunately the production of electricity from the combustion of fuels is not an efficient process. Chapter 4 deals in some detail with the efficient combustion of fuel and the conversion of thermal energy to mechanical energy. The energy conversion diagram, Fig. 4.1, is adapted in Fig. 8.1 to show the main conversion paths for industry with typical values of efficiency marked against each path.

The application of the Carnot Principle to any heat engine cycle shows that however efficient the cycle may be the maximum thermal efficiency is given by

$$\eta_{\text{CARNOT}} = (1 - T_2/T_1)$$

Figure 8.1 Energy conversion diagram

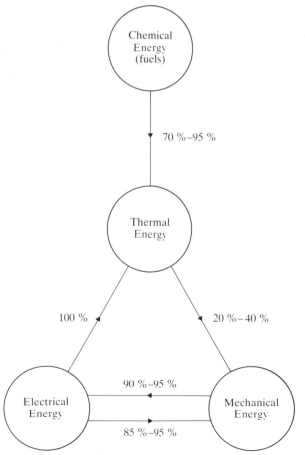

where T_1 is the maximum temperature available (e.g. the metallurgical limit), and T_2 is the lowest temperature available (e.g. cooling water for the condenser). For example, taking a maximum temperature of 1500 K and a cooling water minimum temperature of 280 K, then the maximum possible cycle efficiency is given by $(1 - 280/1500) = 0.813$ or 81.3 %. Due to the high degree of irreversibility in the various processes then the cycle efficiencies in practice are very much lower than the Carnot efficiency. The Central Electricity Generating Board's figures[8.1] in 1988 for the various types of power plant are as shown below.

Power plant	Cycle efficiency
Steam turbine (coal or nuclear fuel)	38 %
Gas turbine	23 %
Diesel engine	41 %

Clearly if a power plant is not converting the input energy into power then it must be rejecting energy in the form of waste heat. To give an idea of the scale of energy loss: the total waste heat from UK power stations is equivalent to the total domestic heat requirement for the UK.[8.2]

For a typical steam turbine power station if the heat rejected in the condenser were utilized the overall efficiency would be increased to about 75 % to 80 %. For existing power stations the scope for using the heat rejected from the condensing steam is limited for two reasons: the temperature of heat rejection is low (about 20 °C); and the power stations are remote from any possible recipients of the thermal energy reclaimed. Possible uses which have been investigated are in greenhouse heating for horticulture, warm water irrigation for frost protection of crops etc.

In some local authorities (e.g. Edinburgh) schemes have been considered for adapting existing power stations; steam can be bled off the steam turbine at a suitable pressure at night when electricity demand is low and used to heat water in a thermal store; hot water is pumped from the thermal store along a pipe line to a second thermal store from which heating and hot water can be supplied to the community when required. Economic considerations have caused some of the proposals such as the above to be shelved mainly because of the large distances involved in transmitting the thermal energy with the consequent large heat losses and the cost of the expensive lagged pipe-work. District heating is considered more fully later in the chapter.

8.2 THE BENEFITS OF CHP

The basic argument in favour of CHP is that it should be possible to obtain a worthwhile overall advantage by local generation of electricity, and at the same time make use of about two-thirds of the waste heat. Other specific benefits can be detailed as follows.

Improved national energy efficiency and the preservation of non-renewable energy reserves

It is particularly important for a nation to reduce its dependence on imported fuels so as to improve the balance of payments. If industry's requirement is reduced then this will stimulate a more competitive position in world markets. The Marshall report[8.3] estimated that if about 25 % of existing domestic and commercial heating requirements were supplied by CHP then the annual national energy savings would be of the order of 15 Mt coal equivalent.

Local generation of electricity

This reduces the cost of transportation of the energy. Long distance electrical transmission is expensive partly due to the need for expensive transformers, partly due to the erection and maintenance costs of pylons and overhead lines. For a cost comparison with other forms of energy transmission, assuming the transmission of the same quantity of energy over the same distance then for a cost index figure of 100 for electrical power transmission by overhead cable the following apply:[8.4]

(a) for oil in an underground pipeline, cost index = 27
(b) for natural gas in an underground pipeline, cost index = 35
(c) for hot water using a flow and return pipeline, cost index = 67.

Reduction in environmental pollution

A reduction in the *amount* of pollution follows from the more efficient conversion of less fuel. Also, the more efficient use of exhaust gases in, say, a waste boiler, (where, if supplementary firing takes place, the fuels are more completely burned), will reduce pollution. This type of combustion system was used in plant in Finland, with the effect that the annual average pollution level in central Helsinki fell by 80 %.

Investment in industry

Atkins[8.2] estimates that around 300 000 jobs would be created in the engineering, construction and service industries by government investment in City-based CHP schemes.

8.3 PROBLEMS ASSOCIATED WITH CHP

CHP is such an attractive proposition thermodynamically that the reader may wonder why it is not used more widely in industrial and commercial premises. The reasons are largely economic and relate to the items discussed in Chapter 2 when assessing the viability of capital investment projects. Detailed arguments are as follows.

(1) If an industry decides to invest in CHP plant then it is committed to a course of action and expenditure for a number of years. The financial appraisal of such a project involves making a number of assumptions on

the future energy demands of the company, fuel prices and availability, taxes, discount rates, maintenance costs etc., which may be accurate only in the short term. It may be safer for the company simply to purchase the electrical power from a central generating utility and generate heat from boilers.

(2) Any change in the manufacturing process may result in a later demand for more heat or power requiring more plant at a cost much higher than the initial installation.

(3) Any change in the manufacturing process may result in a change in the balance of demand for heat and power. This point is covered in more detail in Section 8.4: sufficient to say here that this change will alter the overall efficiency of the plant and hence its economic viability.

(4) Any CHP plant design will include a provision for 'back-up' to ensure security of supply of heat and power. Any support plant can only be regarded as spare capacity which, by definition, will not be frequently used and therefore will not be generating sufficient savings to offset the initial capital costs. Failure to include a back-up system is an unacceptable risk.

(5) Apart from the capital costs of the prime power, there are large costs associated with the distribution of heat in pipelines.[8.5] These costs may be a significant part of the initial capital investment unless the heat user is relatively near the prime mover.

(6) The problems of noise and pollution may involve extra expenditure at a time when regulations are becoming increasingly stringent.

It is unlikely that a given installation would encounter all the above drawbacks. Nevertheless, a true appraisal of costs would include some of the above factors and would obviously reduce the anticipated rate of return on the investment. The case studies which feature later in this chapter show how many of the above difficulties can be overcome and how savings can be made.

8.4 THE BALANCE OF ENERGY DEMAND

CHP, or cogeneration plant, delivers electrical and thermal energy in such a way that much more of the energy content of the input fuel is used. This objective is achieved by utilizing the waste heat. In the case of a diesel engine, for example, heat from the engine's exhaust gases could be recovered by means of a waste heat recovery boiler. After this the gases could then be passed through economizers before exhausting to atmosphere. Heat obtained from the economizers could be used to raise further the temperature of the boiler feed water already preheated by means of heat taken from the engine's cooling water, turbocharger, and lubricating oil. Alternatively, the exhaust gases could be used solely to produce high-grade heat in the form of steam, and the engine cooling water and lubricating oil used to produce low-grade heat in the form of hot water. The input energy to the diesel engine is clearly being used more efficiently than if the engine were being used solely to produce power. The question arises; what is the conversion efficiency of the CHP system and how does it compare with systems producing the heat and power separately?

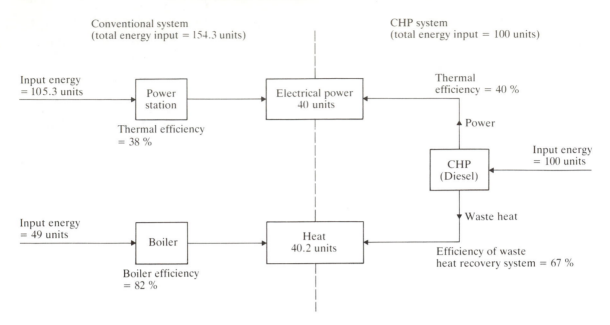

Figure 8.2 Comparison of heat and power generation by conventional and CHP means

Figure 8.2 compares the heat and power generation by CHP and conventional means. In the CHP system, heat and power are produced by a diesel engine. The conventional system generates electrical power from a power station and heat from a separate boiler. As can be seen from the data in Fig. 8.2 the energy input to the conventional system exceeds that to the CHP plant by 54.3 %.

Each prime mover supplies heat and power in a certain ratio. The diesel engine shown in Fig. 8.2 operates with a thermal efficiency of 40 %, so the waste heat is 60 % of the input energy. The heat recovery system is unlikely to be able to recoup more than 67 % of the available waste heat thus giving heat output of $0.6 \times 67\%$ ($=40.2\%$) of the input energy. The heat-to-power (heat/power) ratio will therefore be 40.2:40, which is approximately unity. The diesel engine will therefore be ideally suited to a user who has a demand for equal amounts of power and heat at all times throughout the year.

In practice, the ratio of the demand for heat and power usually fluctuates throughout the year and often throughout the day, making the economics of operation not quite so favourable as at first sight. The easiest way to see how economically the CHP unit performs over a range of heat/power ratios is to compare it with the alternative, which is to purchase the power from a supplier and generate the heat separately from a boiler. Starting with a heat/power ratio of 0, the diesel engine will generate power more efficiently and more cheaply than the generating authority, because the maximum thermal efficiency of a power station is around 38 %. At this condition there will be a slight saving of energy using the CHP plant. Any heat produced by the engine will be allowed to go to waste.

A heat/power ratio of about 1 represents the optimum running condition of the diesel engine which will have a thermal efficiency of about 40 % and a facility to recoup about 67 % of the heat to the cooling oil, cooling water and exhaust (the 67 % figure representing the effectiveness of the heat exchangers used). This point will represent the condition of maximum saving against bought-in power and running a separate boiler to generate heat. The overall efficiency, η, of the

CHP plant can be found as follows:

$$\eta_{CHP} = \left(\frac{\text{net power output} + \text{heat recovered}}{\text{energy input}} \right) \times 100\% \qquad [8.1]$$

For the example given in Fig. 8.2 we have

$$\eta_{CHP} = \frac{40 + \{0.67 \times (100 - 40)\}}{100} \times 100\% = 80.2\%$$

The efficiency with which the bought-in power and boiler heat are generated for a heat/power ratio can be calculated as follows, based on an average generating efficiency of 38 % and a boiler efficiency of 82 %. The required energy input to the power station to generate 40 units of power is

$$\frac{40}{0.38} = 105.3 \text{ units}$$

and the required energy input to the boiler to generate 40.2 units of heat is

$$\frac{40.2}{0.82} = 49.0 \text{ units}$$

The total energy input to generate power and heat separately is thus

$$(105.3 + 49.0) = 154.3 \text{ units}$$

The overall efficiency is given by

$$\eta = \frac{\text{power output} + \text{heat output}}{\text{total energy input}} \times 100\% = \frac{40 + 40.2}{154.3} \times 100\%$$
$$= 52.0\%$$

Not surprisingly, there is a significant difference (about 30 %) in the overall efficiency of the two arrangements. The difference is reduced at increased heat/power ratios.

At a heat/power ratio of 2:1 say, either the engine will have to operate at part load or a back-up boiler will need to be employed to supply the extra heat. Here we shall consider the engine to be operating at full load, with heat being wasted at heat/power ratios of less than unity and extra heat being supplied by boilers at the higher heat/power ratios.

The overall efficiency of the CHP plant plus extra boiler supplying 40 units of power and 80 units of heat is calculated as follows. The heat supplied by the boiler is given by

$$\text{total demand for heat} - \text{heat supplied by CHP}$$

which is equal to

$$80 - 40.2 = 39.8 \text{ units}$$

Therefore the energy input to the boiler is

$$\frac{\text{heat supplied by boiler}}{\eta_B}$$

which is equal to

$$\frac{39.8}{0.82} = 48.5 \text{ units}$$

As previously shown, the energy input to the CHP plant required to supply 40 units of power and 40.2 units of heat is 100 units, and so the total energy input to the CHP plant plus the boiler is

$$100 + 48.5 = 148.5 \text{ units}$$

and the overall efficiency (CHP + boiler) is given by

$$\eta_{CHP+B} = \frac{\text{total heat and power output}}{\text{total energy input}} = \frac{40+80}{148.5} \times 100\%$$
$$= 80.8\%$$

As before, the efficiency with which the bought-in power and boiler heat are generated is based on an average generating efficiency of 38 % and a boiler efficiency of 82 %. The required energy input to the power station to generate 40 units of power is

$$\frac{40}{0.38} = 105.3 \text{ units}$$

and the required energy input to the boiler to generate 80 units of heat is

$$\frac{80}{0.82} = 97.6 \text{ units}$$

The total energy input to generate power and heat separately is thus

$$(105.3 + 97.6) = 202.9 \text{ units}$$

and

$$\eta = \frac{\text{power output} + \text{heat output}}{\text{total energy input}} \times 100\% = \frac{40+80}{202.9} \times 100\%$$
$$= 59.1\%$$

There is now a difference in the overall efficiency of the two arrangements of about 20 %. Both efficiencies are higher than those at a heat/power ratio of 1:1 because of the increased effect of the additional boiler. Ultimately at much higher heat/power ratios the efficiencies of both arrangements will be virtually that of the boiler. Figure 8.3 shows how the energy savings of the (CHP + back-up boilers) scheme over the (bought-in power + boilers) scheme vary with heat/power/ratio for the diesel engine. As a general comment, it can be said that the heat/power ratio alters the energy savings and hence the economics of CHP schemes. It should be pointed out that the arguments in this section have been based on the prime mover giving a constant amount of power at maximum thermal efficiency conditions and the amount of heat varied by either wasting the heat at low heat/power ratios or by using back-up boilers at the higher heat/power ratios. It is possible to achieve some variation in heat/power ratios by operating prime movers at part loads albeit at reduced thermal efficiencies.

Figure 8.3 shows energy savings achieved by a CHP scheme. What of the costs savings? As an example of the relationship between energy and cost savings consider the current domestic tariffs of gas and electricity, representing the price

Figure 8.3 Energy
savings variation with
heat/power ratio

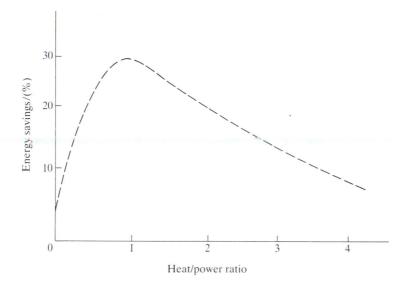

of heat and power respectively:

$$\text{gas tariff} = 38.5 \text{ p/therm} = 0.37 \text{ p/MJ}$$
$$\text{electricity tariff} = 5.97 \text{ p/kWh} = 1.66 \text{ p/MJ}$$

Thus, as a supplied fuel, electricity is 4.6 times as valuable as gas. Even if the gas is burnt in a boiler with an efficiency of 82 %, the factor is around 3.7. The relative value of the two fuels has the effect of increasing the value of energy savings at low heat/power ratios and reducing their value at heat/power ratios beyond the optimum. Bear in mind that industrial tariffs are less than domestic tariffs though the ratio may not be significantly different. This aspect is covered in detailed in Chapter 9.

The ideal situation for a CHP scheme would be where the end-user has a constant demand for heat and power at a heat/power ratio appropriate to a particular prime mover. This is an extremely unlikely scenario for an urban application where the demand for heat is very low in the summer and high in the winter but the demand for electricity is virtually constant. A viable CHP scheme for a typical British town would require customers for the waste heat in the summer, probably industrial users. During the winter months such users would have to operate their own heat raising plant. Such a concept would only work if the heat was competitively priced. During the winter when the CHP plant will be operating at maximum capacity, it is possible that the power developed may be in excess of that required in the locale. Because of the 1983 Energy Act, it is now a requirement for area boards to purchase electricity from a private generator – at a price determined by them – so that any surplus winter generation by the CHP scheme can contribute towards the economic running of the station. This aspect of CHP economics is discussed in the case studies at the end of this chapter. There are urban situations where the demand for heat is roughly constant throughout the year. In North America, for example, the summer fall in the demand for heat is compensated for by the demand for heat to operate air conditioning and refrigeration equipment. In countries which do not have a need for air conditioning during the summer (most Northern European countries, for example) other users for the heat have to be found.

Many process industries have predictable and reasonably constant demands for heat and power, and CHP is a much more economically viable proposition for this type of application.

One solution to the problem of varying demand is to imagine one end-user for the power and another for the heat. The Midlands Electricity Board has developed two CHP stations based on this concept. In both cases the electricity generation is integrated into the local section of the national grid and the heat is supplied to local industries. The design of the stations is based on supplying specific heat rather than electricity demands. These two stations are discussed in detail later in this chapter.

8.5 TYPES OF PRIME MOVERS

There are three main types of prime movers used in CHP plant: steam turbines, gas turbines and internal combustion engines. The basic thermodynamic cycles of these devices have been discussed previously in Chapters 3 and 4 and therefore comments here are restricted to general remarks about applications with simple illustrative examples.

Steam Turbines

Steam turbines are the most commonly used prime mover in CHP installations. Depending on the application the arrangement will be either a back-pressure, passout, or passout condensing turbine. The basic principle of the back-pressure type is: the greater the amount of power desired, the lower the condenser pressure and the less heat generated for process use. Raising the condenser pressure will raise the pressure and temperature of heat supply but will reduce the power output from the turbine. In practice, this type of arrangement is best when operating at a heat/power ratio of around 9:1. A similar argument applies to passout turbines where a variable amount of steam can be extracted from the turbine before full expansion takes place. Here the operational heat/power ratios are around 3:1. A wide variety of fuels are used to raise steam; coal, gas, oil or waste material as described in Chapter 4.

Example 8.1
This example involves a back-pressure turbine, which works with an exhaust pressure appropriate to the process heat requirement. Steam leaves the turbine, is used for heating purposes and normally then passed back to the boiler as condensate. A back-pressure steam turbine unit is installed by a company to supply heat and power. The steam conditions at entry to the turbine are 40 bar and 500 °C. Taking the isentropic efficiency as 0.86,
(i)   calculate the heat/power ratio of the unit when the turbine exhaust pressure is 3 bar. Assume that the steam is returned to the boiler feed pump as a saturated liquid at 1 bar;
(ii) estimate the turbine exhaust pressure which will result in a heat/power ratio of 6:1. Assume that the steam is returned to the boiler feed pump as a saturated liquid at a pressure 2 bar below the turbine exhaust.

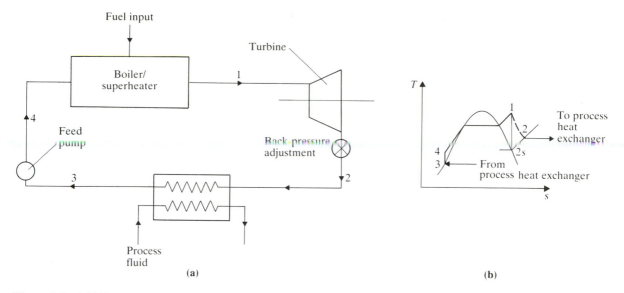

Figure 8.4 (a) Plant schematic for CHP based on back-pressure turbine; (b) $T-s$ diagram for the CHP scheme

Solution

(i)   A plant schematic and the $T-s$ diagram for the processes are shown in Figs 8.4(a) and 8.4(b). From the $h-s$ chart for steam

$$h_1 = 3445 \text{ kJ/kg}$$
$$h_{2S} = 2765 \text{ kJ/kg}$$

Isentropic efficiency is given by

$$\eta_s = 0.86 = \frac{h_1 - h_2}{h_1 - h_{2S}}$$

i.e. $h_2 = 2860.2 \text{ kJ/kg}$

From tables

$$h_3 = h_f \text{ at 1 bar} = 417 \text{ kJ/kg}$$

Therefore

$$\text{heat/power ratio} = \frac{\dot{m}_S(h_2 - h_3)}{\dot{m}_S(h_1 - h_2)} = 4.18:1$$

(ii)   The requirement is that the heat/power ratio is 6:1. This is a higher ratio than obtained when the exhaust pressure was 3 bar in part (i) previously. Therefore the turbine must exhaust at a pressure above 3 bar and it is a matter of trial and error to find the exact value. The calculation procedure will be exactly the same as part (i) previously. As a first guess we will say that the exhaust pressure is 6 bar; the condensate will therefore return as a saturated liquid at 4 bar. From the $h-s$ chart

$$h_1 = 3445 \text{ kJ/kg}$$
$$h_{2S} = 2915 \text{ kJ/kg}$$

Isentropic efficiency is given by

$$\eta_s = 0.86 = \frac{h_1 - h_2}{h_1 - h_{2S}}$$

i.e. $\qquad h_2 = 2989.2 \text{ kJ/kg}$

From tables

$$h_3 = h_f \text{ at } 4 \text{ bar} = 605 \text{ kJ/kg}$$

Therefore

$$\text{heat/power ratio} = \frac{\dot{m}_S(h_2 - h_3)}{\dot{m}_S(h_1 - h_2)} = 5.07 : 1$$

This is below the required ratio of 6:1 and therefore a higher exhaust pressure must be selected. To save unnecessary repetition of the calculation, the reader is left to confirm that the appropriate exhaust pressure is about 8.4 bar.

Example 8.2

This example involves a pass-out turbine, in which steam is bled from the turbine at some point or points between inlet and exhaust. The remainder of the steam is expanded in the turbine and then condensed in the usual way. The process steam is used for heating purposes and normally then passed back to the boiler as condensate. A pass-out steam turbine unit is installed by a company to supply heat and power. The steam conditions at entry to the turbine are 30 bar and 400 °C, and the steam expands with an isentropic efficiency of 0.84 to an exhaust pressure of 0.08 bar. The power output from the turbine is 5 MW. Construct a table which shows how the percentage of steam flow extracted for process heating alters the heat/power ratio based on the steam being extracted at a pressure of 10 bar. Assume that the steam is returned to the boiler feed pump as a saturated liquid at 2 bar, and that the condition line on the h–s diagram is linear. Assume also that the isentropic efficiency remains the same regardless of the steam flow extracted.

Figure 8.5 (a) Plant schematic for CHP scheme based on pass-out turbine; (b) T–s diagram for the CHP scheme

Solution

A plant schematic and the T–s diagram for the processes are shown in Figs 8.5(a) and 8.5(b). From the h–s chart

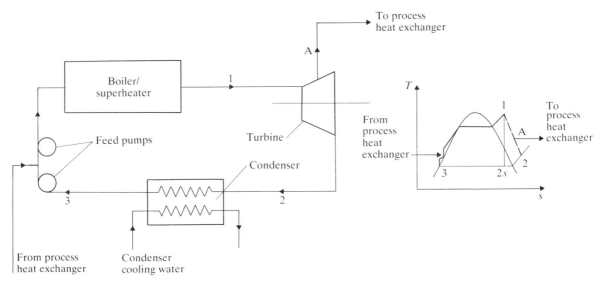

(a)   (b)

$$h_1 = 3230 \text{ kJ/kg}$$
$$h_{2S} = 2175 \text{ kJ/kg}$$

Isentropic efficiency is given by

$$\eta_s = 0.84 = \frac{h_1 - h_2}{h_1 - h_{2S}}$$

i.e.
$$h_2 = 2344 \text{ kJ/kg}$$

From tables

$$h_3 = h_f \text{ at } 1 \text{ bar} = 505 \text{ kJ/kg}$$

Assuming a linear condition line between points 1 and 2, then from the chart

$$h_A = 3015 \text{ kJ/kg}$$

If m_S = the mass flow rate of the steam into the turbine and m_P = the mass flow rate of steam extracted for process heating then the rate of heat delivery for process heating is

$$m_P(h_A - h_2) = m_P(3015 - 505) = 2510 m_P$$

and the power delivered is

$$\dot{m}_S(h_1 - h_A) + (\dot{m}_S - \dot{m}_P)(h_A - h_2) = \dot{m}_S(3230 - 3015)$$
$$+ (\dot{m}_S - \dot{m}_P)(3015 - 2344) = 215\dot{m}_S + 671(\dot{m}_S - \dot{m}_P)$$

Therefore

$$\text{heat/power ratio} = \frac{2510\dot{m}_P}{215\dot{m}_S + 671(\dot{m}_S - \dot{m}_P)}$$

$$= \frac{2510F}{886 - 671F}$$

(where $F = \dot{m}_P/\dot{m}_S$). The values of heat/power ratios generated by different fractions of process steam are shown in the table below.

F	Heat/power ratio
0	0
0.25	0.87
0.50	2.28
0.75	4.92
1.00	11.67

   Clearly the heat/power ratio can be significantly changed by adjusting the balance of steam flows through and from the turbine.

Gas Turbines

Gas turbines are becoming more popular because of low capital cost, reliability, the higher quality of the process heat, but particularly because of their optimum heat/power ratio of around 3:1. The 1983 Energy Act has had the effect of making private generators aware of the benefits of exporting electricity to the national grid at certain times of year. CHP plant can now be run with the philosophy of matching the demand for heat, and exporting excess power to help with the economics of the plant. A prime mover with a lower heat/power ratio will help this concept. A gas turbine used for CHP would normally be fuelled by natural gas, but oil and pulverized coal have been successfully employed.

Example 8.3

A manufacturing company is supplied with 80 000 MW h of electrical energy by the area board each year. The company generates its own heat from a gas-fired boiler. It is proposed to replace this arrangement with a CHP scheme based on a gas turbine. A heat exchanger extracts energy from the turbine exhaust to preheat process air; this energy replaces that formerly supplied by the boiler. Using the data below, confirm that the operating heat/power ratio of the CHP system is about 3:1, and estimate the annual saving in fuel costs of the co-generation scheme over the previous arrangement. A plant schematic and the $T–s$ diagram for the processes are shown in Figs 8.6(a) and 8.6(b).

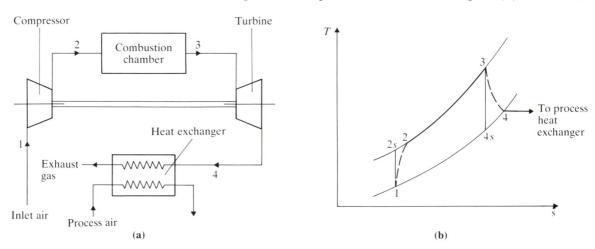

(a) (b)

Figure 8.6 (a) Plant schematic for CHP scheme based on gas turbine; (b) $T–s$ diagram for the CHP scheme

Data

Gas turbine: inlet temperature, 300 K; maximum temperature, 1300 K; pressure ratio, 6:1; all isentropic efficiencies, 0.78; for all the cycle take $\gamma = 1.4$ and $c_p = 1.005$ kJ/kg K.
Generator: efficiency, 0.95.
Fuels: average electricity tariff, 4.2 p/kW h; gas price, 28 p/therm (1 therm = 29.307 kW h).
Boiler: overall efficiency, 0.80.
Heat exchanger: effectiveness, 0.70; inlet temperature of process air, 300 K.
Assume that the thermal capacities of the process air and the turbine exhaust gas are the same. Also assume a combustion efficiency of 100 %.

Solution
For the compressor:

$$\left(\frac{T_{2s}}{T_1}\right)^{\gamma/(\gamma-1)} = \frac{p_2}{p_1} = 6$$

$$\therefore \quad T_{2s} = 300 \times 1.6685 = 500.6 \text{ K}$$

Isentropic efficiency is given by

$$\eta_s = 0.78 = \frac{T_{2s} - T_1}{T_2 - T_1}$$

$$\therefore \quad T_2 = 557.2 \text{ K}$$

For the turbine:
$$\left(\frac{T_3}{T_{4s}}\right)^{\gamma/(\gamma-1)} = \frac{p_3}{p_4} = 6$$

$$\therefore \quad T_{4s} = 1300/1.6685 = 779.1 \text{ K}$$

Isentropic efficiency is given by

$$\eta_s = 0.78 = \frac{T_3 - T_4}{T_3 - T_{4s}}$$

$$\therefore \quad T_4 = 893.7 \text{ K}$$

Therefore, the net power output is

$$\dot{m}c_p(T_3 - T_4) - \dot{m}c_p(T_2 - T_1) = 149.2\dot{m}c_p$$

The heat input is

$$\dot{m}c_p(T_3 - T_2) = 742.8\,\dot{m}c_p$$

Thermal efficiency is given by

$$\frac{\text{net power output}}{\text{heat input}}$$

which is equal to

$$\frac{149.2}{742.8} = 0.201$$

Power delivered is given by

$$\text{net power output} \times \text{generator efficiency}$$

which is equal to

$$149.2\dot{m}c_p \times 0.95 = 141.7\dot{m}c_p$$

For the heat exchanger, effectiveness, E, is given by

$$\frac{\text{actual heat transfer}}{\text{maximum heat transfer}}$$

Since $E = 0.70$ (given) actual heat transfer is

$$0.70\dot{m}c_p(893.7 - 300) = 415.6\dot{m}c_p$$

Therefore

$$\text{heat/power ratio} = 415.6/141.7 = 2.93$$

The thermal requirement is given by

$$(\text{heat/power ratio}) \times \text{electrical energy requirement}$$

which is equal to

$$2.93 \times 80\,000 \text{ MW h} = 234\,600 \text{ MW h}$$

COST CALCULATIONS   For the CHP scheme the fuel cost is given by

$$\frac{\text{electrical energy output (kW h)} \times \text{fuel tariff (£/kW h)}}{\text{thermal efficiency}}$$

$$= \frac{80\,000 \times 1000 \times 0.28}{0.201 \times 29.307} = £3.8026 \text{ M}$$

For the boiler plus bought-in power arrangement the electricity cost is given by

$$\text{number of units (kW h)} \times \text{tariff (£/kW h)} = 80\,000 \times 1000 \times 0.042$$
$$= £3.3600 \text{ M}$$

and the boiler fuel cost by

$$\frac{\text{thermal energy output (kW h)} \times \text{fuel tariff (£/kW h)}}{\text{boiler efficiency}}$$

$$= \frac{234\,600 \times 1000 \times 0.28}{0.8 \times 29.307} = £2.8017 \text{ M}$$

Therefore the total cost is

$$£(2.8017 + 3.3600) \text{ M} = £6.1617 \text{ M}$$

Hence the saving in annual running cost achieved by CHP scheme is

$$£(6.1617 - 3.8026) \text{ M} = £2.36 \text{ M}$$

This saving is of course the major contribution in the financial appraisal of the proposed CHP scheme. The calculation also illustrates the relative value of thermal and electrical energy, in that the cost of the bought-in power exceeds the cost of the boiler fuel, although the thermal requirement is three times bigger than the electrical one. This suggests that schemes with lower heat/power ratios are more likely to be economically viable; a point illustrated by the case studies at the end of the chapter.

Internal Combustion Engines

All internal combustion engines operate in the same way when acting as CHP units; power is used to drive a generator and heat is recovered from the engine exhaust, jacket water and lubricating oil. The size range is enormous; from the 15 kW 'Totem' petrol/gas engine unit for use in the domestic/services sector to the 12 MW diesel unit for use in the industrial sector. The large diesel units operate at heat/power ratios of around 1:1 whereas the smaller petrol/gas units tend to be around 2:1. Case studies of industrial applications of these devices are included later in this chapter. The vast majority of internal combustion engines for CHP applications are fuelled by oil or natural gas for cost and safety with storage reasons.

8.6 THE ECONOMICS OF CHP GENERATION

The general factors which affect the financial appraisal of CHP schemes have already been detailed in Chapter 2. A company considering the use of CHP plant has to be convinced that the cost of producing heat and power by this method is cheaper then by any other method. The most common alternative method involves generating heat from a boiler and purchasing the power from an area electricity board. The savings in costs achieved by a CHP plant have to pay for the capital investment in power generation plant. The main factors which affect the operational economics of CHP plant are as follows.

(1) The heat/power ratio; bearing in mind the higher value of power compared to heat, and how near the operational heat/power ratio is to the 'optimum' design value.

(2) The difference in maintenance costs between a CHP scheme and a 'boiler-only with bought-in power' scheme.

(3) The difference in cost between the fuel used by the CHP plant and that used by the national generating authority. The industrial tariffs for natural gas and fuel oil and the price differential between these two fuels and coal are factors which give the private generator a big advantage.

(4) The extent to which the CHP scheme can export power at the most lucrative times of year and the cost of having the national grid as the back-up system in case of breakdown or maintenance.

(5) The load factor of the CHP plant, since this will dictate the total annual savings and hence the rate of return of the initial investment.

Example 8.4 shows how these factors affect the economic viability of a CHP proposal.

Example 8.4

The Energy Manager of an oil refinery is considering installing a CHP plant comprising two gas turbines and associated site equipment. The electrical power generated would replace that previously purchased from the generating board. The hot exhaust gases would reduce the load on the refinery's furnaces which currently burn oil. Using the data below, calculate the NPV of the project, assuming that the equipment has an expected life of four years, at the end of which it has an expected value of £80 000.

Data
Running costs for gas turbine: fuel, 25 p/therm; operators, maintenance etc., 0.13 p/kW h; overall thermal efficiency, 19 %.
Running costs for furnace: fuel, £55/tonne; calorific value, 40 MJ/kg; furnace efficiency, 78 %; operators, maintenance etc., 0.025 p/kW h.
Exhaust heat, 2.6 kW h of thermal energy is available for each 1 kW h of electrical energy generated.
Electricity consumption: by day, 65 000 MW h; by night, 25 000 MW h.
Maximum demand (kVA), See Chapter 9, Section 9.4; previous December maximum demand, 15 000 kVA.
Negotiated chargeable capacity, 15 000 kVA.

Month	Maximum demand (kVA)	Tariff
January	12 000	£6.50/kVA
February	13 000	£2.10/kVA
March	11 000	nil
April	9 000	nil
May	7 000	nil
June	6 000	nil
July	6 000	nil
August	5 000	nil
September	7 000	nil
October	9 000	nil
November	11 000	£2.10/kVA
December	14 000	£6.50/kVA

Unit charge:
   during the 7-hour night period, 2.06 p/kW h
   outside the 7-hour night period, 4.62 p/kW h
Availability charge:
   £0.96/kVA for the first 50 kVA; £0.79/kVA thereafter

Standby monthly service charge for cogeneration plant = £5000.
Capital equipment/tax/grants: gas turbine generating sets and auxilliaries, £700 000 each; regional development grant (payable in year 2), 20 % of capital cost; corporation tax, 50 %; discount factor after tax, 18 %; tax on profits payable from years 2 to 5 inclusive (i.e. one year in arrears).

Solution
By comparing the running costs of the present system with those using a CHP scheme the annual savings can be estimated.

CURRENT SCHEME   For the current scheme the electricity costs will comprise charges for the number of units consumed and charges associated with maximum demand. (The detailed rationale behind such charges is given in Section 9.4.)
   The units charge is given by

$$\text{no. of units consumed} \times \text{tariff per unit}$$

which is equal to

$$65\,000 \times 1000 \times 0.0462 \qquad \text{for daytime consumption}$$

and

$$25\,000 \times 1000 \times 0.0206 \qquad \text{for night-time consumption}$$

Therefore   units charge is £3.518 M.   [1]

Maximum demand charges comprise availability charges and demand charges as detailed in Section 9.4. Assuming that the company is best suited by the scale 1 tariff then the monthly charges are as follows:

Month	Maximum demand (kVA)	Availability charge (£)	Demand charge (£)
January	12 000	11 859	78 000
February	13 000	11 859	27 300
March	11 000	11 859	nil
April	9 000	11 859	nil
May	7 000	11 859	nil
June	6 000	11 859	nil
July	6 000	11 859	nil
August	5 000	11 859	nil
September	7 000	11 859	nil
October	9 000	11 859	nil
November	11 000	11 859	23 100
December	14 000	11 859	91 000
Totals		142 308	219 400

Therefore the total maximum demand charge is

$$£142\,308 + £219\,400 = £0.3617\text{ M} \qquad [2]$$

The operations/maintenance costs are given by

$$\text{furnace output (kW h)} \times \text{cost per unit output}$$

The furnace output is given by

$$\text{(heat/power ratio)} \times \text{power output}$$

which is equal to

$$2.6 \times (65\,000 + 25\,000) = 234\,000\text{ MW h}$$

Therefore the maintenance cost is

$$234\,000 \times 1000 \times (0.025/100) = £0.058\text{ M} \qquad [3]$$

The furnace fuel cost is given by

$$\frac{\text{furnace output (MJ)} \times \text{fuel tariff (£/kg)}}{\text{furnace efficiency} \times \text{calorific value of fuel (MJ/kg)}}$$

which is equal to

$$\frac{(234\,000 \times 3600) \times (55/1000)}{0.78 \times 40} = £1.485\text{ M} \qquad [4]$$

The total running costs are given by

$$[1] + [2] + [3] + [4] = £5.4227\text{ M} \qquad [5]$$

CHP SCHEME   For the CHP scheme the fuel costs are given by

$$\frac{\text{electrical energy output (kW h)} \times \text{fuel cost (£/kW h)}}{\text{thermal efficiency}}$$

which is equal to

$$\frac{(65\,000 + 25\,000) \times 1000 \times 25}{0.19 \times 29.307 \times 100} = £4.0407 \text{ M} \qquad [6]$$

The standby annual service charge is given by

monthly charge $\times$ 12

which is equal to

$$£5000 \times 12 = £0.06 \text{ M} \qquad [7]$$

The operations/maintenance cost is given by

electrical energy output (kW h) $\times$ cost per unit output (£/kW h)

which is equal to

$$(65\,000 + 25\,000) \times 1000 \times (0.13/100) = £0.116 \text{ M} \qquad [8]$$

The total running costs are given by

$$[6] + [7] + [8] = £4.2177 \text{ M} \qquad [9]$$

$$\begin{aligned}
\text{Saving in running costs} &= [5] - [9] \\
&= £(5.4227 - 4.2177) \text{ M} \\
&= £1.205 \text{ M}
\end{aligned}$$

NPV ASSESSMENT   The discount rate for the calculation is taken as 18 %, based on the prevailing rates of interest, and no inflation. The NPV is calculated as follows:

Year	Capital investment (£M) (A)	Net savings (£M) (B)	Tax incentive (£M) (C)	Tax at 50 % (£M) (D)	Additional grant (£M) (E)	Net after tax (£M) (F)	Discount factor (18 %) (G)	Present value (£M) (H)
0	−1.4	0	0	0	0	−1.4	1.000	−1.4000
1	0	1.205	0	0	0	1.205	0.847	1.0206
2	0	1.205	0	−0.6025	0.28	0.8825	0.718	0.6336
3	0	1.205	0	−0.6025	0	0.6025	0.609	0.3669
4	0.08	1.205	0	−0.6025	0	0.6825	0.516	0.3522
5	0	0	0	−0.6425	0	−0.6425	0.437	−0.2807
								NPV = 0.6512

The column items A to H signify the following:

A   £ − 1.4 M represents the initial capital investment. £0.08 M is the salvage value
B   Net savings on CHP scheme compared to existing scheme
C   Tax incentive (or allowance) − nil in this case
D   Tax for year 2 onwards $= (A + B − C) \times 0.5$
E   Additional untaxable grant of £$(0.2 \times 2 \times 700\,000)$ for year 2 only
F   The residual after tax, which equals $(A + B + D + E)$
G   A column of discount factors (for 18 % in this case)
H   The PV, which equals $(F \times G)$

8.7 CHP IN THE INDUSTRIAL SECTOR

As far as CHP in the industrial sector is concerned, a Department of Energy Survey[8.7] gives a detailed account of the scale of private power generation within industry itself and how much of that is CHP plant. Table 8.1 summarizes the statistics. It can be seen from the table that although electricity consumption in UK industry has more than doubled between 1953 and 1983, production of electricity from CHP plant has increased by only 12 %. This has resulted in the CHP share of electricity consumption in UK industry declining from 15.4 % to 7.5 %. In 1983 four main industrial sectors accounted for 88 % of the CHP generation; chemicals, refineries, paper and board, and the food and drink industries. A more detailed account of industrial sector use is given in Table 8.2. There are a number of reasons why the scale of CHP application in industry has decreased over the years.

Table 8.1 Total consumption and private generation of electrical power by UK industry (1953–1983)

Year	CHP (TW h)	Total private generation (TW h)	Electricity consumption (TW h)	CHP as a percentage of: Private generation (%)	CHP as a percentage of: Electricity consumption (%)
1953	6.01	10.66	38.92	56.4	15.4
1963	7.50	12.95	69.23	57.9	10.8
1973	9.95	15.83	102.50	62.9	9.7
1983	6.74	11.55	90.42	58.4	7.5

Table 8.2 Electrical output/capacity of UK industrial CHP plant in 1983

Industrial sector	Electrical output in CHP plants (TW h)	Electrical capacity of CHP plant (MW)
Chemicals and pharmaceuticals	3.3	962
Refineries	1.6	379
Paper and board	0.9	290
Food and drink	0.5	205
Other industries	0.8	221

(1) The nature of manufacturing industry has changed and there are fewer energy-intensive manufacturing industries with demands for heat and power that are not more economically met by high efficiency packaged boilers and bought-in power.

(2) The advances in heat recovery techniques have reduced the need for CHP plant with a high heat/power ratio; the application of Pinch Technology is an example of this (see Chapter 6). Plant which has been designed to operate at a high heat/power ratio but which is now delivering a lower heat load, can only do so by working at part load. Any extra power required can be obtained from the public supply system. Thus, extra or replacement CHP capacity is not being considered.

(3) Even within the energy-intensive industries, the nature of certain processes has reduced the need for heat and increased the need for power. Virtually all the processes involved in paper manufacture are examples of this.

(4) A number of reasons relate to the investment of capital; they are the same considerations that relate to any investment, namely too long a payback period, capital required elsewhere in the industry, initial capital cost too high.

8.8 CHP IN THE COMMERCIAL SECTOR

CHP in the commercial sector generally uses what are commonly called micro-CHP systems. Micro-CHP is the term used to describe systems which generate heat and electricity simultaneously from reciprocating engines producing an electrical output of up to around 150 kW (see Table 8.3 for manufacturer's performance figures). These engines are derived from suitably adapted automotive units generally fuelled by natural gas but occasionally by LPG or biogas.

In a micro-CHP arrangement the engine is coupled to a generator to produce electricity whilst heat is recovered from the exhaust gas, the cooling water, the lubricating oil and sometimes from the generator itself. The units can work independently of, or in parallel with, the national grid.

The relatively low cost of micro-CHP units, and their heat and electricity outputs make them suitable for use in places such as hotels, hospitals, colleges, residential homes and swimming pools.

The Technology of Micro-CHP Units

The components of a micro-CHP system are as follows:

(a) an engine;
(b) a generator;
(c) a heat recovery system;
(d) a control system.

The engine drives an electrical generator which usually produce a three-phase output at 415 V, and can be synchronous or asynchronous. The former system operates in parallel with the national grid so that any excess demand on a site can be made up by using the national grid. However, it the mains supply is broken then the system shuts down. The asynchronous generator can be stand-alone, providing the total power requirement and not operating in conjunction with the national grid. There are dual systems which operate in parallel with the national grid until there is a power failure on the national grid, at which point the unit disconnects itself from the grid and continues to provide power for essential services. The unit resynchronizes itself to the grid when power is restored. Details of these types of generators are included in Section 4.6, (page 147).

The heat recovery system recovers the waste heat from the engine. All micro-CHP units recover heat from the exhaust gas and engine cooling water using heat exchangers. Additional heat may be recovered from the exhaust gas using a condensing economizer and from the lubricating oil and the generator housing by using additional heat exchangers. In this way, 80 % to 85 % of the heat rejected by the engine is recovered.

The control system, ensures the safe and efficient operation of the installation.

The system should be designed to satisfy criteria in the following areas:

(a) safety;
(b) reliability of supply;
(c) ease of plant operation;
(d) engine efficiency.

Ideally the whole unit should be housed in an acoustic enclosure to reduce noise levels and also provide weather protection. It is common practice for manufacturers to supply a complete unit including all controls and associated gear. It is therefore a relatively simple matter to connect the heat supply to an existing pipe network.

Typical Applications for Micro-CHP

The first micro-CHP units were installed in swimming pools and water treatment plant. In swimming pools and leisure centres there is a continuing demand for heat for the pool water, ventilation air and domestic hot water (for showers etc.), and a fairly constant demand for electricity to operate pumps and lighting. Under such conditions, micro-CHP units will be operating for 12 to 14 hours per day and are therefore likely to be cost effective. Water treatment plants (sewage plants) operate 24 hours per day and therefore have a continuous electrical demand from the pumps. In addition, the engines can run on the methane which is produced during the treatment process; this means that the payback time may be as short as three years.

It is becoming more common to see micro-CHP units in buildings such as hotels, blocks of flats and hospitals. The criteria for the economic operation of

Table 8.3 Typical manufacturer's performance figures for Micro-CHP units

Name of unit	Make	Engine capacity (litres)	Gas input (kW)	Output Heat** (kW)	Power (kW)
TOTEM	Fiat	0.90	60	38	15
CG100	Leyland	2.25	62.5	38	15
CG200	Ford	6.20	163	86	45
32/60	Ford	4.15	102	60	32
40/90	Ford	6.23	153	90	48
Mini CHIP	Waukesha	3.60	107	59	26
		5.40	170	95	40
		13.30	338	180	90
		19.60	495	267	138
NUTEC	Ford	2.80	69	44	18
		4.15	100	55	30
		6.22	150	82	54
		9.90	233	125	75
NUTEC	MAN	11.40	254	143	85
		20.90	463	260	155

** The heat output only includes heat from water at about 85 °C; some engines supply 'low-grade' heat at around 30 °C.

CHP in these situations are just the same as those for the larger units; a sufficiently large and constant heat load throughout a large part of the day and throughout the seasons and a similar pattern of demand for electricity. Given these criteria, a suitable hotel would need to have at least 50 bedrooms with a space-heating requirement for about 18 hours per day extending over at least eight months of the year, together with a high hot water demand for washing and catering needs. At the time of writing there are about 90 sites in the UK with micro-CHP installed. A detailed study of the Castle Hotel in Windsor, UK, where 6 'Totem' units have been installed, is available.[8.9]

In hospitals, there is a demand for space-heating for most of the day and substantial demands for hot water and electricity for up to 20 hours per day. Micro-CHP units can therefore operate for long periods with all the heat and power being used on the site; this makes for short payback periods of around two years.

Example 8.5

A hotel has an electrical load of 48 kW and a demand for hot water which generates a heating load of 90 kW during a year in which the average load factor is 75 %. This demand is currently met from mains electricity and an oil-fired boiler. It is proposed to install a micro-CHP unit which uses natural gas. The installed cost of the unit is £24 000. Using the data below calculate the payback period.

Data

PRESENT SCHEME   Electricity, average unit charge, 4.1 p/kW h. Boiler: efficiency, 70 %; fuel cost, 11 p/litre; GCV of fuel, 40 MJ/litre; annual maintenance costs, £500.

MICRO-CHP SCHEME   Engine: gas input, 153 kW; heat output, 90 kW; power output, 48 kW; fuel cost, 32 p/therm; annual maintenance costs, £1500. (Note: 1 therm is equivalent to 29.307 kW h.)

Solution

The annual number of operating hours is given by

$$\text{load factor} \times \text{total hours per year}$$

which is equal to

$$0.75 \times 8760 = 6570 \text{ hours}$$

PRESENT SCHEME   The electricity cost is given by

$$\text{demand} \times \text{operating hours} \times \text{tariff}$$

which is equal to

$$48 \times 6570 \times 0.041 = £12\,930 \text{ per annum} \qquad [1]$$

The boiler fuel cost is given by

$$\frac{\text{demand} \times \text{operating hours} \times \text{tariff}}{\text{boiler efficiency}}$$

which is equal to

$$\frac{90 \times 6570 \times 3600 \times 0.11}{0.7 \times 40\,000} = £8363 \text{ per annum} \qquad [2]$$

The maintenance cost is given as £500 per annum. [3]
The total is given by

$$[1] + [2] + [3] = £21\,793 \text{ per annum} \qquad [4]$$

PROPOSED SCHEME   The engine fuel cost is given by

gas input × operating hours × tariff

which is equal to

$$153 \times 6570 \times (0.32/29.307) = £10\,976 \text{ per annum} \qquad [5]$$

The maintenance costs are given as £1500 per annum. [6]
The total is given by

$$[5] + [6] = £12\,476 \text{ per annum} \qquad [7]$$

The payback time is given by

capital cost/savings in running cost

Savings in running costs = [4] − [7] = £9317. Therefore

$$\text{payback time} = 24\,000/9317$$
$$= 2.58 \text{ years}$$

Case Study 8.1

As part of the London Borough of Camden's energy conservation policy, a survey was carried out on the Borough's leisure complexes to identify possible uses of CHP units. The Swiss Cottage Swimming Pool and Leisure Complex was selected to receive CHP units as it fitted the basic criteria of high and constant heating and electrical loads. The complex was 20 years old, with gas-fired hot water boilers providing medium temperature hot water for the swimming pool and spaceheating for the buildings. Being so old, the boilers were operating with a low efficiency (around 58 %) and so needed to be replaced with new units. Analysis of the energy consumption indicated that two units, each rated at 90 kWe would be fully employed all year round, and a third unit only 75 % of the year. It was decided to install only two units in the first phase, and, following a monitoring period, a third unit. The predicted energy savings estimated a reduction of 54 % in purchased electricity and a 7 % saving in the use of gas. The CHP units were installed in September 1987, and audited consumption figures indicate that these figures have been achieved. The details of the plant and the associated costs and savings are now discussed.

Figure 8.7 is a schematic layout of the CHP plant. The installation comprises two Waukesha 90 kWe units (see Table 8.3, page 317). Each unit comprises a spark-ignition natural gas engine driving an asynchronous generator. Engine jacket cooling water and initial cooling of the exhaust gas provides 180 kW of high-grade heat. Further cooling of the exhaust gas below the dew point provides

Figure 8.7 Schematic
layout of CHP plant

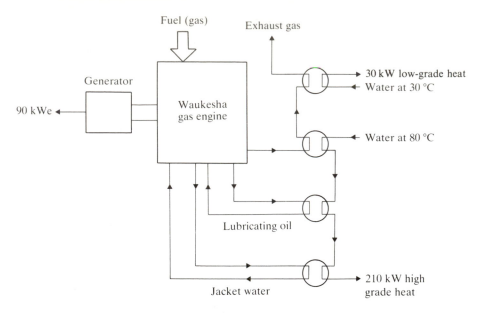

30 kW of low-grade heat. The high grade recovery system comprises a
water-to-water combined shell-and-tube surge-tank heat exchanger which cools
the glycol engine jacket system. The initial cooling of the engine exhaust gases
is achieved by a gas-to-water shell-and-tube heat exchanger. The low-grade
condensing heat recovery system consists of two parallel gas-to-water shell-and-
tube heat exchangers for each CHP unit. The size of the unit was based
on the following rationale.

Electricity consumption

From Fig. 8.8, which shows the patterns of electricity consumption from mains
supply for periods before and after the introduction of the CHP plant, the
previous annual consumption was about 1 500 000 kW h. On average, the leisure
centre usage generated a load factor of about 0.65. The number of kW h generated

Figure 8.8
Pattern of energy
consumption for
Case Study 8.1

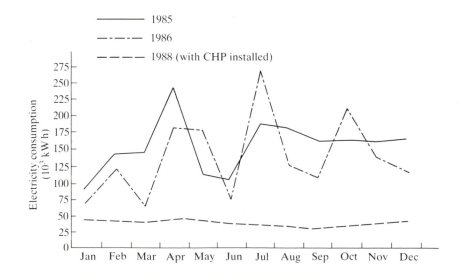

by the CHP units is given by

$$\text{electrical output} \times \text{load factor} \times 8760$$

which is equal to

$$(2 \times 90) \times 0.65 \times 8760 = 1\,024\,920 \text{ kW h}$$

Based on an average tariff (day/night/weekend), the value of this saving is

$$1\,024\,920 \times 3.8 \text{ p/kW h} = £38\,950 \qquad [1]$$

The annual kW h still required from the grid is

$$1\,500\,000 - 1\,024\,920 = 475\,000 \text{ kW h}$$

This figure corresponds to the monthly figure of 50 000 kW h which is indicated on Fig. 8.8 as the continuing consumption after the implementation of the CHP plant.

Gas consumption

From Table 8.3, page 317, the gas input required by each engine = 338 kW. From Fig. 8.7, the heat delivered by each engine = 240 kW. On a load factor of 0.65, the annual number of therms consumed by both engines is

$$2 \times 338 \times 0.65 \times (8760/29.307) = 131\,339 \text{ therms} \qquad [2]$$

The heat supplied by both engines is

$$2 \times 240 = 480 \text{ kW}$$

Previously, this heat was being supplied by boilers of 58 % overall efficiency. The annual energy input to the boilers was given by

$$\frac{\text{output} \times \text{load factor} \times 8760}{\text{boiler efficiency}}$$

which is equal to

$$\frac{480 \times 0.65 \times 8760}{0.58 \times 29.307} = 160\,790 \text{ therms} \qquad [3]$$

The annual saving in gas usage was $[3] - [2] = 29\,451$ therms. The value of this saving was given by

$$\text{saving} \times \text{tariff} = 29\,451 \times 0.34 = £10\,013 \qquad [4]$$

(based on a gas tariff of 34 p/therm). Allocating a cost of £5000 per annum to maintenance, gives the total running cost savings as

$$[1] + [4] - 5000 = £43\,963$$

The installed capital cost of the two 90 kWe units was £90 000. In addition £10 000 was spent on acoustic enclosures, alterations to boiler controls and changes to various building services.

$$\begin{aligned}
\text{Payback time} &= \text{capital costs/cost savings} \\
&= (90\,000 + 10\,000)/(43\,963) \\
&= 2.3 \text{ years}
\end{aligned}$$

This meets the payback criterion acceptable for leisure centres.

Potential for Micro-CHP

The installation of these CHP units in all suitable large buildings in the UK would create about 1 GW of electrical capacity. Micro-CHP units could also supply heat and electricity to domestic dwellings. Several homes would share a single engine, with insulated pipes running underground to distribute the hot water. A study by the Open University's Energy Research Group shows that a system producing 40 kWe could also heat about 50 homes, saving each house about £90 per year in fuel bills at current prices. There are some 22 million dwellings in the UK; allocating each one a 1 kW share in a CHP plant would create an electrical generating capacity of 22 GW. The CEGB produce, on average, about 25 GW.

It is often said that large power stations generate electricity more efficiently than thousands of small machines. A true comparison must include capital costs: micro-CHP costs around £500 to £550 per kWe output, which is slightly less than a coal-fired power station; nuclear stations cost around £1000 to £1500 per kWe output. The precise cost of nuclear-generated electricity depends largely on the rate of interest charged on capital investment and the costs of decomissioning.

8.9 CHP IN THE DOMESTIC SECTOR: DISTRICT HEATING

For domestic use, power plant can be made to provide hot water at temperatures between 80 °C and 140 °C. This can be piped to nearby communities via a well-insulated network of 'heat mains', effectively providing a large-scale central heating system. Clearly the closer the buildings are together, and especially in a compact urban area, the easier it is to operate such a system. The provision of heat to buildings in a particular area is generally known as *district heating* or *whole city heating.*

Although district heating is not the same as CHP, the two are often linked. CHP is a method of producing heat and electricity, whereas district heating is a system for generating and distributing heat. It clearly makes sense to use them together, where they are commonly referred to as CHP/DH.

Present Position

In 1911, probably the first CHP/DH scheme in the UK came into being when the Bloom Street power station in Manchester was modified to supply steam to nearby shops, offices, and factories.[8.11] The plant stopped generating electricity in 1952, although it still provides heat. Until fairly recently, the CEGB operated two small CHP schemes: one heated 11 000 homes in Battersea (London) until it closed in 1983; the other, based on South Denes power station (Yarmouth), provided heat for the needs of the locale. There are no CHP/DH schemes operating in the UK at the present time. It appears that the public generating authorities are committed to the large-scale generation of power and private companies will not accept the long pay-back times which the generation and distribution of heat incur.

Present Energy Policy

It is difficult to identify the place of CHP/DH in the UK national energy policy at the present time. There have been statements like that of the energy under-secretary in 1985, David Hunt, '... our energy policies rest on the belief that market forces are the best way of discovering exactly what the customers want and the best means of finding out whether they can have it at a reasonable price'.[8.12] This view is not exactly in line with lengthy studies which have specified the role which CHP/DH could fulfil and the energy benefits it could bring.

In 1974 the government established the Marshall Combined Heat and Power Group to investigate the economic potential of CHP. A report was published in 1979.[8.3] Its main recommendation was that plans should be drawn up immediately for the development of CHP and 'a start be made on one or more lead cities as soon as possible'. Marshall stressed that '... we need a decision in principle now, so that it is clear that a scheme will be implemented unless the detailed studies reveal unacceptable difficulties. To proceed in the reverse order, that is, to take a decision in principle after completing a detailed study would be a recipe for indefinite delay'. Subsequent events have confirmed this view, in spite of the standpoint in the report that CHP/DH had the potential to meet at least 30 % of the domestic heating needs in the UK.

In 1981, in response to the Marshall report, the government announced that nine urban areas (Belfast, Edinburgh, Glasgow, Leicester, Liverpool, London, Manchester, Sheffield, and Tyneside) would receive first-stage feasibility studies into CHP/DH. The following year, the Atkins report[8.2] was published and was enthusiastic about the prospects for CHP/DH. Little happened until 1984, when the government invited bids for £750 000 which was a contribution towards more detailed site studies in three of the nine lead cities. In 1985, £250 000 was awarded to each of the schemes in Belfast, Edinburgh and Leicester. Work continued on the schemes in Sheffield and Tyneside with local authority support.

Present Situation

More than a decade has passed since the publication of the Marshall report, yet only tentative schemes have emerged from cities like Sheffield, Belfast and Leicester. The reasons are partly to do with the following factors.

(1) In contrast to nuclear power, coal or gas, combined heat and power has no funded bureaucracy dedicated to its promotion. This makes it difficult for CHP to compete in the market place with other energy supply industries.

(2) In 1984, £0.75 million was awarded by central government to support further CHP studies in three lead cities. In the same year, £14 million was allocated to research into renewable resources such as solar, wind and wave power and £220 million was allocated to nuclear power research. This gives some indication of the view of central government.

(3) The requirement to attract private finance is a major problem. Orchard[8.15] considers that the need to produce a return within five years has held back the development of private CHP/DH. In contrast, the CEGB has been

allowed to spread the cost of power stations over 20-year periods; CHP/DH is thus being placed at a disadvantage compared to existing energy supply industries.

European Experience

In contrast to the UK situation, other European countries have greater CHP/DH plant capacity. The most recent international survey published by the OECD highlights the fact that all its member governments are taking a positive approach to CHP except the UK and the USA.[8.14] Tables 8.4 and 8.5 summarize the survey. The central role of CHP/DH in most of these countries is not a result of free market forces. In Denmark, for example, developments are financed, owned and operated by municipal authorities, who are given substantial grants to support the installation and maintenance of the heat distribution networks. In Finland, within three decades, district heating has come to provide over a third of the country's space-heating requirements and CHP plant is meeting over a quarter of the total electricity demand. In Helsinki, about 70 % of the annual heating and electricity demand is met by CHP/DH schemes run by the Helsinki Energy Board, a local government department.

Table 8.4 CHP/DH in European countries

	Installed DH capacity (MWt)	CHP capacity (MWe)
Britain	5 000–6 000	2 500 (solely industrial)
Sweden	20 800	5 200
West Germany	31 000	21 700
Finland	9 127	3 924
Denmark	14 400	1 800

Table 8.5 Proportion of electricity generated by CHP in 1980

	Percentage
Britain	3.5
Austria	7
Sweden	10
Netherlands	10
West Germany	11
Italy	11
Denmark	32

Case Study 8.2

The Hereford CHP station was the first of its kind to be built by the UK electricity supply industry. It was designed by the Midlands Electricity Board (MEB) to provide 13 MW of high-grade heat in the form of steam and hot water to local industry and 15 MW of electrical power for the MEB's 11 kV distribution system in Hereford. The fact that the customers for heat and power are separated shows how the MEB overcame the problems highlighted in

Section 8.4 where a single customer rarely has a constant demand for heat and power or a demand in which the heat and power are in a constant ratio.

The station, which is semi-automatic, is operated by the MEB and was built and commissioned within two years, being completed in 1981. Its overall efficiency is nearly 80 % – over twice that of the most efficient conventional modern power station. It only occupies one acre of ground, and does not need cooling towers or the network of pylon and high voltage overhead transmission lines normally seen at power stations.

The high overall efficiency of the station is due to the fact that the waste heat produced by the generation of electricity is used to provide steam and hot water to two of Hereford's largest companies; H P Bulmer Ltd and Sun Valley Poultry Ltd – both requiring a constant supply of heat throughout the year for food processing operations. If there is any extra demand for heat then this is supplied by back-up boilers. The electrical supply (15 MW) is virtually the base load for Hereford. Extra demand for power will come from the grid in the normal way; there is thus no problem about operating the prime mover at full load and a fixed heat/power ratio. The prime movers are two marine diesel engines which run on specially treated residual oil. Each of the engines drives a 7.5 MW generator. Heat from the exhaust gas of the engines is extracted by means of waste heat recovery boilers fitted to each engine. The gases then pass through economizers before entering the chimney and passing to atmosphere. Heat obtained from the economizers is used to further raise the temperature of the boiler feed water (which is already preheated by means of heat taken from the engine's water jacket, lubricating oil and charge air cooler). Four conventional oil-fired boilers supplement steam provided by the engines when necessary.

As a result of the CHP scheme, the equivalent of 13 MW of heat is obtained as a by-product of electricity generation, thus saving 15 500 tonnes of fuel oil, worth at the current price of 12 p/litre around £2.25 M per year.

Design considerations

Experience led the MEB to look for an industrial process heat load factor in excess of 50 % to provide the necessary economic stability for the project. In Hereford, the MEB sold the heat to two well-established UK companies: H P Bulmer Ltd and Sun Valley Poultry Ltd; two factory sites with substantial heat loads.

It is a basic requirement in the marketing of heat to be in a position to sell heat at a price at least 10 % less than customers can produce it for themselves, since industrialists cannot be expected to sacrifice their independence without some financial benefit. This level of economic performance can only be achieved by extracting maximum thermal efficiency from a station over the maximum time; the station at Hereford was therefore designed to generate the base heat load for the two companies and the base electrical load for the locale.

A major constraint on the design was that the customers of the station required heat for existing factories, and in consequence the station had to be designed not for optimum efficiency but by the dictates of the heat market. This means that it was difficult to design for optimum overall efficiency for a given prime mover. For example, the diesel engine has a surplus of low-grade heat. Figure 8.9 shows the daily variation of heat load for the yearly seasons; the CHP heat recovery system thus has to supply any base load below 20 MW.

Figure 8.9 Heat load at Hereford

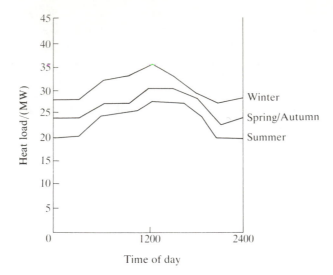

Consideration of the electrical requirements of the design is based on the annual electrical load cycle for Hereford, which is shown in Fig. 8.10. It can be seen that any electrical output below the absolute minimum value of this curve would meet the required base load condition for the CHP station generation.

The next problem was how to meet the peak demand for heat, and design the station to cope with relatively frequent heat load fluctuations. This could have been achieved in one of two ways:

(a) supplementary firing of waste heat boilers;
(b) simple addition of fired packaged boilers.

For Hereford, the decision was taken to install oil-fired boilers, operating on the same fuel as the prime movers, to supplement the heat output.

Prime mover selection was the most crucial decision faced by MEB; three options were considered:

(a) back pressure steam turbine;

Figure 8.10 Electrical load at Hereford

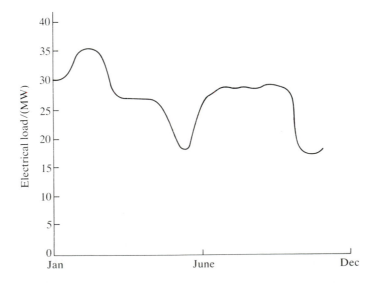

(b) gas turbine;

(c) diesel engine.

The diesel engine was chosen because it had two major advantages; it could run on cheaper fuel (residual oil) and, having the highest thermal efficiency of the three options, it would operate most efficiently at the lowest heat/power ratio – in other words, the energy conversion is biased towards the most valuable product, electricity. More detailed arguments about the characteristics of each prime mover have already appeared in Section 8.5.

The disadvantage of using the diesel engine is that it produces a large quantity of low grade heat in the form of hot water at temperatures between 70 °C and 80 °C. Fortunately, this can be used effectively for both the waste heat recovery boilers and the oil-fired boilers.

At Hereford, using the two 7.5 MW engines as described, the total full-load heat output from the heat recovery systems is 15 MW. If this value is superimposed onto Fig. 8.9 then it can be seen (Fig. 8.11) that the CHP unit supplies heat below the base load condition but that the total station capacity is 45 MW when all four oil-fired boilers are being used. These separate oil-fired boilers give a security of supply to the heat output even when the diesel engine is not operating.

Figure 8.11 Base load output of Hereford CHP station

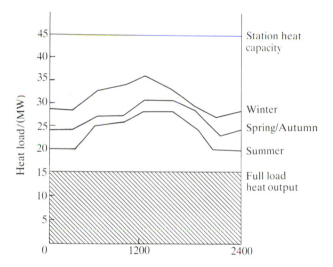

The problem of back-up electrical supply does not arise because the city of Hereford is supplied by the UK national grid in the usual way.

System details of Hereford

A schematic of the system is shown in Fig. 8.12. It can be seen that raw water at approximately 10 °C enters the system from the mains supply. Subsequently the water passes through a water treatment tank, to remove most of the lime deposits in the water so as to avoid the build-up of lime scale in the heat exchanger tubes. After treatment, the water then passes through three heat exchangers:

(1) the charge air cooler, which is used to cool the compressed air after turbocharging to increase the volumetric efficiency of the engine;

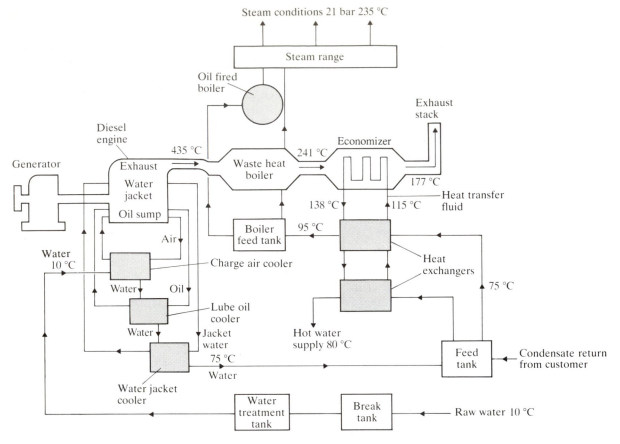

Figure 8.12 Flow diagram for Hereford CHP scheme

(2) the lubricating oil cooler, which cools the engine oil to the required working temperature;

(3) the water jacket cooler, which cools the jacket water to the required working temperature.

These three heat exchangers have by-pass loops to allow the correct working temperatures of the engine to be obtained. For example, if the jacket water is being overcooled then the controller will cause the treated water to by-pass the water jacket cooler until the required working temperature is restored.

The order in which these three heat exchangers are arranged is critical for the efficient heating of the fresh water. The exchangers must be arranged so that the fresh water passes through the heat exchanger system with the temperatures increasing at each stage. This requires the heat exchanger transmitting the lowest temperature to be first and the heat exchanger transmitting the highest temperature to be last. If this order was reversed with the fresh water flowing through the jacket water heat exchanger first, then any subsequent heat exchangers would be unable to transfer any heat to the water as the inlet water temperature would already be higher than the inlet temperature of the charge air.

The fresh water leaves the three heat exchangers at approximately 75 °C and is then pumped to a feed tank, where it is mixed with condensate water returning from the two industrial users: Bulmers and Sun Valley Poultry. From the feed tank, most of the water is directed to another heat exchanger which, using heat from the exhaust gas economizer, further raises the water temperature to 95 °C, at which point it enters the boiler feed tank, ready for conversion to steam at 235 °C and 21 bar by the waste heat boiler. If the demand for steam exceeds the output from the waste heat boiler then the deficit is made up by raising steam in the oil-fired boilers. Not all the water is used in this way; a small proportion is passed through a second exhaust economizer which raises the water temperature to 80 °C for use as a hot water supply.

The exhaust gases from the diesel engine exit the turbocharger at approximately 435 °C and enter the waste heat boiler, generate steam at 21 bar and 235 °C from the feed water at 95 °C, and leave at a temperature of 240 °C. Obviously the exhaust gases still have a considerable amount of energy in them, hence the need for an exhaust economizer. The exhaust gases leave the economizer at 177 °C having raised the temperature of the boiler feed water. The reason for using a heat transfer fluid in the economizer is so that flexibility can be achieved with the boiler feed water temperature, without the need for a complex control system. The temperature of the heat transfer fluid can be allowed to rise to that of the exhaust gases without boiling, and in consequence the heat extracted from the exchanger can remain constant, even if the fresh water flowrate varies; if the demand for steam and hot water decreases then the temperature of the boiler feed water can be allowed to rise.

Leaving the economizer at 177 °C, there is still a considerable amount of energy remaining in the exhaust gas. However, it is necessary to ensure that the exhaust gas temperature does not fall below the acid dew point, to avoid the formation of corrosive acids in the exhaust stack (see Section 4.2). For this reason no further heat is extracted from the exhaust gas.

Environmental considerations

At Hereford considerable research was carried out at the design stage in relation to the reduction of noise and vibration leaving the engine room. The room was designed as an acoustic building using the existing minimum background noise level as the datum. The waste heat recovery boiler was also designed as an acoustic attenuator; as a result, additional exhaust silencing requirements are minimal. The noise levels within the engine room are such as to cause serious hearing loss if there is prolonged exposure without ear protectors. The noise insulation is such that normal speech is quite audible in the adjacent control room. Each engine stands on a concrete bed which is supported by springs. This system significantly reduces the vibration transmitted outside the engine room. However, internal vibration has caused many problems, particularly the cracking of the support bed, and the engine block. Opinions differ on the cause of these cracks as the engines have proven their worth as marine diesel units with years of use and no history of cracks. It is possible that the less rigid structure of a ship allows the engines to flex on their mounting, unlike the more solidly mounted land-based units.

Case Study 8.3

Fort Dunlop was the second CHP station built by the MEB. It was designed to provide high grade heat in the form of steam for local industrial process use in addition to generating 24 MW of electricity which is fed into the MEB's 11 kV distribution system at Fort Dunlop in Birmingham. The station was completed in 1985; it runs almost fully automatically and is operated by the MEB.

The proposed user of heat was to be Dunlop Ltd, tyre manufacturers. The unusual aspect of the heat requirement for Dunlop was the relatively high process steam temperature; to achieve this, two coal-fired boilers were installed in series with the main prime movers, two large marine diesel engines.

Whilst the general concept and design principles remained the same as for the Hereford plant, the Fort Dunlop station has no package oil-fired boilers; instead more use is made of the heat in the exhaust gases by means of 'after-burning' of low-grade fuel in coal-fired boilers.

System details of Fort Dunlop

Figure 8.13 shows a schematic layout of the station. As at Hereford, cold water enters the station and is treated for lime deposits. The cold water then passes through a series of four heat exchangers; the stage two air cooler, the lubricating oil cooler, the jacket water and finally the stage one air cooler. Each of these heat exchangers has a by-pass loop to allow engine temperature control and they are so arranged that the temperature of the fresh water is raised successively

Figure 8.13 Flow diagram for Fort Dunlop CHP station (full load)

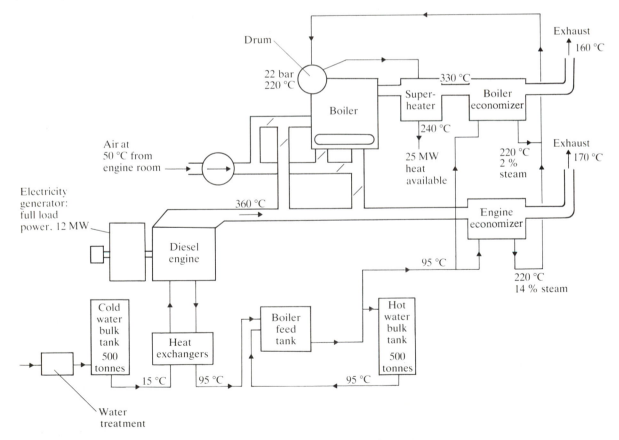

at each stage. The water leaves the heat exchangers at the design temperature of 95 °C and is pumped to either the boiler feed tank or the hot water bulk tanks. There, unlike Hereford (where some water is drawn off for domestic use), all the water passes through either the engine economizer or the boiler economizer, where the water is raised to wet steam at 220 °C. From here it is passed to a boiler and superheater where the steam condition is raised to 22 bar and 240 °C. The total heat output from the station is designed to be 25 MW.

Of the remainder of the heat recovery system, the exhaust gases from the diesel engine are designed to leave the turbocharger at 360 °C and are then distributed between the oil-fired boiler and the engine exhaust economizer. The exhaust gases entering the engine economizer exist at approximately 170 °C and enter the stack to be expelled to atmosphere. The remainder of the exhaust gases enter the coal-fired boiler above the grate and assist in the final combustion of the fuel (called afterburning). The required fresh air is drawn from the engine room itself; such air has been slightly preheated (to 50 °C) by the engine and turbocharger surfaces and is therefore more suitable as combustion air. The combustion gases from the boiler enter the superheater and raise the steam temperature to 240 °C. From there the gases enter the boiler economizer at about 330 °C, heat the hot water from 95 °C to wet steam at 220 °C, and leave at 160 °C, entering the stack to be expelled to atmosphere. The concept of coal-fired supplementary firing of the hot exhaust gas is the fundamental difference between the Fort Dunlop and Hereford schemes. Supplementary firing is necessary because Dunlop requires steam at a higher temperature than can economically be supplied by the diesel engine alone.

As there is still a measure of unused oxygen (about 15 % by volume) in the hot engine exhaust gas, the gas is directed to the coal-fired boiler, together with air from the engine room itself which is slightly preheated. In order to burn the fuel (which is cheap low-grade coal in the form of a dust) efficiently, the boiler is supplied with about 30 % excess air.

If for some reason, no steam output is required by Dunlop, there is no need for hot water to be pumped into the bulk storage tanks. When this situation arises the heat from the engine's heat recovery system has to be dumped. This is done by recirculating the hot water through a unit comprising a dump heat exchanger and a forced draught cooling tower; in this way the water will re-enter the engine heat exchangers at a low enough temperature to maintain the engine cooling systems at their appropriate temperatures.

Energy calculations for Fort Dunlop CHP station

The following calculations are based on flows and temperatures obtained by the station instrumentation when the engine was running at about 85 % of the full load condition.

The data relating to the waste recovery circuit is shown in the table below for part-load running conditions (10.46 MW at 433 rev/min).

For the stage 1 charge air cooler, the heat transferred to the water is given by:

$$Q = \dot{m}_w \times c_p \times (\text{temperature change})$$
$$= (107\,000/3600) \times 4.18 \times (38.1 - 31.9) \text{ kW}$$
$$= 0.77 \text{ MW}$$

Similarly, the heat recovered from the engine lubricating oil is 1.25 MW; the

Flowpath	Inlet temperature (°C)	Outlet temperature (°C)
Water through stage 1 cooler	31.9	38.1
Water through lubricating oil cooler	38.1	48.2
Water through engine jacket	48.2	66.1
Water through stage 2 cooler	66.1	78.6
Water through engine economizer	78.3	Wet steam at 222 °C (dryness fraction = 0.14)
Exhaust gas through turbocharger	501	342.4
Exhaust gas through engine economizer	342.4	170.0
Exhaust gas through boiler economizer	342.4	160.0

Mass flow rates: water through engine economizer = 19 000 kg/h
all other flow rates = 107 000 kg/h
exhaust gas = 95 098 kg/h

heat recovered from the engine cooling water is 2.22 MW; and finally, the heat recovered from the stage 2 charge air cooler is 1.55 MW. Thus the total heat recovered by this part of the system is 5.79 MW.

The data relating to heat recovery from the exhaust gases is also shown in the table above. The specific heat of the exhaust gas will be taken as that of nitrogen at the appropriate mean temperature. The energy recovered to drive the turbocharger is given by

$$\dot{m}_G \times c_p \times \text{temperature change}$$

which is equal to

$$(95\,098/3600) \times 1.098 \times (501 - 342.4)\ \text{kW}$$
$$= 4.60\ \text{MW}$$

The energy extracted from the exhaust gas in the engine economizer is given by

$$\dot{m}_w \times \text{enthalpy change}$$

which is equal to

$$(19\,000/3600) \times (1211 - 327)\ \text{kW} = 4.67\ \text{MW}$$

Not necessarily all the engine exhaust gas flows through the engine economizer. As can be seen from the overall circuit schematic in Fig. 8.13, some of the gas passes to the boiler. The amount directed through the engine economizer can be found by the energy balance for this economizer:

energy gained by water = energy lost by exhaust gases = 4.67 MW

i.e. the mass flow rate of gas to the engine economizer is given by

$$\text{energy transfer}/(c_p \Delta t)$$
$$= 4.67 \times 1000/\{1.065 \times (342.4 - 170)\}$$
$$= 25.44\ \text{kg/s}\ (91\,566\ \text{kg/h})$$

The mass flow rate of exhaust gas to the boiler is given by

total flow − flow to engine economizer
$$= 95\,098 - 91\,566 = 3532\ \text{kg/h}$$

This flow, which is only 3.7% of the total exhaust flow, enters the boiler at 342.4 °C and eventually leaves the boiler economizer at 160 °C. The energy transferred is given by

$$(3532/3600) \times 1.065 \times (342.4 - 160) \text{ kW} = 0.19 \text{ MW}$$

Therefore, the total energy transferred from exhaust gas is equal to

$$4.60 + 4.67 + 0.19 = 9.46 \text{ MW}$$

The overall energy balance for the engine is shown in the table below.

Item	MW	%
Energy input from fuel	29.90	100.00
Energy output as power	10.46	34.98
Energy recovered from exhaust gas	9.46	31.64
Energy recovered by feed water	5.79	19.36
Energy lost in the stack	3.30	11.04
Unaccountable losses	0.89	2.98

It may seem at first sight that the overall efficiency of the unit is remarkably high since only 14% of the input energy is not converted to useful heat or power. It should be remembered that of the 9.46 MW of heat recovered from the exhaust gas, 4.6 MW is used to drive the turbocharger, leaving 4.86 MW. This amount, together with the 5.79 MW of energy recovered by the feed water, gives a total of 10.65 MW of heat recovered for the generation of process steam. The engine delivers the following:

10.46 MW of power (34.98% of the input energy)
plus 10.65 MW of heat (35.59% of the input energy)

giving an overall efficiency as a CHP station of 70.57% and an operating heat/power ratio of 1.02:1.

The above calculations are based on performance figures at a particular time. As it happens, very little (only 3.7%) of the total exhaust flow is directed towards the boiler and therefore the benefits of afterburning are greatly reduced. It is a matter of operational management to split the exhaust flow so as to optimize the overall cost of energy input to the plant, that is to both engine and boiler. Sufficient to say here that the greater the amount of exhaust gas used by the boiler, the less fuel required by the boiler and the less fresh air needs to be pumped from the engine room.

Control systems

The station runs almost fully automatically. The computerized control system is capable of initiating the engine start-up procedures and shutting down the plant. Should a situation arise where a significant fault is detected, then the system is capable of shutting down the whole plant and initiating the alarm signals. Once the design operating temperatures are reached, the control system then maintains these temperatures to ensure efficient running of the prime mover.

The waste heat recovery systems of the plant contain numerous temperature and flow rate sensors which feed information to a central computer; this issues appropriate instructions to the control hardware. If, for example, the jacket water were to return to the engine at a temperature above the design value, then the control system would signal the flow valve to increase the flow of feed water through the heat exchanger (and hence decrease the proportion flowing round the by-pass loop), thus reducing the jacket water temperature. The control and information systems at Fort Dunlop illustrate the recent advances made in information technology; operators can obtain instant detailed reports on any part of the plant which has meant that some problems can be detected earlier, thus enabling preventive maintenance to be carried out. Management confidence in the control system is such that the plant is totally unmanned at certain times of day, thus reducing the station's labour costs.

Fuel prices

Both the Hereford and Fort Dunlop stations were designed in the days before the price of heavy fuel oils began to rise (and more recently fall) significantly. In 1980, soon after the Hereford plant was commissioned, oil prices began to rise thereby significantly increasing the operating costs of the system. Fortunately, 1985 saw a dramatic fall in prices to levels which have not risen significantly in real terms since. The fall in oil price had a knock-on effect on the price of gas; any large industrial consumer could negotiate a tariff for gas based on undercutting the cost of the fuel which the plant would alternatively use. In the case of Hereford, the fuel (residual oil) was one of the cheapest available, so the tariff for gas was agreed to about one-third of the current domestic tariff. The engines at Hereford CHP station are now running very economically on natural gas; in other words, they are producing electricity at a much cheaper rate than the CEGB which uses mainly coal, a fuel which has not reduced in price over the years.

Remember that the point of having a CHP station is to make savings of energy and money against a system of bought-in power and heat from a boiler, so the question now is: having worked out the fuel cost of producing a unit of electrical energy, how much is that unit worth to the customer? Well, the customer in this case is the area board who annually publish a list of tariffs which they will pay for a unit of energy. These tariffs vary with the time of day and the seasons of the year as discussed in Chapter 9. When the fuel cost per unit is less than the income per unit then the station is making an operating profit because there is other income from the sale of heat. Of course, the profit is reduced by maintenance and labour costs.

The foregoing remarks serve to emphasize the points made in Section 8.6 about the factors which affect the economic viability of CHP schemes. Another major factor is the balance of demand for heat and power. In both the Hereford and Fort Dunlop schemes the power is sold to a customer who is, by law, obliged to purchase the power, and who will, if asked, provide a back-up system. The customers for heat are not so constrained and in the case of the Fort Dunlop station, the potential customer for heat, Dunlop Ltd themselves, having been taken over by another company, decided not to use the heat from the CHP station. This left the station without a heat user. The station is running quite profitably at the moment because of the low generating costs and the compensation paid by Dunlop's new owners, but approximately 25 MW of heat is being dissipated to the atmosphere at the time of writing. It is hoped that this heat will be used by large blocks of council houses which are situated relatively nearby. Nevertheless, the situation illustrates how vulnerable CHP stations can be to changes in user operating conditions; a situation always likely in the industrial sector.

8.10 CONCLUSIONS

Combined heat and power is not a new technology; as long ago as 1862 a new sugar beet factory at Lavenham (Suffolk) was using it. So significant an event was the opening of the plant that it warranted a two-page article in *The Engineer* of 5th June 1868. The main thrust of the article was that the plant was the first to make sugar from beet grown in this country. The CHP aspect was dismissed in one line to the effect that 'as usual, exhaust steam from the engine is used in the evaporation of the liquor'.

Surely this is the attitude we should take today over a hundred years later. CHP is no more than a tool to be used 'where appropriate'. It is the 'where appropriate' that is the most important point. CHP will not solve the country's energy problems *per se*, but will, if the circumstances are right, significantly reduce the inefficiency of electrical power generation, saving our valuable fuel resources and money that could be better spent elsewhere. No more, no less.

Problems

8.1   A swimming pool complex uses a CHP system to supply 40 kW of electrical power and 100 kW of heat. The system comprises a gas engine and a gas-fired boiler. Heat is recovered from the engine's cooling water, lubricating oil, and exhaust gas; additional heat is supplied by the boiler. Using the data below, calculate:
(i)   the heat/power ratio of the engine;
(ii)   the total volume flow rate of gas required by the engine and boiler.

Data
GCV of gas, 38 MJ/m^3; brake thermal efficiency of engine, 37 %; mechanical efficiency of alternator, 96 %; boiler efficiency, 82 %; thermal energy recovered from engine, 60 % of the total waste energy from the engine.

(1.02; 17.30 m^3/h)

8.2   A small hotel currently meets its demands for heat and power by purchasing power from the area electricity board and generating heat from a central gas-fired boiler. The hotel management is considering an alternative supply system using a micro-CHP unit also using gas, which will be sized according to the electrical and thermal loads of the summer months; 13 kW and 33 kW respectively. In winter the extra heating requirements will be met by the present gas-fired boiler and extra power purchased from the area board. It is intended that the CHP unit will run at full load for 16 hours per day for all days of the year. Using the data below, calculate:
(i)   the annual savings that could be achieved by installing the CHP plant;
(ii)   the simple payback time for the capital investment required.

Data

Overall efficiency of CHP unit, 85 %; capital cost of CHP unit, £6000; annual maintenance cost of unit, £500; price of gas, 38 p/therm; price of electricity, 4.2 p/kW h; boiler efficiency, 80 %.

(£1714; 3.5 years)

8.3 A plant is designed to use hot industrial waste gases to produce power and to generate steam as shown in Fig. 8.14. Electrical power is produced from a gas turbine unit and the exhaust from the gas turbine is used to generate steam at low pressure while the hot waste gases, as well as providing the heat input to the gas turbine cycle, also generate steam at high pressure. Using the data below and that on the figure, neglecting heat losses and pump work, calculate:

(i)   the mass flow rate of air through the turbine power unit;
(ii) the mass flow rate of LP steam;
(iii) the mass flow rate of HP steam.

Figure 8.14 Plant diagram for Problem 8.3

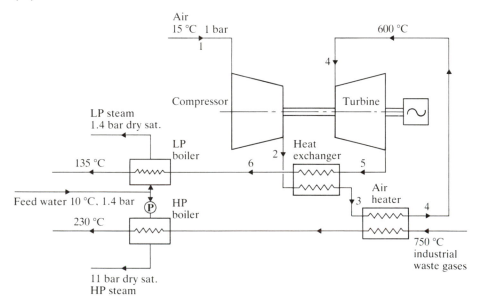

Data

Net power output, 1500 kW; pressure ratio of compressor, 4; isentropic efficiency of compressor, 83 %; turbine pressure ratio, 3.64; isentropic efficiency of turbine, 87 %; mass flow rate of industrial waste gases, 14 kg/s; effectiveness of gas turbine unit heat exchanger, 0.75; for air take c_p and γ as 1.005 kJ/kg K and 1.4; for the waste gases take c_p as 1.15 kJ/kg K; combined mechanical and electrical transmission efficiency of gas turbine unit, 94 %.

(24.1 kg/s; 1.27 kg/s; 0.58 kg/s)

8.4 Low-grade thermal energy from the condenser cooling water of a power station is used to provide the heat source for a heat pump using R11 as shown in Fig. 8.15. The heat pump compressor is driven by a steam turbine. The system is used to heat water in two stages: the first stage in the condenser of the heat pump unit; the second stage in the steam plant condenser. Using the data below, neglecting feed pump work and all losses, calculate:

(i)   the mass flow rate of refrigerant;

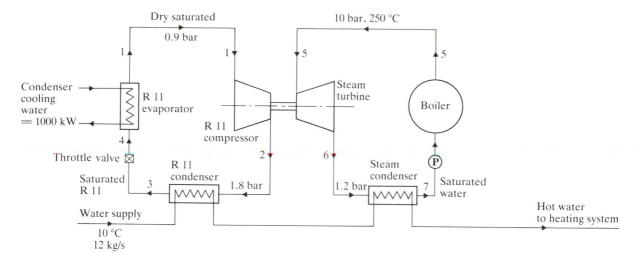

Dry saturated

0.9 bar

10 bar, 250 °C

Condenser cooling water ≡ 1000 kW

R 11 evaporator

Steam turbine

Boiler

R 11 compressor

Throttle valve ⊠

Saturated R 11

R 11 condenser

1.8 bar

Steam condenser

1.2 bar

Saturated water

Water supply

10 °C
12 kg/s

Hot water to heating system

Figure 8.15 Plant diagram for Problem 8.4

(ii) the mass flow rate of steam;

(iii) the temperature of the heated water at exit from the two-stage heating process;

(iv) the ratio of the energy required from the boiler to heat the water by this means to the heat required to heat the water by direct means.

Data

Thermal energy available from power station condenser cooling water, 1000 kW; inlet temperature of water to be heated, 10 °C; mass flow rate of water to be heated, 12 kg/s; isentropic efficiency of heat pump compressor, 80 %; isentropic efficiency of steam turbine, 80 %; heat pump evaporator saturation pressure, 0.9 bar; heat pump condenser saturation pressure, 1.8 bar; steam turbine stop valve pressure and temperature, 10 bar and 250 °C; steam plant condenser pressure, 1.2 bar; vapour entering heat pump compressor, dry saturated; liquid leaving heat pump condenser, saturated; water leaving steam plant condenser, saturated.

(A chart for R11 should be used, for example the CIBSE chart.[4.4])

(6.06 kg/s; 0.306 kg/s; 45.2 °C; 0.434)

8.5 A district heating scheme based on an incineration plant supplies high pressure hot water at 115 °C with a return water temperature of 75 °C. The rate of flow of water is 30 kg/s. The refuse after shredding is mixed with coal before combustion in the boiler; the rate of combustion of coal at full load is 570 kg/h. It may be assumed that during the heating season from October to March inclusive the plant is on full load for seven days per week on a 24-hour basis; supplementary heating is used when the outside temperature falls below a certain value. Using the data below, calculate:

(i) the total quantity of refuse burned during the heating season;

(ii) the annual running cost of the plant compared with that of a gas-fired boiler;

(iii) the break-even point time of the refuse plant compared with the gas-fired boiler, neglecting maintenance costs and interest charges.

Data

Calorific value of coal, 30 000 kJ/kg; calorific value of refuse, 14 000 kJ/kg; calorific value of natural gas, 38 000 kJ/m³; cost of coal, £65 per tonne; cost of gas, 14 p/m³; cost of disposal of customer's refuse, £20 per tonne; overall efficiency of incinerator boiler, 70 %; efficiency of gas boiler, 80 %; capital cost of refuse plant, £650 000; capital cost of equivalent gas boiler, £300 000.

(2766 tonne; £161 830, £420 950; 1.35 years)

8.6 An industrial plant shown in Fig. 8.16 uses a diesel engine and a back-pressure turbine to provide electrical energy and thermal energy for heating. The diesel engine is a slow speed turbo-charged engine burning residual fuel oil with a high air–fuel ratio. There is sufficient oxygen in the exhaust gases to allow further combustion of residual fuel in the boiler without an additional air supply. Air is drawn from atmosphere, compressed in a turbine-driven centrifugal compressor, and passed through an air cooler to the engine inlet manifold. The exhaust from the engine cylinders passes through the turbine which develops just enough work to drive the compressor, the gases then passing to the boiler.

Superheated steam generated in the boiler expands in a steam turbine where at an intermediate pressure some steam is bled off to a process; the condensate from the process is saturated at the bleed pressure and is non-returnable. The steam leaving the condenser is pumped to the hot well where it mixes with the cold make-up water. Feed water drawn from the hot well is pumped through two heat exchangers before entering the boiler. In the first heat exchanger the feed water cools the cooling water from the engine piston, cylinder and lubricating oil system; in the second heat exchanger the feed water cools the cooling water from the air cooler. Neglecting pump work and thermal losses throughout, and using the data below, calculate:

(i) the temperature of the feed water entering the boiler;

(ii) the ratio of the useful energy output to the energy input from the fuel.

Figure 8.16 Plant diagram for Problem 8.6

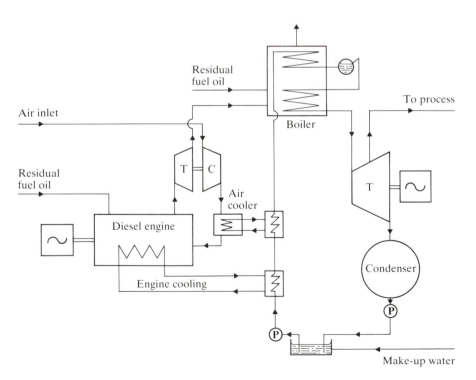

Data

Power generated by the diesel engine, 23 MW; air flow to engine, 213 000 kg/h; air–fuel ratio of engine, 42; atmospheric temperature, 15 °C; pressure ratio of compressor, 2.7; isentropic efficiency of compressor, 0.7; temperature of air leaving air cooler, 40 °C; rate of energy dissipated from engine piston, cylinder and lubricating oil system, 1.7 MW; process steam mass flow rate, 15 kg/s; steam flow leaving boiler, 20 kg/s; steam conditions at turbine entry, 15 bar and 300 °C; pressure of process steam, 1.6 bar; isentropic efficiency of steam turbine, 0.8; condenser pressure, 0.03 bar; temperature of make-up water, 6 °C; specific heat at constant pressure and γ for air, 1.005 kJ/kg K and 1.4; rate of residual fuel oil supplied to the boiler, 3300 kg/h; calorific value of fuel oil, 41 500 kJ/kg.

(109 °C; 67.8 %)

8.7 A large sewage processing plant requires a considerable energy input and it is decided to use the gas products from the plant in a CHP scheme which will also provide the heat input for the sewage processing. An outline of the plant is shown in Fig. 8.17. The steam generator uses a two-pressure system for greater efficiency, and has once-through corrosion-resistant tubing which removes the need for de-aeration of the feed water. After expansion through the high-pressure turbine the high-pressure steam mixes adiabatically with the low-pressure steam from the generator; some steam is bled off at this intermediate pressure for space-heating while the remaining steam expands through the

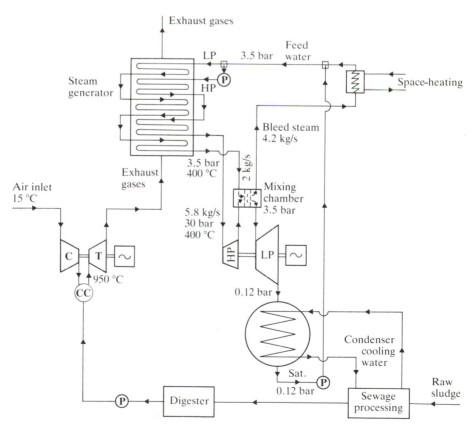

Figure 8.17 Plant diagram for Problem 8.7

low-pressure turbine. Using the $h–s$ chart, the steam tables, and the data given, and neglecting all pressure losses, thermal losses, mechanical drive losses, and pump work, calculate:

(i) the total power output;

(ii) the energy available for space-heating;

(iii) the energy available for the sewage processing plant;

(iv) the volume of sewage sludge treated each day, assuming continuous use.

Data

Gas turbine unit: pressure ratio, 10; air inlet temperature, 15 °C; maximum cycle temperature, 950 °C; isentropic efficiencies of compressor and turbine, 0.82 and 0.95 respectively; combustion efficiency, 100 %; c_P and γ for air, 1.005 kJ/kg K and 1.4; c_p and γ for gases, 1.1 kJ/kg K and 1.333; mass flow rate of exhaust gases, 76 kg/s. Steam system: high-pressure steam, 30 bar and 400 °C; low-pressure steam, 3.5 bar and 400 °C; high-pressure steam mass flow rate, 5.8 kg/s; low-pressure steam mass flow rate, 2.0 kg/s; mass flow rate of steam to space-heating, 4.2 kg/s at 3.5 bar; condenser pressure, 0.12 bar; condensate from condenser, saturated at 0.12 bar; condensate from space-heating heat exchanger, saturated at 3.5 bar; isentropic efficiencies of high-pressure and low-pressure turbines, 0.80 and 0.78 respectively. Sewage system: calorific value of digester gas, 20 500 kJ/m³; production of digester gas at the rate of 0.15 m³ per m³ of raw sludge.

(21.49 MW; 9.89 MW; 8.22 MW; 1.427×10^6 m³ per day)

8.8 A CHP plant is shown diagrammatically in Fig. 8.18. There are two identical boilers using exhaust gases from two identical gas turbine units. The district heating water can be heated by the exhaust gases only or by a combination of the exhaust gases and a heat exchanger in the steam system. In the summer the district heating is off and the power output from the steam turbine is therefore increased. Using the data marked on Fig. 8.8 and neglecting pump work and all losses, calculate:

(i) the power output of the gas turbine units;

(ii) the power output of the steam turbine in the summer;

(iii) the overall system efficiency in the summer;

(iv) the thermal energy available for district heating and the corresponding mass flow rate of water when only the exhaust gas heat exchangers are used;

(v) the thermal energy available for district heating and the corresponding water mass flow rate when the steam heat exchanger is also used;

(vi) the total power output available when the district heating is fully used;

(vii) the overall system efficiency for the fully operational plant in winter.

(50.6 MW; 25.5 MW; 46.5 %; 15.95 MW, 94.9 kg/s; 62.8 MW, 373.8 kg/s; 67.62 MW; 79.8 %)

References

8.1 *CEGB Statistical Yearbook 1987/88* Department of Information and Public Affairs, CEGB

8.2 Atkins G 1986 The advantages of CHP systems. *Proc. IMechE Symposium on CHP* Sheffield 1986

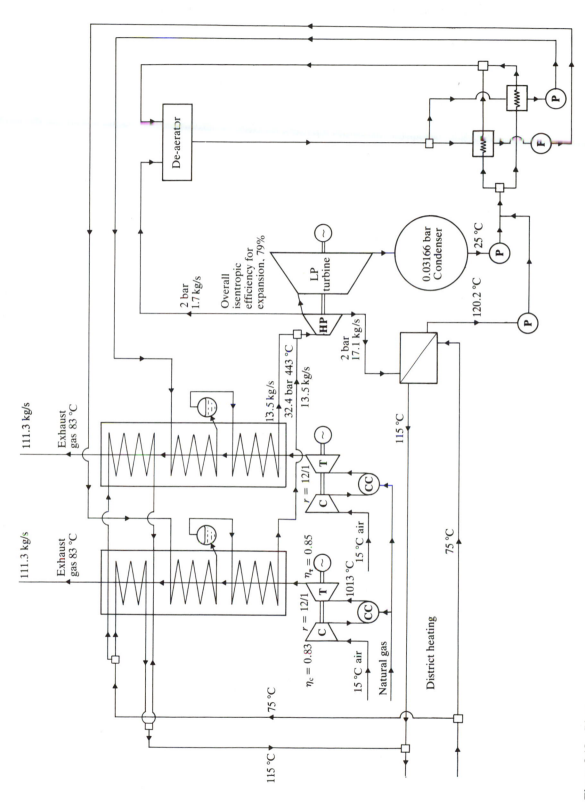

Figure 8.18 Plant diagram for Problem 8.8

8.3   Marshall W 1979 *Combined Heat and Power Generation in the UK* Energy Paper 35 HMSO

8.4   Diamand R M E 1970 *Total Energy* Pergamon Press, Oxford

8.5   Sheffield City Council, Sheffield File 1985

8.6   Forrest R 1985 *Small-Scale Combined Heat and Power* Energy Technology Series 4 Energy Efficiency Office

8.7   Shaffer I R 1986 *Combined Heat and Power and Electricity Generation in British Industry 1983–1988* Energy Efficiency Series HMSO

8.8   Energy Demonstration Schemes involving the installation of CHP. Department of Energy: Project Profile Numbers 161, 188, 208, 219, 227–9, 243, 263–5, 273

8.9   FEC consultants, Totem Total Energy System in a Hotel, Extended Report: Energy Demonstration Scheme, Energy Efficiency Office, 1986

8.10   Andrews D 1986 Power from the people. *New Scientist* **1505**

8.11   Grey R 1985 Bringing heat and light. *Energy in Buildings* Nov/Dec

8.12   Hunt D *Hansard* 25th October 1985

8.13   Garagham H 1984 Britain cold shoulders heat and power. *New Scientist* **1500**

8.14   Orchard W R H 1985 CHP: the nine-star energy option. *Power Generation* Jan

Bibliography

Bonham P 1985 *The Economic Assessment of Proposed CHP Plant* Ruston Gas Turbines Ltd

Meador R 1982 *Cogeneration and District Heating* Ann Arbor Science

Orchard W H R, Sherratt A F C 1981 *Combined Heat and Power – Whole City Heating* Goodwin

Shepherd G T 1980 *CHP – Vision or Delusion?* Paper given at the House of Commons to the Parliamentary Liaison Group for Alternative Energy Strategies

Shepherd G T 1986 Waste not – want not? *Electronics and Power* Oct

9 THE ENERGY MANAGER AT WORK

The techniques used in the management of energy have much in common with those required for the management of any resource in a company or establishment; resources such as materials and manpower. The general concept of monitoring performance and the setting of targets is essential to the control of costs and also the motivation of staff associated with energy usage.

Most companies use energy in one of two major ways:

(a) in the direct manufacturing of a product (for example: in the heating of plastics prior to forming; for powering electric motors; energy to boilers to raise steam);
(b) in the activities which support the manufacturing activity (for example: energy used to maintain the environment of a factory for the benefit of employees; space-heating of offices and warehouses; hot water services; lighting).

9.1 OBJECTIVES OF THE ENERGY MANAGER

The overall objectives of the Energy Manager are to save money. Although it is an over-simplification to imagine that this is just a question of establishing areas of inefficiency and taking appropriate corrective action, it does form a convenient starting point in the description of the activity of the Energy Manager. As a first view, these activities are summarized in Fig. 9.1.

The diagram summarizes the techniques which can save money. Each type of activity will be considered in turn.

Same usage   Reduce costs by tariff negotiation   The fuel and energy costs of a company are related to tariffs which are published by the suppliers. There is often a choice of tariff structure, for example with electricity, and it is a matter of negotiation with the area board as to which is the most advantageous structure to adopt. A similar arrangement applies to tariffs for gas. The tariff structure will not of itself necessarily alter the total consumption of fuel but may suggest a different pattern of usage to reduce costs. This topic is discussed in detail in Sections 9.2 to 9.7.

Less usage   Good housekeeping, by running existing plant in a more effective way   The phrase 'good housekeeping' is often heard when energy management

Figure 9.1
'Engineering' energy
management activity

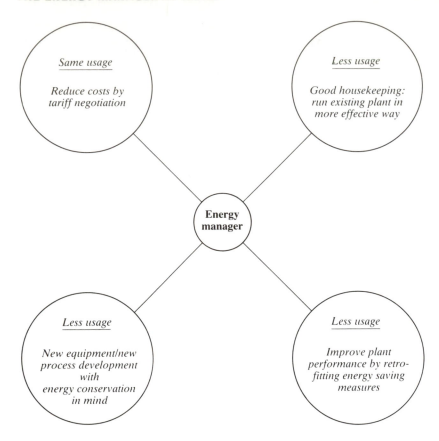

is under discussion. It generally refers to the situation where all personnel within an establishment are constantly aware of the cost of energy and adopt simple measures to save energy. An example of good housekeeping could be something as simple as personnel switching off equipment when it is not being used; lights and portable heaters are examples of this. It could also be a constant awareness of leaks of steam, oil, or more often, compressed air. This aspect of the Energy Manager's job is more to do with the education of personnel, and the success of 'good housekeeping' is as much a matter of good personnel management as the application of engineering principles. This aspect is covered in more detail in Section 9.12.

Less usage   Improve plant performance by retro-fitting energy-saving measures
This is a suitable time to introduce two commonly used phrases: *Energy Monitoring and Targeting* and *Energy Auditing*. Basically, when an Energy Manager collects data about the overall energy consumption of a site from a succession of energy bills, or about a particular piece of plant by instrumenting input and output energy flows then we can say that energy consumption is being monitored. When the data is analysed to show the pattern and efficiency of energy usage this can be said to be an energy audit. Finally, if the audit reveals that either good housekeeping and/or the addition of extra energy-saving devices will improve the energy utilization then targets can be established for future levels of consumption. This aspect will be discussed in more detail in Section 9.11.

Less usage   New equipment/new process development with energy conservation in mind   The situation occurs where the poor condition of the basic plant negates the effectiveness of retro-fitting and requires that the whole is replaced by new plant. It might also be that a new type of system is appropriate, for example in the case of CHP plant replacing a system of bought-in electrical power and heat from a boiler. Clearly with all the experience gained from previous plant, an Energy Manager will have to be satisfied that the value of energy savings will provide an adequate financial basis for capital investment in new plant. This aspect is discussed in detail in Chapter 2 and Section 8.6.

Further considerations

The previous comments on Fig. 9.1 indicate that there are other considerations than those indicated: monitoring, auditing and targeting; staff involvement and motivation; the preparation and presentation of a financial case. There are other factors which bear on the importance of energy management within a company; factors such as the energy-intensiveness of the processes involved, the nature of the production process or service and the previous activity of a company in energy-saving measures. This book is primarily concerned with the 'engineering' aspects of energy management but it should be recognized that all the activities in Fig. 9.1 have other links with other 'engineering' and 'non-engineering' activities. Figure 9.2 attempts to show a wider view.

Figure 9.2   Energy management activities

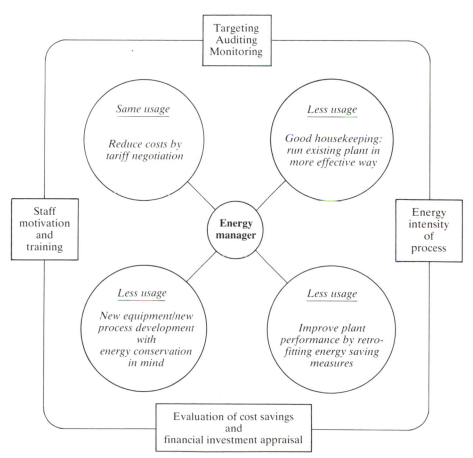

A more detailed discussion of the main features of Fig. 9.2 now follows, with the discussion of tariffs as the starting point. Note: the prices quoted were correct for 1989 and are given to illustrate the tariff systems in general; the *relative* value should not vary significantly; tariffs may also vary slightly from area board to area board.

9.2 DOMESTIC ELECTRICITY TARIFFS

Electricity tariffs are based on the concept that the user will pay not only for the amount of energy consumed but also for the use of the distribution and metering equipment that joins the load to the supply system. Since this equipment is the property of the supplier then the user is effectively paying a rental charge. The detailed structure of electricity tariffs varies from area to area but the overall structure is as follows.

Domestic tariffs are available to private residences where the maximum demand for power does not exceed 25 kW. The standard tariff for domestic consumers comprises a fixed quarterly standing charge and a rate per kW h consumed. All domestic consumers can opt for an 'Economy 7' tariff where consumption is separately metered during the day (17 hours) and night (seven hours); the night-time consumption is charged at a lower rate than the daytime. The seven hours of night period are not necessarily a single continuous period but any seven hours between 2200 and 0900 hours. Typical figures for such tariffs are shown below.

GENERAL DOMESTIC TARIFF
Quarterly fixed charge = £7.50
　　　　　Unit charge = 5.94 p/kW h

ECONOMY SEVEN DOMESTIC TARIFF
Quarterly fixed charge = £9.80
　　　　　Unit charge = 2.07 p/kW h for the night period
　　　　　　　　　　= 6.26 p/kW h for daytime

9.3 BLOCK ELECTRICITY TARIFFS

For non-domestic loads of up to 50 kVA the normal tariff is a block tariff which comprises a fixed quarterly charge, a unit charge for the first 1000 kW h and a lesser charge for additional units consumed. There are variations on this tariff with economy seven rates and other arrangements involving weekend/evening rates. Typical figures for such tariffs are shown below.

GENERAL QUARTERLY BLOCK TARIFF
　Quarterly fixed charge = £7.90
　　　　　Unit charge = 8.13 p/kW h for the first 1000 kW h
　　　　　　　　　　= 6.09 p/kW h for additional kW h

ECONOMY SEVEN TARIFF
　Quarterly fixed charge = £10.20
　Unit charge (daytime) = 8.13 p/kW h for the first 1000 kW h
　　　　　　　　　　= 6.41 p/kW h for additional kW h
Unit charge (night-time) = 2.07 p/kW h

9.4 MAXIMUM DEMAND ELECTRICITY TARIFFS

Large premises are normally supplied under a maximum demand tariff. The structure of these tariffs varies from area to area but typically comprises four items: an availability charge, a demand charge, unit charges, a fuel price adjustment.

The availability charge covers the costs associated with providing the local distribution network to each consumer, and will be expressed as £/kVA of chargeable capacity. In this context chargeable capacity is either a negotiated capacity when the supply is first installed or the highest maximum demand (in kVA) recorded in the 11 immediately preceding months, whichever is the highest.

The demand charge is based on the highest demand (kVA usually) recorded in a month; the charge reflects the seasonal incidence of the domestic and small consumer demand and is only significant in the months November to February inclusive. The maximum demand is defined as the highest average rate of consumption in any half-hour period, or twice the number of units metered in any half-hour period. This means that a single very brief surge of power consumption does not itself establish liability for charges at that instantaneous level of demand. The maximum demand is identified by a device which simply meters consumption for a given half-hour, resets to zero at the beginning of each new half-hour and indicates the maximum reading of all previous half-hour periods since the meter was last used for billing purposes.

The fuel price adjustment is a small correction applied to the unit charge related to the replacement price of fuel burned at power stations from a datum price of coal (£52 per tonne in 1989).

The tariffs vary according to whether the supply is low or high voltage; high voltage is defined as greater than 1000 V. There are also normally three scales of charges.

Scale 1 Separate charges for day and night units apply.
Availability charges and demand charges relate to demands at any time.

Scale 2 Separate charges for day and night units apply.
Availability charges and demand charges only apply to demands outside the seven hour night period.

Scale 3 (Low voltage supplies only) All units are charged at one rate only.
Availability charges and demand charges relate to demands at any time.

AVAILABILITY CHARGE

Low voltage supplies (scales 1, 2 and 3)
For the first 50 kVA of chargeable capacity   £0.96/kVA
For additional kVA of chargeable capacity   £0.79/kVA

High voltage supplies (scales 1 and 2)
For the first 500 kVA of chargeable capacity   £0.72/kVA
For additional kVA of chargeable capacity   £0.60/kVA

DEMAND CHARGE
The following charges apply each month to each kVA of maximum demand recorded in that month.

Low voltage supplies (scales 1, 2 and 3)

In each of the months March to October inclusive	nil
In each of the months November and February	£2.10
In each of the months December and January	£6.70

High voltage supplies (scales 1 and 2)

In each of the months March to October inclusive	nil
In each of the months November and February	£2.05
In each of the months December and January	£6.45

UNIT CHARGE

	Low voltage	High voltage
Scale 1		
For units supplied during the seven hour night period	2.06 p/kW h	1.95 p/kW h
For units supplied outside the seven hour night period	4.62 p/kW h	4.28 p/kW h
Scale 2		
For units supplied during the seven hour night period	2.18 p/kW h	2.07 p/kW h
For units supplied outside the seven hour night period	4.62 p/kW h	4.28 p/kW h
Scale 3		
For units supplied at any time	4.59 p/kW h	—

FUEL ADJUSTMENT

The monthly payment is subject to an addition or reduction at the rate of 0.00035 p (for low voltage supplies) or 0.00033 p (for high voltage supplies) per kW h supplied for each 1 p by which the fuel price is more or less than £52.000 per tonne.

9.5 ELECTRICITY TARIFFS FOR PRIVATE GENERATORS

A number of companies with excess generation capacity sell electrical energy to area boards. As was described in Chapter 8, there are a small number of CHP stations (such as at Hereford) whose total electrical output is exported to the local grid. The tariff paid for units supplied by private generators varies with the time of day, and month of the year. The rationale behind this is that the generating authority's own costs vary with the level of demand and so the value of a private generator's contribution will vary correspondingly. The reader should note that if a private generator has for some reason to cease generation, and because of the continuing demands of a process, has to import electricity, then the tariffs for this energy will be significantly greater than those for the normal consumer.

TARIFFS FOR PRIVATE GENERATORS EXPORTING

	Low voltage	High voltage
For each unit exported between 00.30 and 07.30	1.77 p	1.77 p
For each unit exported between 07.30 and 20.00, Monday to Friday, November and February	5.06 p	5.01 p
For each unit exported between 07.30 and 20.00, Monday to Friday, December and January	7.99 p	7.92 p
At all other times	2.36 p	2.34 p

FUEL ADJUSTMENT   The monthly payment is subject to an addition or reduction at the rate of 0.00 33 p per kW h exported for each 1 p by which the fuel price is more or less than £52.00 per tonne.

TARIFFS FOR PRIVATE GENERATORS IMPORTING

	Low voltage	High voltage
For each unit imported between 00.30 and 07.30	2.06 p	1.95 p
For each unit imported between 07.30 and 20.00, Monday to Friday, November and February	7.60 p	7.13 p
For each unit imported between 07.30 and 20.00, Monday to Friday, December and January	12.00 p	10.83 p
At all other times	3.92 p	3.65 p

FUEL ADJUSTMENT   The monthly payment is subject to an addition or reduction at the rate of 0.00 35 p (for low voltage supplies) or 0.000 33 p (for high voltage supplies) per kW h imported for each 1 p by which the fuel price is more or less than £52.00 per tonne.

AVAILABILITY CHARGE

	Low voltage	High voltage
For each kVA of chargeable capacity imported per month	£0.79	£0.60

9.6 REDUCING ELECTRICITY COSTS

Electricity tariffs may seem complicated!! It is not surprising that companies employ tariff consultants from time to time, although the same advice can be had from the supply industry, free of charge (in the UK). For most consumers, the choice is between a single rate tariff or separate day and night rates; and in some cases, between a block tariff and a maximum demand tariff. The selection that will give the lowest average price is dependent upon the load pattern which the consumer imposes on his electricity supply system throughout the year and in particular upon the proportion of kW h that are taken during the seven night hours and for certain block tariffs, during the night hours and weekends.

Cost reduction can also be achieved by reviewing assessed capacity charges, in particular for premises which have undergone a reduction in activity or changes in electrical utilization. Tariffs should be reviewed at regular intervals particularly when suppliers introduce changes to their tariff structures.

Load Factor

Although the selection of the cheapest tariff is one method of reducing electricity costs, a change of tariff on its own is unlikely to produce large benefits if the load factor is poor.

$$\text{load factor} = \frac{\text{annual consumption in kW h}}{[\text{maximum demand (kW)} \times \text{hours in year (8760)}]}$$

Because maximum demand tariffs comprise a demand and an energy-related component an improvement in the load factor will lead to the demand charge being distributed over a greater number of units and hence lead to a lower average price. Usually an increase in total electricity consumption will also increase the load factor. The maximum demand can often be reduced by energy recovery schemes, or the careful sizing of equipment such as heating elements, or the use of modern methods of lighting and heating buildings.

A technique for improving the overall load factor is demand management; a system which will reduce electricity consumption if the metered demand looks likely to exceed some pre-set limit. The actual value of the maximum demand is taken from a meter which aggregates kW h over a 30-minute period. Systems normally employ microprocessor-based units to predict the half-hour demand and compare this to a reference level. If the forecast demand exceeds the reference level then the management system will either just sound a warning alarm or take action by disconnecting the supply from non-essential usage and later reconnecting when the total demand has reduced. The equipment which is considered non-essential will vary with the type of company or establishment; examples could be space and water heating or air conditioning.

Seasonal, Day and Night Tariff Variations

Since the night tariffs are usually less than half the day tariffs, then night electricity can be the cheapest energy source in terms of useful heat. Off-peak installations which employ a storage system, such as space and water heating; batch processes which can be moved to the night period; a heating system which will use fossil fuels during the day and electricity at night, are all examples of cost savings using night tariffs.

Power Factor

The power factor has been discussed previously in Chapter 4. It is the ratio

$$\frac{\text{useful power (in kW)}}{\text{apparent power (in kVA)}}$$

Certain types of electrical equipment such as resistance heaters and incandescent lamps take only useful power and so the power factor of a system using only these devices would be unity. Other types of equipment such as electric induction motors, transformers and other inductive devices alter the phase angle between the voltage and the current causing the apparent power to be greater than the useful power. The customer will be billed for the apparent power because it is his responsibility to make best use of the energy on his own premises. If a factory has a low power factor then the supplier will have to install a larger supply capacity than is really needed, for which the factory will have to pay via the demand and availability tariffs. There is usually an extra tariff if the power factor falls below about 0.8. The power factor can be improved by installing capacitors (to correct the phase angle) either at the supply point or on individual pieces of equipment (see Section 4.6).

Example 9.1

A company currently uses electric heaters for the daytime space-heating requirements of a large office block. The annual pattern of demand is shown in the table below; demand is during the day only and not at the weekend.

Month	Units consumed (kW h)	Maximum demand (kW)	Maximum demand (kVA) (A)	Demand (£) (B)	Availability charge (£) (C)	Units charge (£) (D)
Jan	6000	75	78.95	513.18	79.60	275.40
Feb	4500	70	73.68	154.73	79.60	206.55
Mar	2500	40	—	nil	79.60	114.75
Apr	1500	25	—	nil	79.60	68.85
May	1000	20	—	nil	79.60	45.90
Jun	500	10	—	nil	79.60	22.95
Jul	500	10	—	nil	79.60	22.95
Aug	1000	20	—	nil	79.60	45.90
Sep	2500	30	—	nil	79.60	114.75
Oct	4500	50	—	nil	79.60	206.55
Nov	5500	75	78.95	165.80	79.60	252.45
Dec	6000	80	84.21	547.32	79.60	275.40
Totals				1381.03	955.20	1652.40

Notes:
(1) Maximum demand in kVA = maximum demand in kW/power factor
(2) Demand charges based on low voltage supply tariffs (Section 9.4)

i.e.,

March to October – nil
January & December – £6.50/kVA
February & November – £2.10/kVA

(3) Availability charges based on low voltage supply tariffs (Section 9.4). Based on negotiated chargeable capacity of 90 kVA,

i.e.,

£0.96 for each of 1st 50 kVA
plus £0.79 for each of remaining 40 kVA

Total = £79.60, which is the same for each month assuming that the maximum demand does not exceed the negotiated chargeable capacity.

(4) Units Charge based on low voltage supply tariffs (Section 9.4) i.e., based on 4.59p/kWh. For this type of tariff, the annual sum could be more simply obtained by multiplying the yearly total of units used (36 000 kW h) by the tariff (4.59 p) to obtain the sum £1652.4.

The Energy Manager decides to install a system of night storage to reduce the maximum demand below 50 kVA and so qualify for a block tariff. Assuming that the night storage capacity is 1500 kW h per month, how much may the Energy Manager spend on the storage system if the company has a limit on payback time of 18 months? Assume that the power factor is 0.95, and ignore the fuel adjustment charge. The availability charge for this company is based on a negotiated chargeable capacity of 90 kVA, which is greater than any maximum demand during the year.

Solution

CURRENT SCHEME　With no night consumption the most economic tariff would be the maximum demand scale 3 tariff, assuming that the supply is low voltage. The various costs are tabulated on the previous page.

$$\text{Overall total} = \pounds(1381.03 + 955.2 + 1652.4)$$
$$= \pounds 3988.63$$

PROPOSED SCHEME　With the night storage capacity, the day/night consumption figures and those for maximum demand are tabulated below.

Month	Quarter	Units consumed (kW h) Day	Quarter	Night	Quarter	Quarterly costs (£) Day	Night	Fixed
Jan		4500		1500				
Feb	1	3000		1500				
Mar		1000	8 500	1500	4500	562.05	93.15	10.20
Apr		nil		1500				
May	2	nil		1000				
Jun		nil	nil	500	3000	nil	62.10	10.20
Jul		nil		500				
Aug	3	nil		1000				
Sep		1000	1 000	1500	3000	81.30	62.10	10.20
Oct		3000		1500				
Nov	4	4000		1500				
Dec		4500	11 500	1500	4500	673.05	93.15	10.20
					Totals	1316.40	310.50	40.80

Notes:
(1)　Day units: 8.03 p/kW h for 1st 1000 units, then 6.41 p/kW h for remainder.
(2)　Night units: 2.07 p/kW h for all units.
(3)　Fixed charge: £10.20 per quarter.

The most appropriate tariff is the economy seven quarterly tariff because of the absence of weekend demand.

$$\text{Overall total} = \pounds(1316.4 + 310.5 + 40.80)$$
$$= \pounds 1667.7$$

The annual savings in electricity costs of the proposed scheme over the current scheme will be £(3988.63 − 1667.7) = £2320.93.

Since the company has a limit on payback time of 18 months, then the maximum amount of capital which may be spent on the night storage scheme will be given by

$$\text{capital} = (\text{savings}) \times (\text{max payback time})$$
$$= 2320.93 \times 1.5 = \pounds 3481$$

If the Energy Manager is able to spend less than this on an installed system, then he may consider increasing the capacity of the night storage to try and reduce the number of daytime units consumed in the first and fourth quarters, as these two items constitute nearly 80 % of the total running cost.

9.7 NATURAL GAS TARIFFS

Natural gas consists almost entirely of methane with traces of other hydrocarbons and nitrogen (see Section 4.1, Table 4.3). The NCV is around 38.6 MJ/m^3 and the density is 0.73 kg/m^3 at standard atmospheric conditions. Traditionally, gas consumption is measured in therms, (where 1 therm is equivalent to 10^5 Btu, 105.506 MJ or 29.307 kW h). The published tariffs for natural gas supplies are very much simpler than those for electricity; there are four basic tariffs.

CREDIT TARIFF (ANNUAL SUPPLIES NOT EXCEEDING 25 000 THERMS)

Standing charge per quarter	£8.70
Charge per therm	39.8 p

This tariff is the one most commonly used in domestic premises where the occupier is billed for the gas already consumed in each quarter. The standing charge is effectively the charge for metering and billing etc.

DOMESTIC PREPAYMENT TARIFF (ANNUAL SUPPLIES NOT EXCEEDING 25 000 THERMS)

Standing charge per quarter	£3.40
Charge per therm:	
for 1st 39 therms per quarter	56.5 p
for remaining therms	42.3 p

This tariff applies only to domestic premises where the occupants pay for gas in advance by a prepayment meter.

TRANSITIONAL TARIFF (ANNUAL SUPPLIES EXCEEDING 25 000 THERMS)

Standing charge per quarter	£8.70
Charge per therm	39.8

This tariff applies to large consumers of gas who will shortly be paying a general tariff (details immediately following) but who have not yet finalized a contract with the gas suppliers.

GENERAL TARIFF (ANUAL SUPPLIES EXCEEDING 25 000 THERMS)

The industrial Energy Manager is highly likely to be concerned with large gas consumption well in excess of 25 000 therms per year and therefore entitled to a tariff below the domestic tariff. The extent of the tariff reduction depends on whether the gas supply is *firm* or *interruptible*.

Table 9.1 Firm gas tariffs

Firm gas – scheduled reference price – p/therm

Volume band	1	2	3	4	5	6	7	8	9	10	11
Nominated consumption therms/annum	25 001 to 50 000	50 001 to 100 000	100 001 to 150 000	150 001 to 250 000	250 001 to 500 000	500 001 to 1 000 000	1 000 001 to 2 000 000	2 000 001 to 5 000 000	5 000 001 to 10 000 000	10 000 001 to 25 000 000	Greater than 25 M
Number of premises											
1	34.0	33.5	33	32	31	30	29	27.5	26	24.5	—
2	—	33.8	33.4	32.5	31.5	30.5	29.5	28.0	26.5	25.0	23.0
3	—	33.9	33.6	33.0	32.0	31.0	30.0	28.5	27.0	25.5	23.5
4–5	—	—	33.7	33.3	32.5	31.5	30.5	29.0	27.5	26.0	24.0
6–10	—	—	—	33.5	32.9	32.0	31.0	29.5	28.0	26.5	24.5
11–20	—	—	—	—	33.2	32.5	31.5	30.0	28.5	27.0	25.0
21–50	—	—	—	—	—	32.8	32.0	30.5	29.0	27.5	25.5
51–100	—	—	—	—	—	—	32.4	31.0	29.5	28.0	26.0
101–400	—	—	—	—	—	—	—	31.5	30.0	28.5	26.5

Firm gas is the name used to describe a supply which is always available for every day of the year (as with the domestic supply). A large user of gas will pay a tariff for this type of supply of between 6 and 14 p less than credit tariff, on the basis that the greater the consumption, the smaller the tariff. Table 9.1 shows typical values for firm gas tariffs. It will be noted that a large company having several separate premises, can negotiate a tariff for the whole group, providing each location uses in excess of 25 000 therms per year. Interruptible gas is the name given to a supply which can be interrupted and which therefore will require the consumer to have an alternative fuel readily available to act as a substitute. Gas will be supplied in this way to customers who use in excess of 250 000 therms per year. There are three types of interruptible contract from which the customer may choose. The periods of interruption, which occur at the supplier's discretion and may or may not be continuous are as follows.

(a) *Short period*   Interruption for a minimum period of three days and up to a maximum of 35 days.

(b) *Medium period*   Interruption for a minimum period of seven days and up to a maximum of 63 days.

(c) *Long period*   Interruption for a minimum period of 21 days and up to a maximum of 90 days.

The current rules which apply are that a customer will be given at least six hours notice that the supply will be terminated for up to a previously agreed number of days. The exact details of the structure for interruptible gas tariffs are given in Table 9.2. Interruptions are most likely to occur during periods of very cold weather when the domestic demand for firm gas rises significantly, and the total demand exceeds the stored and extraction capacity of the supply system. Although the period of notice (six hours) seems rather short, in reality the supplier is rarely so severe and will more likely ask the user to make a specified reduction in demand over a few days. In times of mild winters the supply may only be interrupted for the minimum number of days and therefore the consumer will have the advantage of the lower tariff for virtually the whole year.

Table 9.2 Interruptible gas tariffs

Interruptible gas – scheduled reference price – p/therm

Volume band	1	2	3	4	5	6
Nominated consumption therms/annum	250 001 to 500 000	500 001 to 1 000 000	1 000 001 to 2 000 000	2 000 001 to 5 000 000	5 000 001 to 10 000 000	Greater than 10 000 000
Short period	29.5	28.5	27.0	25.5	24.0	22.5
Medium period	28.0	26.5	25.0	23.5	22.0	20.5
Long period	25.0	23.0	21.0	19.0	17.5	16.0

Given this situation why does the Energy Manager not then opt for a supply totally based on the interruptible tariff? The short answer is that whilst 90 % to 95 % of the usage of a company may be able to switch to an alternative fuel (e.g. oil) by the use of dual-fuel burners etc., some applications such as warehouse heating, canteens, laboratories etc. may not be suitable for any fuel other than gas.

9.8 LIQUEFIED PETROLEUM GAS (LPG)

When crude oil is refined, some hydrocarbon compounds are released in the process which have low boiling points and are gases at normal atmospheric pressure and temperature. Refineries usually produce two main liquefied petroleum gases: propane and butane.

Commercial propane is basically propane (C_3H_8) with very small quantities of hydrocarbons and other substances depending on the site of extraction. Similarly, commercial butane is virtually C_4H_{10}. Both fuels are almost sulphur-free (less than 0.02 % by mass) with no hydrogen sulphide. Traces of unpleasant odour are added to aid leak detection.

The properties of the two LPGs are summarized in Table 9.3

Tariffs: LPG is more expensive than natural gas and is currently 10 % to 15 % higher than the domestic tariff, i.e. about 45 p/therm at current prices (1989).

LPG is normally stored as a liquid in tanks or bottles, and when small amounts are drawn off, the external surface area of the container may be large enough to transfer the heat necessary to vaporize the liquid. At higher rates of extraction, it may be necessary to install a vaporizer to provide the necessary energy. The vaporizer could be a heat exchanger reclaiming heat from waste hot water.

Table 9.3 Properties of commercial propane and butane

	Commercial propane	Commercial butane
Density at NTP/(kg/m^3)	1.8	2.4
Specific heat/(kJ/kg K)	2.3	2.4
Gross-calorific value/(MJ/m^3)	93.0	122.0
Net calorific value/(MJ/m^3)	86.0	113.0
Saturation temperature at atmospheric pressure/($^\circ$C)	−4.0	−2.0
Saturation pressure at 20 $^\circ$C/(bar)	11.72	2.72

Because of its purity, LPG can also be used in a direct-firing mode for process air heaters or for space-heating. This avoids the loss of efficiency inherent in fuels containing sulphur.

9.9 FUEL OIL

Virtually all industrial fuel oils are blends of residual oils originating from the refining process of crude oil. Over a period of time, these oils have acquired commonly used names associated with their viscosity values, although there are nine categories which are signified with a letter.

In Section 4.1 the composition and some properties of oil are given (see Table 4.2) and brief details of combustion methods are also given. Table 9.4 gives the information again with an indicative price per litre added.

Table 9.4 Common
fuel oils and their
properties

	Domestic heating oil	Gas oil	Light fuel oil	Medium fuel oil	Heavy fuel oil
Class	C	D	E	F	G
Kinematic viscosity (cSt at 16 °C)	2.5	4.7	218	1 276	7 424
Density at 16 °C (kg/litre)	0.79	0.85	0.94	0.97	0.98
Gross calorific value (MJ/litre)	35.5	38.7	40.8	41.6	41.7
Typical average price February 1989 (p/litre)	10.50	11.80	11.10	9.90	8.30

Note: Oil prices vary with the size of the order and the region.

$$1 \text{ cSt} \equiv 10^{-6} \text{ m}^2/\text{s}$$

Methods of Conserving Energy in Plant Burning Fuel Oil

Energy-conserving measures centre on either improved control of the combustion process with limitations on the amount of excess air, or the more efficient use of the heat for its intended purpose, or the reduction of energy losses by the use of better insulation or heat recovery applied to the flue gases, although the latter may not be appropriate with the heavier oils on account of the amount of solid particles which are part of the exhaust emissions. Another point which particularly applies to fuel oil is that, because it needs to be stored on site, it will need to be pumped to the burner. Problems arise from the fact that the viscosity of fuel oils varies very widely with temperature. Apart from the difficulties of filtering at low temperature, the oil may be too viscous to pump when the weather is severe and therefore the pipework must be very well insulated and a heater may need to be installed to maintain the oil at the temperatures suggested in Table 9.5.

Table 9.5 Minimum
temperatures for storage
and use of fuel oils

	Domestic heating oil	Gas oil	Light fuel oil	Medium fuel oil	Heavy fuel oil
Class	C	D	E	F	G
Minimum temperatures/(°C)					
for storage	−15	0	10	25	40
for pumping	−5	0	10	30	55
at the burner	−5	0	40	70	90

9.10 COAL

There are a large number of different types of coal; a simplified classification of coal types and their properties and characteristics is given in Section 4.1. Generally speaking, the best quality coals will be found in the deepest mines;

these will be bituminous and anthracite coals. Open cast mines tend to produce brown coal and lignites.

The fact that the physical properties of coal change during the combustion process has an important bearing on the method of handling the residue from combustion in that a powder will require a very different technique to a mass of hard lumps. Apart from coal properties such as volatile matter and caking tendency (see Section 4.1) there are two other factors which relate to the purchase of coal. One is the size of the lumps of coal and the other, the ash content. The size of the coal is generally in one of three categories: smalls, graded, unsized. Graded coal is specified by an upper and lower limit on the size of piece that is acceptable; 'smalls' refers to coal for which only an upper limit on size is specified; unsized can be any size from dust particles to lumps of say 200 mm thickness. The ash content of a coal can be reduced by washing but this will increase the cost by around 15 %. A coal with a large ash content will increase the cost of handling the residue.

In terms of cost, graded coal will be the most expensive, with unsized coal the least. Currently (1989), a typical price for a tonne of smalls of UK Bituminous Coal will be around £65. The price also varies with the calorific value and the scale of distribution required.

A brief discussion of the methods of combustion of coal is given in Section 4.1.

9.11 ENERGY MONITORING, AUDITING AND TARGETING

The techniques used in the management of energy have much in common with those required for the management of any resources in a company or establishment e.g. resources such as materials and manpower. The general concept of monitoring performance and the setting of targets is essential to the control of costs and also the motivation of staff associated with energy usage.

Energy monitoring is the term applied to the regular recording of data relevant to the energy performance of anything from a single piece of plant to an entire site. For example, the electricity consumption of a whole site will effectively be monitored by the meter installed for billing purposes by the area board; a company may wish to have a more detailed knowledge of consumption and will install separate meters at a number of locations of major use. This information can be recorded and analysed to indicate trends in areas of major demand. A similar system can be used to meter the detailed consumption of other consumables such as oil, gas and water. Other data such as temperatures and pressures can be detected by sensors and again recorded (manually or by automatic means).

Monitoring will not of itself lead to energy savings unless the data is analysed and put in context within the company or establishment. Although an Energy Manager will not be pleased if there is an increase in energy consumption, he will not necessarily complain if there is a proportionate increase in output, or in the case of a building, if the number of degree days has significantly increased due to a spell of untypically severe weather (see Section 7.3).

After monitoring, the next step is to analyse the data and carry out an energy audit. A company will pay bills for fuels such as electricity, gas, oil, and coal. The object of the energy audit is to determine how well the company uses the energy and ultimately the energy cost contained in the price of the product. Having then established how the energy is being utilized many industries apply the concept that energy consumption can be decreased by say 5 % per year by the practice of good housekeeping and more systematic maintenance. The starting point for such reductions is usually the present annual consumption related to a unit of production (e.g. MJ/tonne) or to the unit area of a building (e.g. kW h/m² or MJ/m²). This type of approach has two advantages. First, it affords a comparison of one system with another; for example, the daily energy consumption of hospitals is expressed in the unit kW h/m² and Energy Managers try to obtain figures of around 440 to 470 kW h/m². Hospitals with values significantly higher than this can generally employ cost-effective measures to achieve a lower figure, which can be the short-term target. Monitoring the consumption will provide a check on how well the target is met; Section 7.4 discusses targeting in more detail.

The other advantage of expressing energy consumption in terms of unit cost is that changes in output or building size etc., will lead to changes in consumption unrelated to efficiency of energy use.

After targets are established, the monitoring and auditing will allow the Energy Manager to compare actual energy usage and costs with target figures. Even if targets are being met, the Manager may feel that greater reductions are possible by investing in retrofit or replacement plant. He has then to evaluate the potential savings of new or improved systems to assess their economic viability in the ways described in Chapter 2. He may also feel that irrespective of hardware improvements, greater energy awareness through the training of personnel will improve the efficiency of usage and hence reduce costs.

Case Study 9.1

Garforth Fabrications plc has recently decided to review the energy use in its heat treatment plant. Part of this plant is a continuously operated furnace, which has a chain conveyor which carries trays of forgings along the length, initially through an unfired preheat zone and then into the main heating zone which is fired by gas burners.

Basic Data
Furnace: dimensions, 10 m long, 1.50 m wide and 1.00 m high; main heating zone, six burners each nominally rated at 10 therms/hour; usage, continuously for five days/week, 48 weeks per year (four works shutdown weeks); fuel, natural gas – cost 32 p/therm; steel, type EN19.

Monitoring data
Gas consumption, 3500 therm/week (maximum rate = 50 therm/hour, minimum rate = 15 therm/hour); average operating temperature of furnace, 900 °C; flue gas average temperature, 560 °C (flue gas analysis indicates average excess air = 30 %); ambient temperature, 15 °C; throughput, 350 tonne/week.
A visual inspection of the general physical condition of the furnace reveals a damaged front door, some damage to the basic brickwork and severe damage to the furnace lining.

Energy audit of furnace

Using the combustion equations as explained in Chapter 4 we can determine the exhaust gas composition using Fig. 4.13; for an excess air of 30 %, the percentage CO_2 in the dry exhaust products is approximately 9 % by volume.

The percentage heat loss in the flue gas is given by Eqs [4.7] and [4.8]

i.e.
$$\% FGL = \frac{0.35(t_G - t_A)/(K)}{(\% CO_2)} + 0.0083\{1121 + (t_G - t_A)/(K)\}$$

(where $t_G =$ flue gas temperature and $t_A =$ ambient temperature). Using the monitored data:

$$\% FGL = \frac{0.35(560 - 15)}{9} + 0.0083\{1121 + (560 - 15)\}$$
$$= 35 \%$$

The energy absorbed by the metal is given by (tonnage/week) × (change in energy content of metal between 15 °C and 900 °C). The change in energy content of EN19 steel between 15 °C and 900 °C is 602.8 kJ/kg.[9.3] Therefore the energy absorbed by the metal is

$$350 \times 1000 \times 602.8/10^6 = 211 \text{ GJ/week}$$

Since 1 therm = 0.1055 GJ, then the percentage of energy input from the gas which is used to heat the metal is

$$\frac{211 \times 100}{0.1055 \times 3500} = 57.1 \%$$

The final breakdown of energy use will therefore be as shown in the table below.

	Therms/week	%
Input energy	3500	100
Output energy:		
to metal	2000	57.1
to stack	1225	35.0
unaccounted for and radiation losses	225	7.9

Energy saving options

The Energy Manager reviews the potential energy-saving measures and considers that the following options are available, each with their potential saving and cost:

(1)   Adjustment of air–gas ratio control; can be adjusted with furnace in service. Should be carried out every four weeks. Savings of the order of 5 % at a nominal cost of £40/adjustment.

Comment   this option is a simple maintenance task and its effect would be monitored by the exhaust gas analyser.

Cost   40 × 48/4 = £480

(2)   New air–gas ratio controller; savings typically 12 % at a cost of £700. The controller requires eight weeks delivery but can be installed in a weekend.

Comment   this option might be necessary if the burner adjustment is so worn that the air–fuel ratio changes soon after option 1 has been implemented.

Cost   £700

(3)   Structural maintenance; including replacement of damaged doors and repair of broken brickwork. The damaged door could be replaced to give a 4 % saving at a cost of £400. A new door requires five weeks delivery but can be installed in a weekend.

Comment   this option is a basic maintenance task which will reduce unwanted excess air and heat losses from the structure.

Cost   £400

(4)   Complete reline with ceramic fibre; savings typically 4 % at a cost of £75 per square metre of wall and roof area. The fibre requires 10 weeks delivery but can be installed only in a works shutdown period.

Comment   this option will reduce the heating-up time of the furnace because of the low thermal inertia of ceramic linings; it will also reduce the heat losses from the furnace because of its low thermal conductivity compared to conventional refractory linings. In addition, the faster cooling down time will mean less shutdown time for maintenance. Also, ceramic fibres do not suffer, as conventional linings do from rapid temperature changes.

Cost   (area of roof and walls) × cost per unit area = 35 × 75 = £2625

(5)   Recuperative burners; savings typically 25 % at this level of exhaust temperature at a cost depending on the size chosen:

Rating (therms/hour)	6	7.5	10
Price installed (£)	1500	1700	1900

The recuperative burners require 10 weeks delivery and can be installed only in the works shutdown period.

Comment   this option will significantly reduce the gas consumption by the amount of pre-heat transferred to the incoming combustion air. The high flue gas temperature makes this option an attractive one depending on the cost and payback time involved.

Cost   (By using recuperation, the maximum demand is reduced by 25 % from 10 to 7.5 therm/hour)

$$\therefore \quad \text{Cost} = \text{cost per burner} \times \text{number of burners}$$
$$= 1700 \times 6 = £10\,200$$

Cost benefits

The total annual cost of gas at present is given by

$$\text{therms consumed per week} \times \text{number of weeks} \times \text{tariff}$$
$$= 3500 \times 48 \times 0.32$$
$$= £53\,760 \text{ per year}$$

The percentage saving of each option can be calculated by multiplying the total cost by the appropriate fraction. This will be a saving based on that option alone being implemented. The following table indicates the annual savings.

Option	% saving	Annual saving (£)	Cost (£)
1	5	2688	480
2	12	6451	700
3	4	2150	400
4	4	2150	2625
5	25	13440	10200

It should be remembered that when implementing more than one option, the second option will have a smaller total consumption on which to operate, and therefore the value of the savings will be less than when that option alone is applied. Also, most options require a few weeks of ordering time and some can only be implemented in works shutdown periods. Thus, in the first year of operation, most options will not generate the full year's savings indicated in the table. When energy-saving measures have been implemented, the monitoring and auditing will reveal whether or not the claimed percentage savings have been achieved.

Case Study 9.2

This case study concerns an Energy Manager newly appointed by a company anxious to reduce its energy costs but not quite sure how to do it. The company manufactures confectionery and basically has three main product lines which are housed in separate buildings. The person previously responsible for energy management within the company, the works services engineer, was not able to devote much time to a detailed assessment of the energy use within the company and simply kept a record of the monthly fuel bills and the levels of production. The details of the monthly gas and electricity bills and tonnages of production are detailed below and on pages 363 and 364, along with some analysis by the works services engineer.

Month	Production figures (tonnes)			
	Line 1	Line 2	Line 3	Total
April	1399	1618	92	3109
May	1112	1654	84	2850
June	1066	1446	89	2601
July	1208	1547	121	2876
August	661	835	74	1570
September	1147	1229	105	2481
October	1075	1673	128	2876
November	795	850	75	1720
December	1186	1242	102	2530
January	857	1140	97	2094
February	858	1167	119	2144
March	882	1167	104	2153

Gas consumption –
interruptible and firm

Month	Days	Interruptible gas consumption			Firm gas consumption		
		Therms	Tariff (p/therm)	Cost (£)	Therms	Tariff (p/therm)	Cost (£)
	(A)	(B)	(C)	(D)	(E)	(F)	(G)
Apr	30	118 158	27.00	31 903	7 424	33.50	2 487
May	32	100 704	27.00	27 190	5 571	33.50	1 866
Jun	28	107 671	27.00	29 071	1 899	33.50	636
Jul	31	114 469	27.00	30 907	2 541	33.50	851
Aug	29	79 908	27.00	21 575	1 034	33.50	346
Sep	32	130 313	27.00	35 185	1 128	33.50	379
Oct	31	131 233	27.00	35 433	4 501	33.50	1 508
Nov	28	137 701	27.00	37 179	10 944	33.50	3 666
Dec	34	110 982	27.00	29 965	9 227	33.50	3 091
Jan	32	105 621	27.00	28 518	13 159	33.50	4 408
Feb	28	88 093	27.00	23 785	13 463	33.50	4 510
Mar	30	120 452	27.00	32 522	10 372	33.50	3 475
Totals		1 345 305		£363 233	81 263		£27 223

Overall total = £390 456

Notes: (see Section 9.7, Tables 9.1 and 9.2, for the supply tariffs)
A   The number of days between meter readings
B   Interruptible gas consumption in therms
C   Gas tariff from Table 9.2; short period interruption for a total annual consumption of 1.345 million therms (total obtained by summing column B for the year). The appropriate tariff is 27.00 p/therm
D   Cost = B × C
E   Firm gas consumption in therms
F   Gas tariff from Table 9.1, based on total annual consumption of 81 263 therms (obtained by summing column E for the year). The appropriate tariff for the one set of premises is 33.50 p/therm
G   Cost = E × F

After perusing the figures, the new Energy Manager decides on some short- and long-term courses of action. These are presented after the consumption and production charts.

There seems to be a lot of information!! At first sight the two most important figures seem to be the *total* electricity and gas bills, which are £587 943 and £390 456 respectively. The Energy Manager reviewed the figures and discussed certain matters with colleagues and the previous Energy Manager. Drawing the following conclusions, he decided on the courses of action indicated.

(1)   The maximum demand for electricity is controlled by an electronic energy management system which monitors the half-hour maximum demand clock and preferentially switches off certain power-consuming equipment for short periods. This system should ensure that the maximum demand (Column B) never exceeds the negotiated chargeable capacity (Column C), thereby maintaining the level of the availability charge. Why then, in December, did the demand rise to such a high level (4450 kVA)? The reason was that during the summer months, certain pieces of equipment were disconnected from the monitor on account of changes in production priorities. This was meant to be a temporary measure. Unfortunately, the equipment was not reconnected until after the maximum

Electricity consumption
(maximum demand
tariff – Scale 2
(Section 9.4))

| Mon | Days | M.D. (kVA) | S.C. (kVA) | Unit consumption | | | Load factor (%) | Payments | | | | | Avge. cost (p/kW h) | Average consumption | |
| | | | | Day (MW h) | Night (MW h) | Total (MW h) | | M.D. (£) | S.C. (£) | Units (£) | Fuel (£) | Total (£) | | Day (MW h) | Night (MW h) |
	(A)	(B)	(C)	(D)	(E)	(F)	(G)	(H)	(I)	(J)	(K)	(L)	(M)	(N)	(O)
Apr	25	2700	2950	584.2	168.4	752.6	46.5	nil	2339	30459	2054	34852	4.63	23.4	6.7
May	30	2850	2950	750.1	218.4	968.5	47.2	nil	2339	39154	2644	44137	4.55	25.0	7.3
Jun	32	2750	2950	637.6	182.8	820.4	38.8	nil	2339	33223	2240	37802	4.61	19.9	5.7
Jul	29	3000	3000	837.9	235.6	1073.5	51.4	nil	2379	43564	2818	48761	4.54	28.9	8.1
Aug	35	2850	3000	639.9	183.8	823.7	34.4	nil	2379	33850	2162	37891	4.60	18.3	5.3
Sep	30	2850	3000	735.6	210.8	946.4	46.1	nil	2379	38327	2186	42892	4.58	24.5	7.0
Oct	31	2950	3000	895.9	263.8	1159.7	52.8	nil	2379	46825	2679	51883	4.47	28.9	8.5
Nov	32	2850	3000	803.5	209.2	1012.7	46.3	5985	2379	43095	3190	54649	5.40	25.1	6.5
Dec	24	4450	3000	679.9	173.1	853.0	33.3	28925	3524	34977	2777	70203	8.23	28.3	7.2
Jan	33	2850	3000	709.4	194.5	903.9	40.1	18525	3524	36781	2847	61677	6.82	21.5	5.9
Feb	34	2800	3000	850.8	227.4	1078.2	47.2	5880	3524	43991	2830	56225	5.11	25.0	6.7
Mar	30	2750	3000	793.0	203.6	996.6	50.3	nil	3524	40831	2616	46971	4.71	26.4	6.7

Total £587 942

Notes: (see Section 9.4 for maximum tariff details for low voltage supplies)
A   The number of days between meter readings
B   Maximum demand in kVA
C   Chargeable capacity in kVA
D   Day units consumed in MW h
E   Night units consumed in MW h
F   Total = D + E
G   Load factor = F × 1000/[B × A × 24]
H   Maximum demand charge = A × £2.10 for November and February
    = A × £6.50 for December and January
I   Availability charge = £0.96/kVA for 1st 50 kVA and £0.79/kVA for the remainder, based on maximum demand or chargeable capacity, whichever is the greater (Section 9.4)
J   Units charge: (Scale 1) = 2.06 p/kW h for night period and 4.62 p/kW h for day period
K   Fuel adjustment charge (according to prevailing prices of coal)
L   Total = H + I + J + K
M   Average cost per unit = L × 100/[F × 1000]
N   Average consumption (day) = D/A
O   Average consumption (night) = E/A

demand tariff began to apply. The effect was not only to increase the maximum demand charge but also to increase the availability charge which, for the next 11 months will be based on 4450 kVA and not the negotiated capacity of 3000 kVA (see Section 9.4). The total effect will be to add about £24 000 to the total electricity bill over the next 11-month period.

Action Check to see if all the high electrical energy users are connected to the electrical management system prior to the autumn period.

(2) Some of the months include holiday periods; also some have a five-day working week, some a six-day week. This leads to misleading figures of load factor (e.g. in the months of June and August) and average daily consumption.

Action Obtain a detailed account of the pattern of working hours per month.

(3) The single electrical meter reading does not indicate where the energy is being used. Similar comments apply for the gas consumption figures. Whilst the production figures indicate the tonnage per month, it is not clear which lines are consuming how much and what type of energy (electricity or gas).

Action Investigate the capital costs of installing electricity and gas meters on each of the production lines and other energy-using equipment such as boilers, to facilitate some calculation of the energy cost in the product of each line.

(4) The electrical power factor is not known.

Action Arrange for it to be measured. Assess the capital cost of raising the power factor if it is less than 0.95, given that the company has a policy of not investing in capital if the payback time exceeds 18 months. Invite quotations from manufacturers of power factor correction equipment.

(5) The gas figures contain no indication of efficiency of usage, nor the period of interruption of supply.

Action Check on the maintenance procedures, particularly for the boiler, and after monitoring the boiler performance, carry out an energy audit to assess the energy balance of the unit. The boiler is not fitted with an economizer; the energy-cost savings of retrofitting this need to be evaluated, again given the constraint of an 18-month payback time. Obtain the invoices for all the fuel oil purchased in the last two years and check on the usage to see if it has been used as boiler fuel. (Fuel oil is the alternative to gas when the latter supply is interrupted.)

(6) At present the energy bills are monitored and the results noted on pieces of paper and graphs distributed around the walls of the office. The application of a simple computer-based spreadsheet would give instant recall of energy consumption figures and permit analysis of patterns of usage. Another scheme would be to relay all the meter readings to a central computer, thereby continuously monitoring relevant information.

Action Determine the cost of purchasing a database package and a suitable microcomputer system. Determine the cost of the instrumentation and data logging system required for continuous monitoring.

After a few weeks in the post, the Energy Manager attends the monthly meeting of the area Energy Managers and obtains the following checklist for energy savings from Dr Nigel Gwyther who was the guest speaker.[9.6]

9.12 CHECK LIST OF FUEL-SAVING IDEAS

Each section of the check list commences with the more obvious and simple checks by which energy savings can be made and progresses towards those which are more obscure, require capital expenditure or management policy decisions, or are applicable only in special cases.

The words 'consider' and 'investigate' in the list imply that there may be other conflicting interests, financial or otherwise, that require a decision to be made concerning more factors than just the saving of energy.

An asterisk * has been placed at those items of advice where there is a danger in taking a proposal to an extreme value. For example, too much thermal insulation applied to a cold face of a furnace may cause a structural failure.

There are three basic methods of saving energy:

(a) by developing more efficient processes for its use, which takes time and money;

(b) by rationing or reducing demand for it through pricing so that only a few can afford to buy it;

(c) by operating existing processes more efficiently, which requires the dedicated enthusiasm of all.

The check list and commentary describe suggestions for fuel saving derived from large organizations which have teams of energy specialists able to take measurements and thus prove the effectiveness of the suggestions. Small businesses may not have these specialists and hence may need help from outside bodies in applying some of the suggestions and in identifying their problems.

Lighting

Lighting usually represents a small fraction of the total energy requirement of a factory, but economies are easy for all to see and can give impetus for conservation of less tangible items. Good lighting design improves productivity, but glare can lead to accidents. Modern building design has created problems in certain cases by demanding permanent artificial lighting through lack of sufficient windows and, in other cases, in the summer by calling for refrigeration from the air conditioning system because of the solar gain from too much window space.

1   Make the best use of daylight by keeping windows and roof lights clean, also by suitable arrangement of working places near windows.

2   Keep lamps and fittings clean.

3   Replace lamps when their efficiency drops through ageing.

4   Use suitable reflectors and diffusers which transmit light in the desired direction.

5   Avoid dark background colours which absorb light.

6   Have separate switches to control lights near windows.

7   Make sure that light is adequate but that it is switched off when not required. (Frequent switching reduces the life of lamps.)

8   Consider automatic switching of lighting.

9   Use fluorescent or discharge lamps rather than filament lamps.

Space-heating

Temperatures above 19 °C for space-heating are illegal in the UK following the fuel crisis of the 1970's; currently this limit appears to be generally ignored. Occupants could be encouraged to wear heavier clothing at lower internal temperatures. Excessive temperature variation with hot and cold spots needs to be controlled by strategically placed thermostats.

10   Block off unoccupied working areas and do not heat them.

11   Limit maximum temperature to the legal limit of 19 °C. Check the accuracy of temperature control. Minimize variation in temperature.

12   Use warm air curtains in conjunction with automatic door closing where possible to improve comfort without draught or excessive loss of hot air. Despatch bays are often the cause of excessive heat loss.

Ventilation

A ventilation system which includes heating, humidification and filtration is expensive to install, but a good system will improve the working environment as well as productivity. The correct positioning of louvres is essential. A system which consumes power is abused by unnecessary opening of doors and windows resulting in loss of heat. Despatch bays are often windswept because the doors are never closed; a situation which often brings discomfort to adjacent working rooms in cold weather. The proper siting of doors, their closure and the fitting of warm air curtains in exterior walls will help improve the general working conditions.

13   Avoid draughts by a properly sealed system, sealing doors, windows etc.

14   Doors and windows should normally be closed in cold weather.

15   Excessive ventilation caused by leaving windows open involves excessive space-heating, consequently the number of changes per hour in a room should be restricted.

16   In air-conditioned buildings ensure that the controls for moisture content, temperature and direction of air flow are effective.

17   Avoid air leaks in ducts by sealing them.

18   On non-working days do not switch on air heating and ventilation too soon but maintain a level just sufficient to give frost protection.

19   Switch off the air conditioning system up to one hour before the building is due to be vacated for a long period.

Electrical Equipment

Electricity costs are based on the rate as well as the amount of consumption. This requires a company to balance the costs of the maximum demand and the load factor. Maintaining a high power factor is important in reducing costs by increasing efficiency of usage.

Certain items of plant are likely to have a low power factor involving the installation of larger transformers, switchgear and cables to carry the additional current: electric arc furnaces, induction motors and furnaces, power transformers and voltage regulators, welding machines, choke coils and magnetic systems, neon signs and discharge lamps. The power factor with these can be improved by the installation of the correct size of capacitor which may be rotary or stationary. A works with its own alternators can use them to make the necessary correction. Transmission losses can be minimized by the use of higher voltages and the correct equipment.

20   Switch off equipment which is not required for any prolonged period.

21   Lower the maximum demand by regulating the intermittent use of equipment.

22   Match electric motors to their required duties.

23   Use higher voltage where possible to reduce transmission losses.

24   Match the size of transformers to their load requirements.

25   Make the best use possible of the three-phase system for power distribution.

26   Obtain advice on power factor control and most advantageous tariff.

Thermal Insulation

The material chosen has to be of low thermal conductivity and suitable for the temperature of operation so that it does not shrink, melt or otherwise deteriorate. When applied to an existing structure an insulating material should not cause overheating and failure of the structure. Applied to the outside wall of a furnace it is called cold-face insulation and when installed inside a furnace is called hot-face insulation. The latter method is preferred for a batch reheating furnace where the heat stored in the structure can be significantly reduced, while the former tends to be used for continuously-heated furnaces where the continuous heat loss is more important than the heat stored.

Thermal insulation is normally provided in the walls of a building by a cavity of air about 50 mm wide, and it may be advantageous to fill this with loose fibre or polyurethane foam, providing that the building is not subject to damp effects. 100 mm of insulation is recommended for loft or roof insulation of a building. Double glazing is less attractive for its thermal saving but it can reduce draughts from ill-fitting windows and eliminate noise.

There is an economic thickness for each application, offset by the value of the heat saved had there been no insulation. Some insulation has poor mechanical strength and needs protection from damage and inclement weather by the aid of wire netting or sheet metal cladding.

27 Except for heat transfer units such as space-heaters and refrigerators, heated and cooled pipes should be lagged.

28 *Any exposed surface not at room temperature should be thermally insulated. Included should be valves, flanges, flues and chimneys.

29 Protect thermal insulation from damp, inclement weather, and mechanical damage.

30 Use cavity insulation in buildings.

31 Apply thermal insulation to lofts or roofs (100 mm recommended).

32 Double glazing provides thermal saving, also reduces draughts and noise levels.

33 Make exterior doors self-closing.

34 Consider the use of double doors or revolving doors in the entrances to buildings.

35 Consider the replacement or enhancement of thermal insulation for low, medium and high temperature using material of low thermal conductivity.

36 Cover the surface of a hot liquid with a lid or a floating insulator to reduce the heat loss.

37 *Apply cold-face insulation to the exteriors of furnaces and ovens.

38 Use where possible hot-face insulation inside ovens and furnaces which are intermittely heated so as to reduce the heat storage loss.

39 Prevent radiation escape by closing apertures such as inspection holes and doors.

40 Use polished metal exterior surfaces to minimize radiation loss.

Steam, Compressed Air and Other Services

Oversized steam and other heated pipes, even if lagged, may have a low delivery temperature because of heat losses. Undersized pipes also have losses, but in this case are due to friction which has to be overcome by a higher than normal pumping pressure or by a low delivery pressure which may fluctuate excessively with flow changes. All common services should be regularly inspected for leaks since they are often not attended for long periods due to poor accessibility. Air is considered by many to be cheap, but its leakage from a pressurized system can involve a surprisingly large expenditure.

41 Take particular care to avoid leaks. Check thoroughly for leaks and repair where necessary.

42 Inspect and maintain stream traps.

43 Seal off redundant pipework and ducts.

44 Ensure that steam mains are properly sized.

45 Recover steam condensate and return it to the boiler if not contaminated.

46   Avoid the use of steam-reducing valves for low-pressure steam. Back-pressure engines or calorifiers should be considered.

47   Switch off compressed air services when they are not required to run continuously.

48   To reduce pressure losses clean air filters and renew packing.

49   Mechanical efficiency falls with use, hence renew values, springs rings, glands, etc.

50   Avoid overheating air in a compressor by returning it through a by-pass to the inlet.

51   Multiple-stage compressors are more efficient with intercooling between stages; regularly clean the heat exchangers to ensure minimum pumping cost.

52   Ensure that fan rotating parts are in balance.

53   Fan impellers often become dirty and corroded. Surface cleaning and polishing will improve efficiency.

Boilers for Steam and Hot Water

Where more than one boiler is in use, it may be possible to shut one down for part of the time. Hot water storage of a boiler will vary with the load on it and is greatly affected by the efficiency of combustion and by the transfer of heat from the waste gases to the water or steam, hence the need for regular cleaning and setting of controls.

The treatment of feed water protects a boiler from the effects of corrosion, reducing the risk of failure by overheating and loss of efficiency which would otherwise result from the build-up of deposits. Carried to excess, blowdown of the dissolved solids in boiler water can cause up to 10 % of the energy supplied as heat to be lost. Too little blowdown can be a danger to a boiler since it can result in priming, foaming and carry-over of solids with the steam to the steam-using equipment.

54   Stagger the demands for steam as much as possible to give a more even loading on the boilers.

55   Avoid boiler safety valves blowing unnecessarily by proper control of firing.

56   Controls should be serviced and adjusted to maintain optimum efficiency, temperature, pressure, etc.

57   Is the treatment of feedwater appropriate?

58   Do not have excessive blowdown.

59   Can the temperature of hot water be reduced with an improvement of efficiency?

60   Where superheated steam is appropriate, this should be provided at a high steady temperature.

Combustion

The efficient combustion of a fuel demands that the combustion apparatus be suitable for the fuel concerned. Solid fuel has to be of the correct size and hardness, and have suitable moisture, ash, volatile matter and sulphur contents. The ash fusion temperature of coal may be critical, leading to loss of unburnt fuel and problems with fly-ash, grit and permeability of the fuel bed. Variable quality of fuel poses additional problems. High volatility in fuel can lead to incomplete combustion of waste gases unless the supply of combustion air is suitably distributed.

Most heavy liquid fuels are atomised so as to produce a fine spray of uniform-sized droplets before combustion. Atomization can be carried out by mechanical means with a spinning cup, or pressure through a jet, or by pressure energy in steam or compressed air (see Section 4.1). The shape and size of the spray depends partly on the pressure of the oil and partly on the size and shape of the burner nozzle, but the viscosity of the oil plays a significant role. Oil of uniform quality gives poor atomization if it is too cold, but gives carbon deposits which upset the spray pattern and leads to heating problems if it is too hot. Proper temperature control of the oil is therefore important, as is the ratio of atomizing fluid (air or steam) to oil.

Gases are easier to burn more efficiently than solid or liquid fuels, but burners for them must be carefully selected and require a steady gas pressure.

All burners for pulverized fuel, oil or gas need regular maintenance and should not be allowed to overheat. The air–fuel ratio for a given fuel should be constant for optimum conditions, a slight excess of air being desirable for complete combustion (50% to 100% for coal, 15% to 20% for liquid, and 10% for gaseous fuels). Automatic proportioning with a variable fuel flow is desirable.

To ensure that complete combustion is maintained it is necessary to analyse waste gases regularly to detect the percentage carbon dioxide in them. Furthermore it is desirable that oxygen, carbon monoxide and water vapour should be measured to verify that the correct air–fuel ratio is being obtained. Waste gas analysis may indicate air leaking into a flue system due to faults in the brickwork or badly fitting doors and dampers, or stratification of the productions of combustion through bad mixing of fuel and air.

Furnace draught is necessary to cause the flow of products of combustion through a furnace, to promote good mixing of the fuel and air, and to improve convective heat transfer from the gases to the charge. Natural draught requires large diameter, tall chimneys, carrying waste gases at a high temperature. Fans to supply air (forced draught) and fans to suck waste gases from the combustion chamber (induced draught) enable better heating efficiency with smaller chimneys and lower stack losses than is possible with natural draught. They also give better control of furnace pressure which, when balanced, can reduce air leaking in or waste gases escaping. Cladding the furnace exterior with metal sheeting is more effective than grouting to seal the developing cracks in the brickwork. Thermal insulation of metal chimneys is often desirable.

61   Switch off unwanted burners.

62   Excess air in fuel-fired equipment is probably the chief cause of energy waste in industry. Therefore the control of air–fuel ratio should provide

only sufficient of each for complete combustion consistent with proper mixing.

63 The percentage carbon dioxide etc., in waste gases should be monitored regularly to check that combustion is efficient.

64 Air leaks in brickwork and flues and around dampers should be stopped.

65 Furnace draughting should be balanced; fit flue dampers to isolate boilers not on line to prevent a natural draught pulling in cold air.

66 Fuel supplied for combustion should be in the proper condition to achieve optimum heat release.
Coal – uniform mix of suitable size range, volatile matter, ash and moisture content, hardness and reactivity.
Oil – constant temperature to control viscosity, constant pressure.
Gas – constant pressure.

67 A combination of different fuels calls for mixing control.

68 Burners should not overheat. Their efficiency is maintained by adjustment, cleaning and replacement of worn parts.

69 Adequate mixing of fuel and air at the burner is necessary for proper direction of the flame and completion of heat release from the fuel within the combustion chamber.

70 Multiple burners and multiple zone heating need balancing to produce even heating.

Furnaces and Ovens

For heat transfer efficiency the hearth area of a furnace needs to be completely covered with the charge, but overloading is undesirable as it slows down the rate of heating. Control of temperature to give uniform heating aids productivity and minimizes rejects. Machinery associated with heating process is itself subject to overheating, and this is often overlooked.

A change in the method of heating may sometimes be desirable. The use of muffle furnaces for controlled atmosphere is thermally inefficient and may sometimes be avoided. When many components of the same dimensions need identical treatment it may be worth adopting electric induction heating. Ovens using microwave and infra-red methods of fast electric heating are useful in cooking and curing. Dielectric heating can be applied to non-conducting solids.

71 * Hearth coverage is increased by proper loading.

72 Doors should be left open for as short a time possible for loading and unloading and should be well-maintained so as to minimize radiant heat loss and air in-leakage.

73 The mass of carriers, trays etc., can often be reduced without loss of control.

74 Steady operation by planning throughput is desirable.

75   Heating cycles, particularly in batch furnaces, should be short but consistent with the maintenance of product quality.

76   Consider the possibility of changing to a more efficient form of heating by changing the burners, or fuel, or by using an electrical method.

Reduction of Waste

Hot water wasted means energy wasted. Is the water temperature higher than necessary? Besides placing notices to discourage waste on wash room doors, the fitting of spring-loaded taps and of low throughput showers, together with a high standard of insulation and maintenance can be of assistance.

Heat dissipation is sometimes necessary where there is no case to incorporate waste heat recovery. Traditionally, in a number of industries, a water cooling tower and draught system have been used to produce cooling by partial evaporation of the water. Where water is scarce and the attendant noise is not likely to be offensive, a closed circuit of water can be cooled by air from a fan.

77   Closer specification in the use of all materials saves in many ways. Energy is directly saved with those involving a high consumption of energy in their production, such as metals, rubber, plastics, paper, glass, cement and refractories.

78   Sort waste materials at source ready for recycling.

79   Incinerate refuse and recover its heat, e.g. for space-heating.

80   Consider CHP schemes where power and heat are produced together with minimum overall losses.

81   Fermentation of organic waste could provide methane gas as a fuel to drive stationary engines for power or perhaps as part of a CHP scheme.

Waste Heat Recovery

When energy is rejected in the form of pressure, temperature of waste gases and other products, or unburnt fuel from one process, it may be possible to harness it to reduce the energy needs of the same or some other process. An investigation is necessary to establish the facts before recovery of energy in the form of preheat of charge, combustion air, fuel, hot water, steam or unburnt fuel is attempted. Waste heat recovery involves capital investment. Criteria on which this investment could be justified are as follows.

(a)   The source must provide sufficient energy at a high enough temperature or pressure, or in the case of unburnt fuel, at a high enough calorific value.

(b)   The quantity of energy saved must be worth more than the equivalent cost of fuel which would otherwise be burned directly.

(c)   The energy in the form recovered must be needed for some processes or adaptable for that need.

(d)   The pattern of energy demand for the waste heat recovery must comply with the pattern of availability (storage of energy in a tank or accumulator, standby plant or auxillary fuel firing may be possible).

(e) There must be space available for the recovery equipment.
(f) The return on capital must be acceptable.
(g) A loan for the capital has to be available. Government grants and short-term loans are available for approved energy-saving projects.

General Comments

An Energy Management Department is usually to be found in a large company; its role will be to promote efficiency in the use of heat and power. This it does by initiating heat and power surveys, determining plant efficiencies, giving advice to higher management on fuel costs and the economical operation of fuel-using equipment, and ensuring compliance with safe procedures and legal requirements involving the use of fuel. Amongst other things it has technical responsibility in the purchase of fuels and fuel-burning equipment, executive responsibility for operatives engaged in handling fuel, the generation of power, steam and compressed air, water and similar services, instrumentation, heating, ventilation, refrigeration, and for the safety, training and education of personnel within the department.

A small firm requiring external advice may obtain it from appropriate Government Departments, advisory councils, energy and fuel suppliers, insulation manufacturers, polytechnics and universities, research organizations and trade organizations.

82   Investigate the various tariffs available for the supply of fuels to ensure that they are those which best meet the needs of the process.

83   Keep a regular record of stocks, purchases and consumption of energy.

84   Provide instrumentation to measure and control the consumption of energy. Regularly maintain and check the calibration of the instruments.

85   Examine the records to pinpoint any change in the energy consumption, making use where appropriate of computer database packages to relate energy consumption to production.

86   Carry out efficiency tests on machines and other plant to verify that they are not deteriorating.

87   Institute or revise planned engineering maintenance to reduce energy waste, improve safety and general working conditions.

88   Ensure that energy conservation has due consideration in production planning.

89   Encourage all personnel to save energy by investigating any suggestion which might reduce energy consumption in the long term.

90   Be prepared to argue a case for the installation of capital equipment which can reduce energy consumption; investigate what grants are available from regional and government sources.

91   Use persuasion and incentives to achieve desired results, and above all maintain the pressure for fuel economy.

9.13 SUMMARY

The techniques used in the management of energy have much in common with those required for the management of any resource in a company or establishment. The general concept of monitoring performance and the setting of targets is essential to the control of costs and also the motivation of staff associated with energy usage.

Whilst this chapter has concentrated on some aspects of energy management all the previous chapters have in effect done the same by indicating likely sources of inefficient use of energy and the various design techniques and types of equipment that can be used to reduce energy wastage.

Energy costs will continue to rise in the future, and so will have a direct bearing on the profitability of many industries. The Energy Manager will therefore have an increasingly important role to play.

Problems

9.1   The electrical energy use by a company on low voltage supply is as follows.

$$\text{annual unit consumption} = 72\,000 \text{ kW h}$$
$$\text{maximum demand (every month)} = 40 \text{ kW}$$
$$\text{measured power factor} = 0.95$$

There is no significant night or weekend use. The present tariff is the General Quarterly Block Tariff. Calculate the annual savings, if any, of changing to a Maximum Demand Tariff. First decide which of the scales 1, 2 or 3 is most appropriate. Assume that consumption is the same in every month and that the availability charge is based on the maximum demand. Ignore the fuel clause charge.

(Annual costs: block tariff, £4498; MD tariff (scale 3), £4514)

9.2   The annual electrical energy use by a company on low voltage supply is 95 000 kW h. At present, there is no significant night or weekend use. The tariff is the General Quarterly Block Tariff. Some of the space-heating load can be transferred to the economy seven tariff by using night storage heaters. The effect will be to reduce the annual daytime consumption of energy by 25 000 kW h. Calculate the annual savings achieved by this measure. Assume that consumption is the same in every month, and ignore the fuel clause charge.

(£785)

9.3   In the previous example, the tariff is the General Quarterly Block Tariff. What is the saving achieved by using the appropriate Maximum Demand Tariff, if the storage heater scheme reduces the maximum demand from 40 kW to 26 kW? Use the following data:

$$\text{measured power factor} = 0.93$$
$$\text{chargeable capacity} = 50 \text{ kVA}$$

(£1093)

9.4   The space-heating of a warehouse is supplied from radiators using hot water delivered from a gas-fired boiler. The boiler consumes 75 000 therms per year, operating with an overall efficiency of 72 %. During a period of extreme demand, the boiler breaks down and the heating load has to be met from electrical heaters. By the time the boiler is repaired and back in service these heaters have supplied 20 % of the annual heating load normally supplied by the gas-fired boiler. Calculate the extra cost incurred by the boiler breakdown. Assume that the energy tariffs are simply:

$$gas = 33 \text{ p/therm}$$
$$electricity = 4.6 \text{ p/kW h}$$

(£9610)

9.5 A company on a high voltage supply (scale 2 MD tariff) has a monthly maximum demand of 490 kW, and operates with a power factor of 0.7. The chargeable capacity is 750 kVA. The Energy Manager decides to invest in measures to increase the power factor to 0.95. The cost is £60 per kVA installed. Calculate the payback time on the investment, assuming that the chargeable capacity is renegotiated to 550 kVA.

(4.45 years)

9.6 A hospital generator system can deliver 10 MW of low voltage power during the day. The cost of generation is 2.91 p/kW h; this figure includes all relevant costs. By examining the tariff structure in Section 9.5 determine which months of the year offer a profit-making export tariff and hence the annual profit to be made by exporting electricity to the grid.

(November, December, January and February. Annual profit = £0.393 M)

9.7 A company currently uses electric heaters for the daytime spaceheating requirements of a large office block. The annual pattern of demand is shown in the following table; demand is during the day only and not at the weekend. The Energy Manager decides to install a system of night storage to reduce the maximum demand below 50 kVA and so qualify for a block tariff. Assuming that the night storage capacity is 1200 kW h per month, how much may the Energy Manager spend on the storage system if the company has a limit on payback time of two years? Assume that the power factor is 0.93, and ignore the fuel adjustment charge. The availability charge for this company is based on a negotiated chargeable capacity of 90 kVA, which is greater than any maximum demand during the year.

(£4050)

Month	Units consumed (kW h)	Maximum demand (kW)
Jan	5750	70
Feb	4650	65
Mar	2950	40
Apr	1750	30
May	1000	25
Jun	300	10
Jul	250	5
Aug	1500	15
Sep	3050	20
Oct	4700	40
Nov	5600	70
Dec	6000	75

References

9.1 Fuel Efficiency Booklets Nos 1–18. Energy Efficiency Office, Thames House South, London SW1P 4QJ

9.2 *Energy Manager's Workbook* Volumes 1 and 2 Energy Publications, Cambridge 1982

9.3 BISRA (ed.) 1953 *Physical Constants of Some Commercial Steels at Elevated Temperatures* Butterworths

9.4 Murphy W R, McKay G 1982 *Energy Management* Butterworths

9.5 Ottaviano V B 1983 *Energy Management* Fairmont Press

9.6 Gwyther N *150 Ways of Saving Energy* Private Publication. Dr Gwyther is available at: Dept of Mechanical and Production Eng., Teesside Polytechnic, Teesside, UK

9.7 Focus on Monitoring and Targeting. *Energy Management* Department of Energy Feb. 1989

INDEX